Soviet Robots in the Solar System

Mission Technologies and Discoveries

Wesley T. Huntress, Jr. and Mikhail Ya. Marov

Soviet Robots in the Solar System

Mission Technologies and Discoveries

 Springer

Published in association with
Praxis Publishing
Chichester, UK

Dr Wesley T. Huntress, Jr.
Geophysical Laboratory
Carnegie Institution of Washington
Washington DC
USA

Professor Mikhail Ya. Marov
Vernadsky Institute of Geochemistry and
Analytical Chemistry
Russian Academy of Sciences
Moscow
Russian Federation

SPRINGER–PRAXIS BOOKS IN SPACE EXPLORATION

ISBN 978-1-4419-7897-4   e-ISBN 978-1-4419-7898-1
DOI 10.1007/978-1-4419-7898-1
Springer New York Dordrecht Heidelberg London

Library of Congress Control Number: 2011921306

Cover design: Jim Wilkie
Project copy editor: David M. Harland
Typesetting: BookEns, Royston, Herts., UK

Printed on acid-free paper

Springer is part of Springer Science+Business Media (www.springer.com)

This book is dedicated to all those men and women around the world in the 20th Century who dreamed of going to the Moon and planets, and did something about it. They may not have traveled themselves, but they built machines – robots – to carry their eyes, ears, nose, hands, arms and legs to places where they could not go. By their work, the world experienced space travel through the spacecraft that they built and launched on bold, risky, and dangerous missions to the Moon, Venus, Mars, and the far reaches of the solar system. The ingenuity of these engineers and scientists was beyond measure, and they created a magnificent spectacle for the rest of us.

Contents

x   **Contents**

Illustrations

Authors' preface

The competition between the United States and the Soviet Union in the Cold War produced one of the greatest adventures of exploration in the history of humankind. As a by-product of military competition between the two countries in weapon delivery systems and laying claim to the propaganda 'high ground', both countries applied themselves to the conquest of space by attaching civil payloads to their rockets in order to conduct both human missions in Earth orbit (and to the Moon in the case of the US Apollo program) and robotic missions beyond Earth orbit to the Moon and planets.

This book describes the 20th Century history of the Soviet adventure in robotic exploration of the Moon and planets. Our chronicle includes just those missions launched by the Soviets into deep space whose objective was to explore the Moon or planets. It does not include missions sent into deep space to study the Sun or Earth-Moon space environment. Test missions launched beyond low Earth orbit with operating lunar or planetary spacecraft, such as the Zond series, are included. Launch tests carrying non-operating model spacecraft are not included. We have endeavored to provide a comprehensive and accurate account of all relevant missions conducted between the year 1958, the date of the first Soviet spacecraft launch attempt to the Moon, and 1996, the date of the last Russian deep space mission to be launched in the 20th Century. All missions that were assembled on the launch pad with intent to fly are included. Some launch attempts suffered explosions on the pad, or shortly after booster ignition, or at some point during the flight of the launch vehicle. The Russians were particularly beset by launch vehicle failures, most often involving the upper stages.

There are inconsistencies in the data reported both in Western and Russian sources on Soviet lunar and planetary missions. We have attempted to provide the best possible information based on the published data and on interviews conducted with Russian participants in the former Soviet space program. In some cases we have made judgments to select what appears be the most accurate.

Wesley T. Huntress, Jr.
Mikhail Marov
January 31, 2011

Acknowledgments

Wesley Huntress sincerely thanks the Geophysical Laboratory of the Carnegie Institution of Washington for his emeritus position, and also the Jet Propulsion Laboratory of the California Institute of Technology, Director Charles Elachi and Chief Scientist Moustafa Chahine, for their support. Much of this book was written during the time spent at JPL as a Distinguished Visiting Scientist. I would also like to thank several friends who provided assistance, including Viktor Kerzhanovich, Sasha Zakharov and particularly my co-author Mikhail Marov. Most importantly, I acknowledge the patience and understanding of my wife Roseann while I worked on this manuscript.

Mikhail Marov expresses his thanks to the M.V. Keldysh Institute of Applied Mathematics where he has worked for nearly 50 years as an Institute staff member involved essentially in all major endeavors of the Soviet robotic and human space program, and where he served as Scientific Secretary of the distinguished Space Research Council of the Soviet Academy of Sciences (MNTS KI) that was hosted by the Keldysh Institute while Mstislav Keldysh, its Director, and President of the Academy of Sciences, was Chairman of the Council. I would like to thank all my colleagues in the organizations of industry and Academy of Sciences, in particular in the S.P. Korolev Rocket-Space Corporation Energiya and Scientific-Industrial (NPO) Lavochkin Enterprise with whom I worked on space activities resolving numerous problems. I also thank my Russian friends who helped to find out and/or clarify historical data for this book, including Victor Legostaev, Vladimir Efanov, Igor Shevalev, Yury Logachev, Arnold Selivanov and Sasha Zakharov. Special thanks to Olga Devina who assisted me in data compilation and cross-examination. Finally, I appreciate Wesley Huntress's kind invitation to participate in this project and co-author the book, as well as friendly cooperation and mutual understanding while working on the manuscript.

The illustrations in this book are mainly from governmental sources in the US, including NASA, and in Russia including Energiya, NPO-Lavochkin, the Institute for Space Science and the Institute of Geochemistry and Analytical Chemistry. We found the books '*S.P. Korolev Rocket-Space Corporation Energiya. 1946-1996*' Volume 1, published by Menonsovpoligraph (1996), '*S.P. Korolev Rocket-Space*

Corporation Energiya at the Boundary of Two Centuries. 1996-2001' Volume 2 (ed. Yu.P. Semenov), OOO Regent Print (2001), and '*Automatic Space Vehicles for Fundamental and Applied Studies*' published by NPO-Lavochkin, MAI PRINT Moscow (2010) particularly useful. We appreciate the kind permission of V.P. Legostaev, the First Deputy of the President and General Designer of Energiya, and also of V.V. Khartov, the General Designer of NPO-Lavochkin, to reproduce photographs, diagrams and drawings from their organizations either published in their publications or placed on internet sites.

For non-governmental sources every effort has be made to trace the original copyright holders and seek formal permission for all figures that have appeared in previously published works. A number of these images are from older and out-of-print books, and due to mergers and acquisitions in the publishing industry it has not been possible to track down all potential original copyright holders. We offer our apologies to any that we may have inadvertently overlooked. In all such cases we have cited the publication, author, or artist if known. We have used several unattributed drawings from the 1981 '*Space Travel Encyclopedia*' (in Hungarian) by I. Almas and A. Horvath, others from the 1972 '*Robot Explorers*' by Kenneth Gatland with art by John Wood and others, several by Ralph F. Gibbons from the '*Soviet Year in Space*' series published by the American Astronautical Society, some drawings by Peter Gorin in Asif Siddiqi's '*Challenge to Apollo*', and by an unattributed artist in the NASA '*Pioneering Venus*' publication. Many thanks to Asif Siddiqi and Don Mitchell for permission to use material from their print and web publications. Special thanks to James Gary for permission to use his art work and to Ted Stryk for images from the Soviet program that he has reprocessed with modern methods. Unfortunately, many of the older diagrams and images from the Soviet program are not the best quality for modern publication, but are illustrative of the times. Finally, thanks to David M. Harland for his diligent job of editing and his many improvements to the manuscript.

Part I

The pieces: people, institutions, rockets and spacecraft

1

Space race

FIRST ON THE MOON, FIRST ON VENUS, AND FIRST ON MARS

The latter half of the 20th Century will forever be known as the time when the human race broke its Earth-bound chains and began to explore the boundless reaches of interplanetary space. The Soviet Union initiated this enterprise with its Earth orbiting satellite 'Sputnik', meaning 'fellow traveler', in 1957, and shortly thereafter Soviet scientists made the first attempts to send spacecraft to the Moon and to the planets. What followed were 38 years of triumph and tragedy in one of the most exciting adventures in recent human history.

The first pioneers of space flight lived in the first half of the 20th Century. Tsiolkovsky, Tsander and Kondratyuk in Russia, Oberth in Germany, Goddard in the US, and later Korolev and Glushko in the Soviet Union, von Braun in the US, and Esnault-Pelterie in France, all believed that humankind could travel to other planets in the Solar System using new developments in rocket propulsion. These early visionaries established the notion that it was in fact possible to fly to the planets, but their dreams became reality only after the intervention of World War II created the technological catalyst for accomplishing deep space propulsion. By the end of the 20th Century, humans had set foot on the Moon and had sent robotic spacecraft to most of the planets, as well as to some comets and asteroids.

Most of the history of space exploration in the 20th Century is characterized by intense competition for dominance between the USSR and USA. At the dawn of the 'space age' the two nations were developing ICBMs to drop nuclear warheads on each other's cities. Europe and Japan were preoccupied with rebuilding after the devastation of World War II. The USSR launched the first artificial satellite, Sputnik on October 4, 1957, using a modified version of their first operational ICBM. The first human space flight was also Soviet; Yuri Gagarin's orbital flight of April 12, 1961. These events shocked Americans, who had difficulty imagining how they could not have been first into space. The Americans also immediately recognized the implications of these events for their national defense. The USA mobilized a massive space development program of its own in 1958, and on May 25, 1961 President

Kennedy formulated a national goal to land a man on the Moon before the decade was out, implicitly meaning that this man should be American, not Russian.

The Soviet Union was slow to respond to the challenge, but in 1964 initiated a national program to send a cosmonaut to the Moon ahead of the Americans. The Americans won the 'race' on July 20, 1969, with Apollo 11's touchdown on the Sea of Tranquility. The Soviet program stalled after a series of failures of the N-1 heavy rocket, their equivalent to the American Saturn V, but the USSR produced dramatic results with robotic lunar rover and sample return missions through 1976. The Americans shut down their Apollo lunar program in 1972 after six successful flights to the lunar surface.

The 'space race' was a Cold War phenomenon, but just like the international air races in the first half of the 20th Century, the space race resulted in an explosion of research and technological development. While competition in space exploration between the USSR and the USA originally focused principally on flying humans to the Moon, there was also competition to fly robotic spacecraft to the Moon and beyond, and this yielded remarkable feats of engineering and enormous scientific progress in understanding our Solar System and technological progress for Earth applications. If it had not been for the political imperatives of the Cold War, it is highly unlikely that the national investments required for this progress would have been made. After the fall of the Soviet Union in 1991, the Russian robotic space exploration program withered away.

This book provides the technical details of the Soviet Union's robotic space exploration missions, beginning with the attempted launch of a lunar impactor on September 23, 1958, and concluding with the final launch in the Russian national scientific space program in the 20th Century, the Mars-96 mission, on November 16, 1996. Each flight campaign is placed into the political and historical context in which the entire endeavor occurred, chronicling the boldness of the program, the daring spirit of its creators, the genius of its implementation, and the successes and in some cases the tragic failures in its execution. The book is in two parts. Part I describes the pieces that must be combined to make up a space program: the key players who make things happen; the institutions that design, build and operate the hardware; the rockets that offer access to space; and the spacecraft that carry out the enterprise. Part II is a chronological account of how the pieces are put together to undertake space flight and mission campaigns. Each chapter covers a particular period, usually several years, when specific mission campaigns were undertaken during launch windows determined by celestial mechanics. Each chapter in Part II gives a short overview of the flight missions that occurred during the time period and the political and historical context for the flight mission campaigns, including what the Americans were doing at the time. The bulk of each chapter is devoted to the scientific and engineering details of each flight campaign, and in each case the spacecraft and payloads are described in as much technical detail as is available at the time of writing this book, the progress of the flight is described, and a synopsis of the scientific results is given.

The Soviet robotic space program was dramatic, and was driven by a thirst for technological achievement and a desire for international recognition and respect. It achieved all these things. Soviet robotic spacecraft were first on the Moon, first on Venus, and first on Mars.

2

Key players

INTRODUCTION

Any great enterprise is the product of people. It is people who make things happen. Institutions are the means by which great enterprises are realized, but it is the people in these institutions, and in particular the leaders of these institutions, that drive the mechanisms to create great products. And so it is in the space exploration enterprise. We begin the story of the Soviet Union's space exploration program in the 20th Century with a description of the people who led the development of this great enterprise. While there were many administrators, engineers and scientists who were essential, we have room here only to describe those at the top of the enterprise, those whose personal and institutional power created the USSR's space program. At the top are the Communist Party leaders and government ministers who had control over selecting and funding national projects; second, and most particularly, the individual Chief Designers of the space program who proposed the projects; third the directors of the design bureaus which were responsible for building rockets and spacecraft for the projects; and finally the President of the Soviet Academy of Sciences, who besides his own leadership of the space program provided academic resources via the directors of the Academy's research institutes where space mission goals were developed using the rockets and spacecraft built by the design bureaus.

The single most important individual in the development of the Soviet space program after WW-II was Sergey Pavlovich Korolev. After Joseph Stalin decided to make rocket development a national priority at the end of the war, Korolev was retrieved from exile in a labor camp, together with others from his small band of engineers that built research rockets before the war. They started with the V-2 and a group of captured German engineers, just as occurred in the US. During the 1940s and 1950s Korolev's design bureau developed the USSR's first long range rockets using the German rocket engineers' expertise to build their own design skills. By the mid-1950s the German engineers had been generally dismissed, and the enterprise was entirely Russian. Korolev began testing his R-7 ICBM in the spring of 1957, the rocket that would launch not only Sputnik and other early Earth satellites, but almost all of the Soviet lunar and planetary missions throughout the 1960s and all Soviet cosmonauts. In upgraded and modified versions, this venerable rocket has become the core of the Soyuz launcher that is used commercially today for both manned and unmanned missions.

Korolev was an excellent engineer and designer, with considerable leadership and political skills. These qualities and his mission successes made him the darling of the Soviet space program. His identity was kept secret and he became known as 'Chief Designer', a term invented for the titular head of the Soviet space program. There were only two others that followed him after his death in 1966, but neither man had the full measure of qualities possessed by Korolev and the program seemed to lose much of its driving force. Had Korolev remained in charge, the USSR may have landed a cosmonaut on the Moon – even if later than planned and after the Americans. The Chief Designer of the Soviet space program, the de-facto leader inside Kremlin circles, was at the same time a director of one of the implementing design bureaus. There was no equivalent in the US: Wernher von Braun had a similar leadership role but was not at the same time the Administrator of NASA. In the USSR, there was no equivalent of NASA. The space enterprise was only a portion of the government's Ministry of General Machine Building, which had wide control over all of Soviet space industry and the design bureaus that implemented the policies of the ministry.

The design bureaus and research institutes were the places where all the hardware

Figure 2.1 Korolev's Council of Chief Designers in 1959. Left to right: A.F. Bogomolov, M.S. Ryazansky, N.A. Pilyugin, S.P. Korolev, V.P. Glushko, V.P. Barmin, V.I. Kuznetsov.

was developed and built to execute the Soviet space program, except for the science instruments supplied by the Soviet Academy of Sciences. The directors, also known as 'Chief Designers', of the several design bureaus and research institutes were the key 'movers and shakers' of the program. At the beginning of the Soviet rocket and space enterprise, Korolev established a Council of Chief Designers to coordinate all efforts in rocket development and space exploration. The members of the Council are shown in Figure 2.1. Council member Academician Valentin Petrovich Glushko (1908–1989) was an early colleague of Korolev's before WW-II and supplied the rocket engines for the R-7, but later he became a dedicated rival to Korolev. He was one of the most important figures in the history of the Soviet program, and his role following Korolev's death is described later in this chapter. Academician Nikolay Alexeevich Pilyugin (1908–1982) was Chief Designer of NIIP and responsible for autonomous control systems (avionics) for rockets and spacecraft. Pilyugin was one of Korolev's closest colleagues and pioneered the development of flight computers and precision avionics for autonomous navigation. Corresponding member Mikhail Sergeevich Ryazansky (1909–1987) was Director and Chief Designer of NII-885 and developed radio systems including on board transmitters, receivers, radio command links and terrestrial antennas for rockets and deep space missions. In particular, he pioneered the study of radio systems to facilitate autonomous navigation by vehicles in deep space and the development of imaging systems for spacecraft. Academician Alexey Fedorovich Bogomolov (1913–2009) was Director of Design Bureau OKB MEI (until 1989) and principally responsible for the development of on board radio telemetry and trajectory tracking, in addition to terrestrial antennas for rockets and spacecraft. He also greatly contributed to radar remote-sensing techniques including the instrument for mapping of Venus by Venera 15 and 16. Academician Vladimir Pavlovich Barmin (1909–1993) was Chief Designer of all ground complexes for ballistic missiles and space launchers. He also contributed to the development of soil-sample devices for Luna and Venera missions. Academician Victor Ivanovich Kuznetsov (1913–1991) was Chief Designer and Director of NII-10, and as such he developed gyroscopes for rockets and spacecraft and pioneered inertial navigation systems in the USSR.

The design bureaus were all in competition with one another. One or the other of the directors, such as Korolev, was by force of personality and political connection the 'Chief Designer' of the whole space program. With no dedicated governmental space administration to marshal the competition between design bureaus, the Soviet space program was rife with rivalry, animosity and political intrigue. The resulting inefficiencies were wasteful of resources and a cause for much delay and many a failure. After Korolev died, there was no one with all the personal skills necessary to hold it all in check.

Almost equivalent in stature to Korolev was Mstislav Vsevolodovich Keldysh, head of the Institute of Applied Mathematics and after 1961 President of the Soviet Academy of Sciences. While Korolev was the 'Chief Designer' of the Soviet space program, Keldysh was 'Chief Theoretician'. They worked together both to advocate and implement the space exploration program. From 1956 until his death in 1978, Keldysh was the Chair of the highly recognized Inter-Departmental Scientific and

Figure 2.2 Sergey Pavlovich Korolev (left) and Mstislav Vsevolodovich Keldysh (right).

Technical Council on Space Research (MNTS KI; Mezhduvedomstvennyi Nauchno-Tekhnicheskii Soviet po Kosmicheskim Issledovaniyam) which was responsible for space science and technology development in the Soviet Union. The Council and the Academy determined the objectives for the space program, advised the government and recommended individual projects, provided expertise in space navigation, and supplied scientific investigations for flight missions. Acting together, Korolev and Keldysh were responsible for many of the achievements of the space program.

The final highly influential group were the directors of research institutes of the Soviet Academy of Sciences. The two leading space science organizations were the Vernadsky Institute of Geochemistry and Analytical Chemistry established in 1947 and the Institute for Space Research set up in 1965. The Academy's science institutes devised the science objectives and instruments for space missions. The leading design bureau and science institute directors were strong individuals who advised Korolev and Keldysh on which missions to fly and determined what science investigations would be carried.

MINISTER

Afanasyev, Sergey Aleksandrovich
1918–2001
First Minister of General Machine Building
1965–1983

Afanasyev

Sergey Afanasyev's organization managed the institutions and workforce that built ballistic missiles and satellites vital to the defense of the Soviet Union, as well as the spacecraft and launch vehicles for their politically important space exploration program. Leonid Brezhnev once told him, "We believe in you, but if you fail we will put you against a brick wall and shoot you." Known as "the big hammer", he could be a very rude and intimidating man but he had a talent for orchestrating immense projects. He was among the most powerful people involved in the USSR's space program, which included Korolev and his rival Glushko. His criticism of Korolev's management of the manned space program resulted in the separation of the robotic program from Korolev's bailiwick to that of Georgi Babakin in 1965. He oversaw the Soviet Union's response to the Apollo project and was ultimately responsible for canceling it after many setbacks.

FOUNDER AND CHIEF DESIGNER OF THE SOVIET SPACE PROGRAM

Korolev, Sergey Pavlovich
1907–1966
Founder of the Soviet Space Program
Chief Designer OKB-1 1946–1966

Korolev

Chief Designer Sergey Korolev (this common spelling is not phonetically correct, Korolyov is proper) was the behind-the-scenes Soviet equivalent of von Braun in the US. His Experimental Design Bureau No.1 (OKB-1) led the development of first military and, shortly thereafter, peaceful applications of rocketry in the USSR. His identity was a state secret known only to an inner circle; to others he was simply the 'Chief Designer'. While von Braun was openly engaged with the public and served as an enthusiastic communicator on the American civilian space program, Korolev worked under heavy state security. He was not even allowed to wear his medals. His identity was not made public until after his death.

A passionate advocate of space exploration, Korolev began as a young engineer leading a research group, GIRD, that built small rockets in the 1930s at the same time as Robert Goddard was making his rockets in the US. Korolev became a victim of one of Stalin's purges in the late 1930s, confessing to trumped up charges under duress. He was sent initially to a gulag before being transferred to the 'sharashkas', slave labor camps for scientists and engineers, where he could continue to work on rockets in exile for the military. As a result of the considerable hardship he endured, he developed health issues that would persist for the rest of his life. He was released near the end of WW-II to evaluate the captured German V-2 missile and build a Soviet rocket capability.

In 1946 Korolev was appointed Chief Designer of a new department in Scientific Research Institute No.88 (NII-88) to develop long-range missiles. The R-1, which was basically a Soviet-built V-2, led to a succession of ever more powerful rockets named R-2, R-3 and R-5. He proved himself to be a very talented technical designer and manager, and in 1950 his department was upgraded to a design bureau, and then in 1956 was separated from NII-88 to become OKB-1. He began work in 1953 on an ICBM to deliver the heavy 5-ton nuclear warhead. This would require a rocket of unprecedented size and power. The resulting massive multi-stage R-7 (which NATO referred to as the SS-6 Sapwood) was first tested in the spring of 1957, long after technology had reduced the size of the warheads. It was overly large and awkward as a weapon, taking 20 hours to prepare for launch, and only a few were deployed before more practical delivery systems were produced by competing organizations. However, the R-7's lifting power allowed Korolev to adapt it for space exploration purposes, including Sputnik, which proved to a reluctant Kremlin the political value of non-military uses for large missiles. Like von Braun, Korolev's passion was the exploration of space, but he needed the military business to build his rockets. Thus his designs owed as much to his dreams as to hard military requirements. Korolev's lobbying to use the R-7 for space exploration, and his insistence on the large and militarily impractical cryogenic rockets best suited to this role, drew impatience from the military, which reacted by placing contracts with competitors, in particular Mikhail Yangel's OKB-586 and Vladimir Chelomey's OKB-52.

Korolev was a charismatic man who through sheer perseverance, political savvy, technical expertise, and talent for leadership established the Soviet space exploration program on the backs of the military, with consequent resentment. Nevertheless, he triumphed because his space spectaculars won him the support of the Soviet political hierarchy and in particular Nikita Khrushchev. The R-7 in its various incarnations became the most reliable and most used space exploration launch vehicle in the 20th Century. The Soyuz version continues in use today to launch cosmonauts into low Earth orbit. The Molniya version launched all of the early Soviet lunar and planetary missions until the more powerful Proton developed by Chelomey became available, and later versions are still used for this purpose. His sudden death in January 1966 was a severe shock, and without his leadership the Soviet lunar program devolved into rivalry between factions, impeding progress and dashing any chance the Soviet Union may have had after their late start.

PRESIDENT OF THE SOVIET ACADEMY OF SCIENCES

Keldysh, Mstislav Vsevolodovich
1911–1978
President, Soviet Academy of Sciences 1961–75

Keldysh

While Sergey Korolev was the engineering genius behind the Soviet space program, Mstislav Keldysh was its scientific genius and his eager partner. There was no single person equivalent to Keldysh in the US space program. As a brilliant and elegant mathematician, he was particularly adept at applying mathematics to complex practical problems, with a special interest in aerodynamic engineering. From 1946 to 1961 he was head of the research organization NII-1, which is now the Keldysh Research Center. NII-1 was originally Korolev and Glushko's rocket research group prior to their arrest in the purges. In 1953 Keldysh was named head of the Division of Steklov's Mathematical Institute which in 1966 became the Institute of Applied Mathematics and now bears his own name. In 1961 he was elected President of the Soviet Academy of Sciences.

Keldysh's involvement in space research began in 1954 when he co-chaired with Korolev the committee that designed the scientific spacecraft that ultimately became Sputnik 3. Beginning in 1956 he chaired the Academy's powerful MNTS committee and was regarded as the 'Chief Theoretician' of the space program, in charge of the scientific aspect of space including military applications in computers and nuclear weapons design. He and the Academy's science institutions provided the theoretical basis for space exploration, rocket design, mission design and navigation in space. Unlike in the US, the Soviet Academy of Sciences was charged with developing the mathematical and scientific tools, including instruments, for space exploration, and as head of the Academy Keldysh was a major force in the development of lunar and planetary exploration in the USSR. The government often had the Academy assess the merits of projects proposed by the various design bureaus. Also, the government presented Keldysh to the international community as the face of the Soviet space exploration program, representing it abroad and to the media. His prominence went hand in glove with Korolev's obscurity.

CHIEF DESIGNERS AND DIRECTORS OF THE DESIGN BUREAUS

Tikhonravov, Mikhail Klavdievich
1900–1974
Deputy Chief Designer OKB-1 1956–1974

Tikhonravov

Although not chief of a design bureau, Mikhail Tikhonravov was a key member of Korolev's team in the early days of OKB-1, and one of the pioneers of the Soviet space program. He was an early glider enthusiast and worked with N.N. Polikarpov in the 1920s developing aircraft. In 1932 he joined GIRD and became interested in the theory of rocket flight and space technology, working with Korolev to build the first Soviet liquid propellant rocket. Tikhonravov escaped the terror of the late 1930s and during WW-II worked on Katyusha rockets and a rocket-powered fighter. After the war, he was fascinated with the German V-2 rocket and designed his own high-altitude rocket for carrying a pilot into space. In late 1946 he became Deputy Chief of NII-4 in Moscow to manage research into ballistic missile development. There he began a pioneering study into multistage rockets and orbital flight that would later be applied in launch vehicle and spacecraft development. Following Tsiolkovsky, he originated the concept of 'packet' design for multistage rockets adopted by Korolev for the R-7. On November 1, 1956, he was transferred to OKB-1 where he worked hand-in-hand with Korolev in developing robotic spacecraft for flights to the Moon, Venus and Mars, and spacecraft for OKB-1's manned spaceflight program.

Mishin, Vasily Pavlovich
1917–2001
Chief Designer OKB-1 1966–1974

Mishin

As Korolev's deputy and protégé, Vasily Mishin took over management of OKB-1 after his mentor's unfortunate death during surgery in 1966. It was during Mishin's tenure that OKB-1 attempted to develop Korolev's giant N-1 Moon rocket and the Soyuz spacecraft to send cosmonauts to the Moon. When he took over, the project was plagued with technical problems and unrealistic schedules. Mishin was a well-regarded engineer and a kindly man, but did not possess Korolev's leadership talent, nor the charisma and connections that Korolev used to mobilize the massive Soviet political and industrial machine and to thwart his enemies. While NASA succeeded with Apollo, Mishin oversaw four disastrous N-1 launch attempts, failures in lunar Soyuz test flights, failures in three space station missions, and the deaths of the pilot of Soyuz 1 in 1967 and the three-man crew of Soyuz 11 in 1971. He was deposed in 1974 by a coup orchestrated by Korolev's bitter rival, Valentin Glushko. Two years later any further attempts to send cosmonauts to the Moon were terminated.

Mishin was exiled to the Moscow Aviation Institute and blamed as "the man who lost the Moon race". He was unfortunate to have been the man in charge when the ambitious technological challenges began to crumble in the face of the relentless American Apollo juggernaut; he just didn't have the 'right stuff' to overcome them. Although many in the West thought that he had been executed, Mishin resurfaced in the late 1980s and published a number of controversial accounts of the history of the Soviet space program.

Glushko, Valentin Petrovich
1908–1989
Chief Designer OKB-456 1946–1974
Chief Designer NPO-Energiya 1974–1989

Glushko

A contemporary of Korolev, Valentin Glushko began working on rocket engines in the 1920s and became head of the Gas Dynamics Laboratory. The military merged it with Korolev's GIRD rocket research group in the 1930s. Like Korolev, Glushko was a victim of the purges. After WW-II he was made head of Design Bureau OKB-456 to develop rocket engines for missiles designed by Korolev's OKB-1, Chelomey's OKB-52 and Yangel's OKB-586. When Korolev began to design a successor to the R-7 and ignored Glushko's advice to use hypergolic propellants they became bitter enemies. In fact, the animosity between the two harked back to the purges. Korolev was convinced that Glushko was responsible for his internment. Glushko was arrested first, and there is a story that under duress he denounced Korolev for undermining progress by preferring liquid rather than solid fuel rockets, and shortly thereafter Korolev was arrested. Glushko criticized Korolev's plans for the Moon program and impeded Korolev's progress by refusing to build the engines for the N-1, forcing Korolev to resort to an inexperienced supplier.

In 1974, with the N-1 suffering spectacular failures, OKB-1's enemies, including Glushko and Chelomey, convinced Brezhnev to fire Mishin. Glushko was appointed in Mishin's place. His first act was to precipitously cancel the N-1 program. He then absorbed OKB-1 into his own design bureau OKB-456. On gaining membership of the Central Committee of the Communist Party he also absorbed Chelomey's design bureau to create a massive rocket engineering empire named NPO-Energiya. Then, having defeated the legacy of Korolev, Glushko focused on building a new rocket and reusable spacecraft system in his own image – the Energiya and Buran – to replace the Soyuz system and compete with the US Space Shuttle. The Energiya rocket flew twice in the late 1980s and Buran once, unmanned, and were promptly canceled as unaffordable. They are now only silent monuments to a man described by his critics as vain, stubborn, petty and manipulative. Nevertheless, the Energiya-Buran project is a monument to the skilled people in the Soviet Union who made this ambitious and complex project possible. By supreme irony, today Korolev's Soyuz rocket and spacecraft are still in front line service and the conglomerate that Glushko built bears Korolev's name as the S.P. Korolev Rocket and Space Corporation Energiya.

Glushko was a superb engineer and designer of rocket engines and his OKB-456 created some of the most efficient engines ever produced. He managed to build closed-cycle engines that eluded the skills of American rocket engine makers. At the same time he was a stubborn critic of cryogenics, even though he built engines using liquid oxygen, and insisted that hydrogen was not a suitable rocket fuel while the US was using it for the upper stages of its most powerful launch vehicle, the Saturn V. Unable to eliminate combustion instability in large single-chamber engines, Glushko devised an ingenious solution using four smaller combustion chamber/nozzles which shared a common fuel/oxidizer feed. The four-chamber RD-107 and 108 engines he built for the R-7 are still in use today with the Soyuz launcher. In one of the ironies of the Cold War, the very powerful four-chamber RD-170 engine that he made for the Energiya rocket was split in two and the two-chamber variant, the RD-180, is now in service powering the latest model of the US Atlas launch vehicle!

Chelomey, Vladimir Nikolaevich
1914–1984
Chief Designer OKB-52 1955–1984

Chelomey

Vladimir Chelomey, a mathematician dealing with non-linear wave dynamics, began his career working on cruise missiles. In 1955 he became head of OKB-52, and in 1958 began work on his first ICBM, the UR-100 (NATO designation SS-11), which became the Soviet Union's answer to the US Minuteman. While Korolev never lost his preference for cryogenics, both Chelomey and Mikhail Yangel opted for storable propellants and their missiles were better suited to military requirements. This led Korolev to focus on the politically-supported lunar cosmonaut program. Chelomey's attention to military requirements gained him respect in the military establishment and access to far greater resources than Korolev.

In the early 1960s, Chelomey began development of the UR-500 Proton rocket intended to be a heavy lift ICBM. When the military canceled it, Chelomey, with

Keldysh's support, used his political connections to save it for the Moon program. Chelomey had a rival plan to Korolev's for development of rockets and spacecraft to take cosmonauts to the Moon. He proposed his plan in competition to Korolev when the USSR finally made its decision in 1964 to compete with the US Apollo program. Khrushchev (whose son was an engineer at OKB-52) was indebted to Chelomey for providing practical and vital military ICBMs, and so Chelomey managed to have his UR-500 chosen in preference to Korolev's new design for the test and circumlunar phases of the manned lunar program. However, the spacecraft would be the lunar Soyuz that Korolev proposed, and Korolev's massive N-1 Moon rocket was selected over Chelomey's even larger UR-700 for the lunar landing missions. The Chelomey-Korolev rivalry continued as both programs were separately managed and funded by Khrushchev and later by Brezhnev in a process that divided the backing required for an efficient and timely outcome. After a long run of early failures, the Proton was used to launch an automated Soyuz test spacecraft under the cover name of Zond on flights which looped around the Moon and returned to Earth. It went on to launch heavy satellites and modules for the Salyut and Mir space stations. Georgi Babakin at the Lavochkin Design Bureau, who had inherited Korolev's robotic exploration program, recognized that the Proton was well suited to launch the heavy spacecraft that he was designing and, with upper stage modifications which included using one of the stages from Korolev's N-1 rocket, the Proton became the launcher of choice for the Soviet lunar and planetary spacecraft of the 1970s and beyond. It is today a world standard for commercial heavy launch services.

Babakin, Georgi Nikolayevich
1914–1971
General Designer NPO-Lavochkin 1965–1971

As a self-taught engineer, Georgi Babakin did not gain a college degree until the age of forty-three. He worked on rocket control systems at NII-88 from 1949 to 1951, where he first met Korolev, and then designed military missile systems at OKB-301 for Chief Designer Semyon A. Lavochkin, where he rose to become a deputy chief designer and then General Designer (Director) of OKB-301, now renamed NPO-Lavochkin. Meanwhile, OKB-1 had become overwhelmed with responsibility for both manned and unmanned programs, and was suffering a run of failures. Trusting Babakin implicitly, Korolev trans-

Babakin

ferred all robotic lunar and planetary space probes to Lavochkin. Subsequently, Babakin solved the quality control problems plaguing the Luna Ye-6 and 3MV planetary spacecraft, leading to a long run of successes at the Moon and Venus. The heavy Proton-launched spacecraft were developed under his direction and he experienced their initial success with the Luna 16 sample return and Luna 17 rover.

He was a worthy successor to Korolev, but died suddenly at the early age of fifty-seven in August 1971 before his new Mars spacecraft reached their destinations.

Kryukov, Sergey Sergeyevich
1918–2005
General Designer NPO-Lavochkin 1971–1977

Sergey Kryukov worked with Korolev, Tikhonra-vov and Mishin on the development of the R series of rockets, and rose to deputy chief designer to Korolev along with Mishin and others at OKB-1. He had a falling out with Mishin over development of the Block D upper stage for the N-1 (also used on the Proton) and transferred to Lavochkin. After less than a year, he became General Designer when Babakin died. He inherited the problems that would plague the Mars program and the successes that would come in the Venus program. After the 1973 Mars fleet disaster, he was tasked by Afanasyev to design new and even larger Mars missions to send rovers to the surface and to return samples. These

Kryukov

missions turned out to be too complex and costly for the traumatized post-Apollo Soviet space program and were canceled in 1977 in favor of the rather less ambitious Phobos mission. Kryukov was replaced by Vyacheslav Kovtunenko and transferred to Glushko's organization, where he worked until retirement in 1982.

Kovtunenko, Vyacheslav Mikhailovich
1921–1995
General Designer NPO-Lavochkin 1977–1995

While working for Yangel's design bureau, Vya-cheslav Kovtunenko designed the Cosmos and Tsyklon rockets and was responsible for the Intercosmos series of small science satellites. On succeeding Kryukov as Director of Lavochkin, he developed the new generation Universal Mars Venus Luna spacecraft, which was essentially a renovation and upgrade of the heavy Venera spacecraft. He encountered obstacles to funding, not faring well against industry heavyweights such as Glushko, and the first of the new spacecraft was unable to be launched until 1988, as the Phobos mission. Kovtunenko would guide Lavochkin through the successes of Venera 11 to 16 and Vega

Kovtunenko

1 and 2, and the partial failures of Phobos 1 and 2, and the transition from the USSR to Russia leading up to the final Mars-96 debacle. He died in office in 1995.

DIRECTORS OF THE SCIENCE INSTITUTIONS

Petrov, Georgi Ivanovich
1912–1987
Director of the Institute for Space Research (IKI)
1965–1973

Petrov

A brilliant aerodynamics engineer having contributed significantly to ICBM design, Georgi Petrov was selected in 1965 by Keldysh to be the first Director of the newly formed Institute for Space Research. Petrov worked hard to establish his institute in the panoply of scientific communities, all of which were scrambling for funding in the new scientific space program. It was several years before IKI developed into a world-class institute for space research and the building of scientific instruments for space science missions. He established highly capable teams of space scientists and engineers and successfully motivated them to explore near-Earth space, the Moon, and the planets. IKI benefited immensely from his leadership, and mirrored his style of creativity and open discussion.

Sagdeev, Roald Zinnurovich
1932–present
Director of the Institute for Space Research 1973–1988

Sagdeev

Roald Sagdeev was a nuclear physicist working in the remote 'science city' of Akademgorodok when, at the advice of the distinguished physicist Leo Artsimovich, he was tapped by Keldysh to replace Petrov at IKI. He took leadership of IKI as the second generation of heavy Venus spacecraft was being introduced by Lavochkin, and shared in its success. He reassigned planetary geology to the Vernadsky Institute and focused his own institute's scientific efforts on planetary atmospheres and space plasma. These two institutes became dominant and competitive centers for planetary science. IKI remained the center for space astronomy.

A hallmark of Sagdeev's experience in a 'science city' far from the Kremlin was a culture of open, questioning discussion with promotion on the basis of merit rather than on political connection. Although upon becoming Director and a member of the Communist Party he initially conformed to the Soviet system, he later imported the Akademgorodok attitudes to IKI, bringing *perestroika* (transformation) and *glasnost* (openness) to his institute before Mikhail Gorbachov introduced it to the USSR. His most remarkable and enduring achievement was the opening of the Soviet planetary exploration program to international participation, leading his country into an era of scientific mission cooperation with the West as *perestroika* was driving the Soviet Union. Succeeding through charm, patience and shrewd political judgment, first the Vega Venus-Halley mission and then the Phobos Mars mission were approved as progressively more open to international scientific participation. He was aided by the mass and size of Soviet spacecraft, which were able to accommodate a large number of foreign instruments to undertake comprehensive scientific missions. The new policy was highly successful at the outset, catching the US in the doldrums after its successes of the 1970s, and the Soviets overtook the US as international leader of planetary exploration in the 1980s.

After the success of the Vega missions in 1986, Sagdeev became a local hero and international celebrity. But the joy was short lived. The loss of the Phobos missions in 1988 raised an international furor in the space science community. This was not a comfortable situation for Sagdeev and he left IKI in 1988, married the daughter of Dwight Eisenhower, and moved to the US to become a Professor at the University of Maryland. He remained a force in international space science and exploration for a time, but his influence on space policy decreased as he focused his efforts more on East-West relations. The high level of international participation in the Vega and Phobos missions, and the ensuing Mars-96 mission, has never been equaled.

Vinogradov, Aleksander Pavlovich
1895–1975
Director of the Vernadsky Institute of Geochemical and Analytical Chemistry 1947–1975

Alexander Vinogradov was the Soviet Union's leading geochemist, head of the Vernadsky Institute and Vice President of the Soviet Academy of Sciences at the opening of the 'space age', and Chairman of the Moon and Planets Section of the Space Council MNTS KI. He was a pioneer in using chemical and isotope analysis to study the formation of minerals in Earth and meteoritic materials. He developed the use of gamma-ray spectroscopy to study the composition of planetary surfaces, and analyzed samples returned from the Moon. Under his leadership, the Vernadsky Institute developed many of the geochemistry instruments flown on missions to the Moon, Venus and Mars.

Vinogradov

Barsukov, Valery Leonidovich
1928–1992
Director of the Vernadsky Institute 1976–1992

Valery Barsukov was a geologist experienced in field work. After taking over the Vernadsky Institute and its new role in planetary geology in 1976, he promoted missions and flight experiments with geochemical goals. He assumed leadership at a time when Mars exploration was in decline and Venus exploration was dominating the planetary program. He was an effective lobbyist for planetary geology missions and proved an effective rival to the Institute for Space Research led by Sagdeev. Both Barsukov and Sagdeev were well connected and fought, sometimes bitterly, to establish their own space science missions.

Barsukov

With Sagdeev's departure in 1988, Barsukov and the Vernadsky Institute assumed effective leadership of the Soviet planetary exploration program. Until his death in 1992, Barsukov pursued a complex Mars exploration plan even more international in scope than Sagdeev's Phobos mission, with a particular focus on US involvement. Under the joint leadership of Barsukov from Vernadsky and Professor James Head from Brown University, the Vernadsky-Brown Symposium on Cosmochemistry was organized. This continues to function as a forum for Russian-American cooperative research in lunar and planetary science.

3

Key institutions

PARTY, GOVERNMENT AND MILITARY

In the Soviet Union there were three separate organizations that ran the country – the Communist Party, the government and the military. The Party was in overall charge through its Central Committee and the executive Politburo. The Party business was managed by a Secretariat that included a Secretary of Defense Industry and Space, and a Department of Sciences. The Soviet Academy of Sciences, while claiming to be independent, was implicitly part of the Department of Sciences and hence a Party organization. It ran the Inter-Department Scientific and Technical Council on Space Exploration (MNTS KI; Mezhduvedomstvennyi Nauchno-Tekhnicheskii Soviet po Kosmicheskim Issledovaniyam) which was nominally responsible for specifying the national policy and strategy for the space program.

The Soviet government comprised the Council of Ministers and its executive the Presidium. The Presidium had a Commission for Military Industry (VPK; Voenno-Promyshlennaya Komissiya) which included the various ministers controlling the defense industries. The Ministry of General Machine Building (MOM; Ministerstvo Obshchego Mashinostroenija) controlled the planning and budget for the Soviet space program. MOM was the closest equivalent to NASA in the USSR but it had a much wider remit, including the design and production of rockets and space systems for the military. It established, controlled and funded the various design bureaus (OKBs) that developed the rocket and space systems, and the scientific research institutes (NIIs) that provided the science and technical support required for space projects. There was no separation of civilian and military space programs. MOM was the focal point of the powerful Soviet military industrial complex. It controlled a massive industrial system, providing funding enormous in scale, and operating in complete secrecy.

How this dual system functioned between Party and government, sitting atop a large and convoluted system of industry, design bureaus and research institutes, is somewhat mystical. It was complicated even more by the third force in the Soviet system, the military. The Armed Service's Strategic Missile Forces was the heavyweight competing for funding from MOM, and the military controlled the launch pads, the launch ranges, and the tracking stations. In reality, the personal power and political influence of the Chief Designers of the design bureaus such as Korolev, Glushko, Chelomey, Yangel and Babakin were most influential for planning and execution, especially in the early years. These powerful men competed mightily with one another for influence and funding, sometimes bitterly, as between Korolev and Glushko or between Korolev and Chelomey.

DESIGN BUREAUS

OKB-1

The founding space exploration enterprise in the Soviet Union was Experimental Design Bureau No.1 (OKB-1). It had its beginnings in the Scientific Research Institute No.88 (NII-88). A new design section, Department No.3, was set up by the government in May 1946 for the dozens of engineers who had just returned from over a year of investigating the German rocket industry. Sergey Korolev headed the department as Chief Designer. It comprised almost 150 engineers and technicians, and its task, Stalin stated, was to build a Soviet version of the V-2. After succeeding with the R-1, and proceeding to design new rockets of its own, the department was restructured into a larger design bureau OKB-1 in the early 1950s and then separated from NII-88 in 1956. OKB-1 built the first Soviet ballistic missile to carry a nuclear warhead, the intermediate range R-5M, and the first submarine-launched ballistic missile, the R-11FM. Korolev's proposal to build the first intercontinental ballistic missile, the R-7, was approved by the government in 1954. The first successful test of the missile was carried out in August 1957 and on October 4, 1957 it was used to launch Sputnik. The R-7 has been modified, augmented and upgraded in various forms to become the most prolific and successful space launch vehicle in history.

While building for the military, Korolev's real passion was for space exploration. OKB-1 would eventually lose the military rocket business to rivals, but it achieved great success in space exploration, along with frustrating failure, before Korolev's death in 1966. After Sputnik, Korolev and OKB-1 pursued more ambitious goals – robotic flights to the Moon and planets, and manned flights into Earth orbit. OKB-1 built the first spacecraft to impact the Moon, Luna 2, the first to photograph the far side of the Moon, Luna 3, and the first interplanetary spacecraft intended for Mars and Venus, but the failure rate was terrific. From 1958 through 1965, only four of 21 robotic flights to the Moon were successful (Luna 1, 2 and 3, and Zond 3); none of eleven attempts at Venus and none of the seven attempts at Mars were successful. On the other hand, OKB-1 had a singularly excellent record in manned spaceflight, launching the first man into space in 1961, the first woman into space in 1963, the first multi-person spacecraft in 1964, and the first spacewalker in 1965.

There were other design bureaus critical to the space program in the mid-1960s. Valentin Glushko's OKB-456 was the premier developer of rocket engines. Glushko supplied engines for Korolev's early rockets as well as other military rocket builders such as Chelomey. Chelomey's OKB-52 built the Proton rocket which became the staple heavy launcher for Soviet lunar and planetary spacecraft. In 1964 the Soviet Union made the late decision to compete with the US and send cosmonauts to the Moon. Korolev, Glushko and Chelomey each presented plans to the government for building the necessary rockets and spacecraft. After considerable wrangling, OKB-1 won on the basis of its head start in the manned program and long-standing work on the design of a Moon rocket. Chelomey did save his Proton rocket from the military scrapheap for the precursor manned circumlunar flights, but OKB-1 was to provide the final upper stage and the spacecraft.

During the battle for control of the manned lunar program, while still conducting both manned and robotic flight programs, succeeding with one and struggling with the other, Korolev realized that OKB-1 had taken on too much. It was essentially responsible for the entire Soviet space effort including communications satellites, reconnaissance satellites, robotic and manned space exploration programs. OKB-1 had to offload something in order to relieve the pressure on his organization, so in March 1965 Korolev reluctantly transferred the robotic program to NPO-Lavochkin. Keldysh played a significant role in this decision. If any comparison to the US could be made at this point, it would be that the USSR had two NASAs – one for manned missions (OKB-1) and another for robotic missions (NPO-Lavochkin). This is not a perfect comparison, however, since neither had full control of its own funding or its suppliers; that came from MOM.

After Korolev died in January 1966, OKB-1 was renamed the Central Design Bureau of Experimental Machine Building (TsKBEM) and his deputy Vasily Mishin took over. But unlike Korolev, Mishin was not a charismatic and politically savvy leader and he immediately ran into trouble. He introduced Korolev's three-person Soyuz spacecraft into service for the first time in April 1967 with tragic results, killing the test pilot Vladimir Komarov when the parachute failed to deploy properly as he returned to Earth. He then presided over the repeated failure of the N-1 rocket, which would have launched the Soviet Union's challenge to Apollo. In 1974 he was replaced by Glushko, who merged the organization with his OKB-456, and then with Chelomey's OKB-52, to form the giant NPO-Energiya. This organization went on to produce the Energiya heavy lift rocket, the Buran space shuttle, and the Salyut and Mir space stations. Now known as the S.P. Korolev Rocket and Space Corporation Energiya (RRK Energiya) it dominates the Russian manned space flight enterprise, having operated the Mir space station for almost 15 years, supplied the Zvezda habitat module for the International Space Station, and a decade of flights of the Soyuz and Progress spacecraft to service the ISS.

NPO-Lavochkin

The Scientific Production Organization NPO-Lavochkin was originally founded in 1937 as the Lavochkin Aircraft Design Bureau, OKB-301, named for its Chief Designer. Lavochkin produced a number distinguished fighter aircraft during WW-II

Figure 3.1 Scientific Production Association Lavochkin.

and then surface-to-air missile designs after the war, producing the first operational system for the defense of Moscow. In 1953 the SAM business was transferred to a new design bureau and OKB-301 pursued ramjet intercontinental cruise missiles as a hedge against problems with ICBM development. But the successful introduction of ICBMs in the late 1950s left OKB-301 without work. Semyon Lavochkin died in 1960 and the organization transferred in 1962 to Chelomey's OKB-52. The factory was closed, but reopened in 1965 as NPO-Lavochkin under the steady and capable leadership of Georgi Nikolayevich Babakin specifically to take responsibility for the robotic lunar and planetary spacecraft programs transferred from OKB-1.

The new NPO-Lavochkin realized immediate success using its inheritance from OKB-1 augmented by a history of great skill and experience in aviation technology. Luna 9 soft landed on the Moon's surface in January 1966, and before the year was over there were three successful lunar orbiters and a second soft lander. The first successful Venus entry probe, Venera 4, followed in 1967. NPO-Lavochkin went on to continue this highly successful series of spacecraft at Venus, a successful series of lunar orbiters, rovers and sample return missions, and the singularly complex and successful Vega missions which delivered landers and balloons to Venus enroute to a flyby of Halley's comet. Unfortunately, NPO-Lavochkin had no success at Mars: their campaigns in 1969, 1971, and 1973 were riddled with failures, and worse was to come in 1988 and 1996. Their astronomy missions have met with better success, in particular the Granat and Astron space observatories. Today NPO-Lavochkin is the single engineering center for the production of robotic scientific spacecraft.

THE ACADEMY OF SCIENCES AND ITS RESEARCH INSTITUTIONS

Whereas in the US the National Academy of Sciences is an advisory body to the government, the Soviet Academy of Sciences had governmental and implementation roles. It was integral to the Party, and made decisions on the worthiness of proposed space projects, approving those to be undertaken. However, the Ministry of Machine Building allocated the funding for these projects. The President of the Academy, Mstislav Keldysh, was a powerful and highly influential figure in the Soviet space program during his tenure. Before he became President of the Academy in 1961, he was head of the Institute of Applied Mathematics (IPM) and kept this position until his death in 1978, when IPM was named the Keldysh Institute. It played a major role in space navigation and mission design.

Also unlike in the US, where university laboratories and NASA's various field centers prepared scientific experiments for planetary missions, in the USSR the research institutes of the Academy of Sciences filled this role. These institutes were established by the Academy, but were funded through MOM. In the early years, the leading player was the Vernadsky Institute, more formally known as the Institute of Geochemistry and Analytical Chemistry. In 1965, at about the same time as Korolev transferred the robotic program to NPO-Lavochkin, the Soviet Academy of Sciences under Keldysh's initiative established the Institute for Space Research (IKI; Institut Kosmicheskikh Issledovanii), which gradually built up its role in scientific missions, including providing flight instruments, and by the 1970s was a fierce competitor to Vernadsky. With Roald Sagdeev's appointment as Director in 1973, IKI assumed scientific leadership of deep space missions. After Sagdeev quit in 1988, Vernadsky shared leadership under Valery Barsukov until the latter died in 1992. Today IKI is the leading space science institution. The institutes develop flight instruments and NPO-Lavochkin is responsible for the spacecraft and operations.

Figure 3.2 Institute for Space Research.

Figure 3.3 R-7 Pad 1 at Baikonur today and as photographed by the U-2 (NASA & Bill Ingalls).

LAUNCH COMPLEX

The USSR's first missile test range was established after WW-II at Kapustin Yar near Volgograd, formerly Stalingrad. Throughout the 1950s this was used to test the early short and intermediate range Soviet rockets, and later for launching the smaller Cosmos satellites. As Korolev worked on his first ICBM design, the R-7, it became clear that a new launch site would be required to accommodate radio guidance and tracking stations along a much longer range within Soviet territory, and to move the work beyond range of US tracking stations in Turkey. Tyuratam in Kazakhstan was selected for the R-7 launch complex. The site was called Baikonur, after a railhead some 270 km to the northeast, in an attempt to deceive the Americans in targeting their missiles. Construction started in 1955, and over the years the site has become an immense facility some 85 km by 125 km in extent including dozens of assembly and launch complexes, numerous control centers and tracking stations, work areas for tens of thousands of workers, the town of Leninsk to house them, and a 1,500 km test range.

The first launch complex to be built was the one for the R-7, and it is still in use today. It is part of the 'Center' or 'Korolev' area that includes the N-1 assembly and launch complex that was later converted for Energiya and Buran. The 'Left Flank' or 'Chelomey Arm' to the northwest has assembly and launch complexes for the Proton, Tsiklon and Rokot. The 'Right Flank' or 'Yangel Arm' to the northeast has a backup R-7 pad and facilities for Zenit and Cosmos.

COMMUNICATION AND TRACKING FACILITIES

Lunar and interplanetary missions required facilities for tracking spacecraft on their journeys through deep space and to communicate with them for navigation, control and data acquisition. Korolev chose Yevpatoria in 1957 as the site for these facilities because it offered a southerly latitude near the plane of planetary orbits. It was also conveniently close to the Black Sea and Crimean resorts.

Known as the Center for Long Range Space Communications (TsDUC), its first facility was a 22 meter antenna built in 1958 for lunar missions. The first phase of construction for planetary missions was ready in 1960. Korolev built his receiving stations by scavenging old naval parts – using the hull of a scrapped submarine, a revolving turret from an old battleship, and a railway bridge on top of the turret to hold the antenna array. Each array consisted of eight antennas in two rows of four all of which moved in unison. There were two sites, one to the north for the receivers and the other to the south for the transmitters. The receiving station had two such antenna arrays, each using eight 15.8 meter dishes. They operated in the meter band at 183.6 MHz, in the decimeter band at 922.763 and 928.429 MHz (32 cm), and in the centimeter band at 3.7 GHz (8 cm) and 5.8 GHz (5 cm). The transmitter station had one array of eight 8 meter dishes. This 'Pluton' transmitter was rated at 120 kW and operated at 768.6 MHz (39 cm). A ground-link microwave station was set up for transmitting data to a second station at Simferopol and then on to other locations in the USSR. The TsDUC facilities went online September 27, 1960, only a day before

Figure 3.4 Receiving station at north facility (left) and transmitting station at south facility (right), Yevpatoria.

Figure 3.5 *Cosmonaut Yuri Gagarin*, king of the Soviet tracking ship fleet.

the optimal launch date for Mars, although the first Mars launch did not occur until October 10. Between 1963 and 1968 Yevpatoria and Simferopol each received a 32 meter 'Saturn' dish, and five others were installed at Baikonur in Kazakhstan, Sary Shagan in Balkash, Shelkovo near Moscow, and Yeniseiesk in Siberia. In 1979 a 70 meter 'Kvant' dish was built at Yevpatoria. TsDUC now also has a 64 meter antenna at Bear Lake near Moscow, and a 70 meter dish in Ussuriisk near Vladivostok. All deep space missions were operated from Yevpatoria until a new control facility was opened in Moscow in 1974.

The USSR did not have a worldwide network of tracking stations like NASA's Deep Space Network, with serious consequences for deep space mission operations. Critical operations such as planetary encounters had to be planned for times when the spacecraft could communicate. And since signals could not be picked up when a spacecraft was below the horizon, spacecraft were designed to transmit only when Yevpatoria had a line of sight. This system required carefully controlled timing of spacecraft operations and reorientation of the spacecraft for high-gain operations. To provide a measure of relief from the limitations on spacecraft operations imposed by a single ground station, the Soviets deployed tracking ships in the world's oceans. These ships also tracked missile tests, covered manned space missions, and tracked interplanetary missions making the second firing of their upper stage to escape Earth orbit into interplanetary space. The ships were not a wholly satisfactory solution for deep space tracking, as only small dishes could be mounted on the ships and weather conditions could severely hamper operations. The first ships deployed in 1960 were the *Illchevsk*, *Krasnodar* and *Dolinsk*. In 1965/6 the *Illchevsk* and *Krasnodar* were

replaced by the *Ristna* and *Bezhitsa*. A third generation consisting of the *Borovichi,
Kegostrov, Morzhovets* and *Nevel* were deployed in 1967. These were all converted
merchant ships of about 6,100 tons displacement with crews of 36. In May 1967 the
first purpose-built tracking ship was introduced, the *Cosmonaut Vladimir Komarov*
(17,000 tons). The *Cosmonaut Yuri Gagarin* (45,000 tons) and *Academician Sergey
Korolev* (21,250 tons) joined the fleet in 1970. In addition, a number of smaller
tracking vessels were deployed: the *Cosmonaut Pavel Belyayev, Cosmonaut Georgi
Dobrovolskii, Cosmonaut Viktor Patsayev* and *Cosmonaut Vladislav Volkov*.

4

Rockets

EARLY SOVIET ROCKET DEVELOPMENT

The enabling technological step towards lunar and planetary space flight was the development of the military intercontinental ballistic missile (ICBM). From this, it is only a small incremental step to the development of a rocket capable of launching Earth-orbiting satellites, and then only another small step to one capable of sending spacecraft on trajectories to the Moon and beyond. The developers of ICBMs in both the US and USSR dreamed about space flight from the very beginning, and always in the back of their minds knew that the weapons on which they were working could ultimately be used for space exploration. This was as true for Sergey Korolev in the Soviet Union as for Wernher von Braun both in wartime Germany and later in the US. Each rapidly adapted their large rockets for flights to Earth orbit and beyond. The launch of Sputnik and the first Soviet launches to the Moon were made during the initial months of testing the R-7, the Soviet Union's first ICBM. Subsequently, various versions of the R-7 became standard launchers for both military and civilian Soviet space missions. The 'space race' in the 1960s between these two nations was essentially defined by the development of ever more powerful rockets on both sides. The first intercontinental rockets developed in the US were the Atlas and Titan, and both were used in the civilian program for manned and robotic missions. However, the giant Soviet N-1 and American Saturn V rockets were developed to land men on the Moon, and hence were far larger than required for military applications. Military rockets were modified by both nations to send spacecraft to the Moon and planets by adding upper stages for the extra boost required to achieve interplanetary velocities. Without these military rockets and the development of their associated upper stages, there would have been no access to space for interplanetary missions.

The history of rocketry in Russia can be traced back to their use by the military in the 13th Century – the same time that rockets made their appearance as a weapon in western Europe. A Rocket Enterprise was founded in Moscow in the 1780s, and in 1817 the Russian engineer Alexander Zasyadko wrote a manual on the production of

Figure 4.1 Early GIRD rocket and team in the 1930s.

rockets and their use for artillery bombardment. By the beginning of WW-I, Russia had developed the artillery rocket into a significant weapon with a range of almost 10 km. This development gained momentum after the Russian revolution in 1917, as the newly established Soviet Union became an industrial state with a large military force. The establishment of the Gas Dynamics Laboratory in Leningrad in 1928 for development of military missiles marked the beginning of the later powerful Soviet military rocket design bureaus.

The first consideration of the rocket for use other than as a military weapon was by the Russian visionary Konstantin Tsiolkovskiy, whose book '*The Exploration of the World's Space with Jet-Propulsion Instrument*' was published in 1903; the same year as the Wright brothers' first powered flight. Tsiolkovskiy, a schoolteacher, laid the theoretical foundation for space flight and interplanetary space travel using the rocket. In the 1930s, his work led a number of enthusiasts to found an organization called the Group of Research in Jet Propulsion (GIRD) whose first project was to construct a rocket-powered airplane. Sergey Korolev, the famed 'Chief Designer' of the Soviet space program in the 1960s, was a founding member. The government began to sponsor the organization in 1932, and the group launched both a hybrid engine rocket and a liquid-fueled rocket in 1933. They were merged with the Gas Dynamics Laboratory in September 1933 as the Jet Propulsion Scientific Research Institute (RNII).

Progress was slow and resources very limited for these amateur rocketry pioneers in the 1930s. At that time, no government was interested in supporting a program to develop peaceful exploration of space. Military applications were the only hope for obtaining state budgetary support, and this happened first and most successfully in Germany during WW-II.

THE COLD WAR RACE TO BUILD AN ICBM

At the end of WW-II the US and USSR each captured German rocket scientists, V-2 rockets, and rocket development equipment. The captured engineers and technology enabled both nations to vastly accelerate their own rocket development work. The V-2 was flown many times in Russia and America. This famous rocket became the springboard for initiating a race between the two post-war superpowers to be first to build an ICBM capable of dropping a nuclear warhead on the other side. Building rocket defenses to counteract US strategic bombers, in the 1950s the USSR initially appeared to have the edge in this competition, which led in turn to the perception in America of a 'missile gap'. This term was applied in several ways. Technically, it meant a gap in the range and 'throw weight' of a missile, but in some cases it was simply a measure of how many operational weapons each side had, and the Americans had an exaggerated view of the number of missiles actually pointed at them from inside the Soviet Union – a misperception that delighted the Soviet government.

Despite starting at the same point in the late 1940s with captured V-2 rockets, the Soviet and American development programs took different paths towards an ICBM. In the early 1950s, the US had substantial advantages both in electronics technology and the ability to construct smaller, high-yield weapons. The fact that Soviet atomic devices were much heavier led to more powerful rockets than those required by the Americans. The Soviets led in rocket mobility and deployment, in part because they assembled their rockets horizontally in production line fashion, and rolled them fully assembled on a railcar to the launching facility. The Americans built their rockets in sections and assembled them slowly on the pad by stacking them vertically one stage at a time.

The Soviet program culminated in the versatile R-7 two-stage rocket, which had its first successful test in August 1957. The kerosene and liquid oxygen propellants imposed a lengthy loading procedure prior to launch. The R-7 reached operational capability, but only five were ever deployed because by then the Soviets had learned how to build smaller warheads and were developing a more suitable missile. It was rapidly replaced by smaller rockets that could be placed in hardened silos and loaded with storable propellants. The R-7 survived to serve as a space launcher because of its large 'throw weight', the mass that it could launch, and its versatility to use upper stages for various military and civilian missions. It was called the 'Vostok' launcher after the Gagarin flight, and became the base vehicle for Soviet lunar and planetary missions until superseded by the larger Proton vehicle in the 1970s. It is still in use today as the core vehicle for the Soyuz family of Russian launch vehicles.

R-7 ICBMS AND SPUTNIK

The first ICBM built in the USSR was the R-7, affectionately named 'Semyorka' by its makers. It was designed and built by Sergey Korolev's design bureau, OKB-1, in great secrecy in the 1950s. Its multi-stage design was quite different than the scheme used in the US where the stages were stacked on top of one another. The R-7 used a 'packet' design in which identical propulsion units were clustered around a central core unit and dropped from the core after burnout. The core continued to burn as the second stage. This concept was suggested by Tsiolkovsky, and championed by M.K. Tikhonravov working at the Defense Scientific Research Institute (NII-4) starting in the late 1940s. Korolev adopted the idea at OKB-1, and in the early 1950s directed feasibility studies by Keldysh's Department at the Mathematics Institute (MIAN) of the Soviet Academy of Sciences to examine the utility of variants of the scheme. In 1952, work at these three institutes resulted in a preliminary design that evolved two years later into the definitive design of the R-7. The Soviet government approved construction of the R-7 on May 20, 1954, with the project designation 8K71.

The R-7 launcher had a central core propulsion unit 33 meters tall, including the warhead, with four identical strap-on booster propulsion units around the bottom 20 meters. Each strap-on unit was an integral propulsion stage with an RD-107 engine and its own tanks for kerosene and liquid oxygen. The central core was powered by a nearly identical RD-108 engine delivering somewhat less thrust but sustained over a longer time and optimized for high altitudes. Each RD-107 had a pair of gimbaled vernier engines for steering and trim, and the RD-108 had a set of four such engines. The main engines were built by Valentin Glushko's OKB-456, and each had a cluster of four combustion chambers fed by a single turbopump. All engines, boosters and core, operated for lift-off. The four boosters would burn for about 2 minutes before dropping off, leaving the central core as the second stage sustainer, which continued for several minutes until it had achieved the required velocity and altitude.

The first model of the R-7 was delivered in December 1956 and used for captive tests. The first flight model followed in March 1957. The first three launch attempts failed. On the first, on May 15, 1957, the flight was cut short after 103 seconds when a booster engine failed. The second vehicle was removed from the pad on June 11 after three aborted launch attempts. The third attempt on July 12 failed when the vehicle began to rotate rapidly and shed its boosters. The fourth attempt on August 21 was a qualified success, with the rocket delivering the payload along the desired trajectory, but the payload disintegrated during re-entry. A fifth test on September 7 led to the same result. In these latter tests, however, the rocket itself had performed satisfactorily.

Figure 4.2 R-7 test vehicle on the stand in the late 1950s (courtesy Energiya Corp).

From the very beginning of ICBM development in the late 1940s, Korolev clearly had in mind using his rocket to access space. He repeatedly lobbied the government for support and on January 30, 1956, in response to a letter to the Politburo signed by Korolev, Keldysh and Tikhonravov, a decree was passed for the development of an artificial satellite designated Object D and a special version of the R-7 to launch it. With Object D development proceeding rather slowly, Korolev, fearing that von Braun in America might place the world's first satellite into orbit, was eager to try to do so while his R-7 was undergoing its early test flights, and he decided to launch a very simple satellite, essentially a small sphere containing a radio transmitter, after the first successful test flights. Launch vehicle 8K71PS serial number M1-1PS (PS for 'Prosteishyi Sputnik', meaning provisional satellite) was modified by removing unnecessary warhead targeting equipment and test instrumentation, reprogramming the burn sequence, and replacing the dummy warhead with the satellite and shroud. It lifted off on October 4, 1957 and opened the 'space age' by placing Sputnik into a slightly lower orbit than planned after the sustainer shut down 1 second early.

R-7E AND THE EARLY LUNA PROBES

On March 8, 1957 Korolev's OKB-1 founded a new department to develop manned satellites and spacecraft for lunar exploration. It was headed by Tikhonravov, and in April he submitted the first plans. These required that the basic 8K71 R-7 rocket be fitted with a third stage, and by the summer of 1957 technical plans were completed. The third stage would be mounted on top of the sustainer using an open truss so that its engine could be started before the sustainer was shut down – a measure designed to prevent the cavitation in the propellant tanks that would occur in zero-G if ignition were to be delayed until after the core stage had shut down.

Work began on two different vacuum-performance engines for the third stage, one by Glushko's design bureau, OKB-456, and the other by OKB-1 itself, working with Semyon A. Kosberg's OKB-154 in Voronezh. The decision to build two versions of the third stage originated in a clash between Korolev and Glushko concerning the development of Glushko's engine, resulting in a bitter rivalry that persisted through the development of the ill-fated N-1 Moon rocket and contributed to the ultimate failure of the Soviet manned lunar program. Glushko wanted to develop a powerful 10-ton thrust engine using new hypergolic fuels, unsymmetrical dimethlyhydrazine (UDMH) and nitric acid, instead of the standard kerosene fuel used by Korolev and Kosberg's 5-ton thrust engine. However, Korolev was wedded to the LOX-kerosene combination and disliked Glushko's toxic fuel, calling it "the devil's own venom". He doubted that such a new engine could be developed in time for his schedule, which called for the first test of the three-stage rocket in June or July 1958, the launch of a lunar impact probe in August or September, and a flyby mission to photograph the far side of the Moon in October or November. He ordered the development of a 5-ton thrust engine at OKB-1 based on the R-7's verniers. He was aware of developments at Kosberg's aviation design bureau, which was working on a restartable LOX-kerosene engine using a new turbopump based on jet engine designs. To speed his own development at OKB-1, Korolev engaged Kosberg. As a neophyte in the rocket engine business, Kosberg initially demurred, but Korolev persuaded him to collaborate on an engine that could operate in vacuum. For his part, Glushko was not happy with this parallel work, especially since OKB-154 was outside the circle of space developers. Glushko felt that he was due deference from Korolev, and he considered Korolev's overtures to Kosberg an insult. But Korolev's instincts proved correct. Glushko was struggling with problems when Kosberg's engine became available for use in August 1958. It used a higher density kerosene to yield the needed thrust levels. The development of a third stage based on Glushko's engine was finally canceled in 1959.

Korolev's ambitious schedule had to take second priority to developmental tests required to make the basic R-7 an operational ICBM. During the first half of 1958, his lunar plans were constantly threatened by difficulties with numerous changes to the engines and failures in development flight tests. A prototype of the lunar rocket with a dummy third stage equipped with avionics and telemetry, but no propulsion, was launched on July 10, 1958, powered by an improved set of booster and sustainer engines but these failed a few seconds into the flight, bringing down both the rocket and the timetable. Korolev shot for the Moon at the earliest opportunity on the very first flight of the new third stage on September 23, 1958. This and a second attempt on October 12 failed when the boosters fell apart and all flights had to be suspended until the problem was analyzed and fixed. Frustratingly, the cause turned out to be longitudinal vibrations in the strap-on boosters caused by the addition of the third stage. With this problem fixed, a third attempt failed on December 4, 1958, when the second stage engine shut down prematurely. On January 2, 1959, the rocket worked properly and although Luna 1 did not impact the Moon, the 6,000 km flyby was a sufficiently impressive achievement for the Soviets to declare that this had been the objective.

This R-7E, which was an 8K71 with a Block E third stage, was designated 8K72 and informally known as the 'Luna' launcher. It could put 6 tons into low Earth orbit and send 1.5 tons into deep space. It was used exclusively for the first generation of Luna probes in 1958–1960 including the successful Luna 1, 2 and 3. It came to an ignominious end with an explosive failure less than one second into the flight of the final such probe on April 19, 1960. The booster and core stages were upgraded and the third stage improved, including upgrading its engine, to produce the 8K72K three-stage heavy lift orbital version of the R-7. This was used to launch the manned Vostok orbital spacecraft and the first Soviet photoreconnaissance satellites, known as the Zenit 2 series.

R-7M: THE 'MOLNIYA' LUNAR AND PLANETARY LAUNCH VEHICLE

In early 1958 Korolev began planning for planetary missions. His original intention was to use the 8K73, a version of the 8K72 with a more capable third stage. During that summer OKB-1 began work on spacecraft for launch to Venus in June 1959 and Mars in September 1960. However, the 8K73 project and the 1959 Venus mission were abandoned when Glushko's engines for the new third stage had development problems. Korolev turned to Kosberg again and decided to adapt the second stage of the new silo-based ICBM under development, the R-9A, also known as the 8K75, as the third stage for his planetary launcher. Kosberg fitted the stage with larger tanks to sustain the longer engine burn times. In the meantime, the R-7 was still in its final development phases in preparation for operational deployment.

During 1958 an improved version of the basic two-stage R-7, the R-7A or 8K74, was being developed for easier operational servicing and greater performance. The 8K74 had all-inertial guidance with the original radio guidance system retained only as a backup, improved engines for reliability, redesigned verniers for simpler control and increased performance, a new ignition system, and some portions of the engines were moved nearer to service hatches. The first launch of the 8K74 on December 24, 1959, was a success. The 8K74 became the basic two-stage booster for generations of launchers to the present day. The only 8K71-based vehicles used after this 8K74 test were two 8K72 Luna launches on April 12 and 18, 1960, both of which failed.

In all this rush of rocket development in 1958–59, Keldysh's mathematicians had determined that continuous burn of all stages was an inefficient use of energy to reach interplanetary velocities. Continuous burn also required precise timing without margin for launch delays. Instead they recommended a scheme in which the booster placed an escape stage into low Earth orbit. This would be ignited when the orbital phasing was optimum for launch towards the Moon or planetary target, and once on course it would release its payload.

Abandoning the three-stage approach for lunar and planetary launches, in early 1959 Korolev began work on a four-stage approach. The airframe of the 8K74 core vehicle was strengthened to support the mass of the new upper stages, modifications were made to the operating pressures and burn programs to increase the thrust of the core vehicle, a stronger open truss was provided between the sustainer and third stage, and new guidance and control systems were supplied for the upper stages. The Kosberg third stage was modified further with an increased propellant load and an upgraded 8D715K four-chamber engine and designated Block I. The 8K74/III two-stage core vehicle with the new Block I third stage and a first burn by a new fourth stage, Block L, built by OKB-1, would put the Block L and spacecraft combination in Earth orbit. The Block L was made restartable, so that its second burn would put the spacecraft on an interplanetary trajectory. It would be capable of sending 1,600 kg to the Moon or 1,200 kg to either Venus or Mars. This four-stage 8K78 is known as the 'Molniya' launcher. A prototype with a dummy fourth stage was successfully tested on January 20, 1960, with a second successful test 10 days later. The Block L completed its ground tests in the summer of 1960, and Korolev rushed preparations for three Mars launches on the first tests of this new launcher. The spacecraft were also built in a great rush before the launch window closed in mid-October. Only two rockets made it to the launch pad on time.

The first flight test of the complete 8K78 occurred on October 10, 1960, with a 1M Mars spacecraft at the top of the stack. The spacecraft had to be stripped down in order to provide sufficient mass for rocket test instrumentation. The launch failed when resonant vibrations in the upper stages damaged the avionics during third stage burn and the rocket veered off course. A second attempt on October 14 also failed when the third stage engine did not ignite because a LOX leak on the pad had frozen kerosene in the fuel lines. The first test of the Block L did not come until the third flight on February 4, 1961, which attempted to launch a 1VA Venus spacecraft. The first three stages performed perfectly, but the Block L was stranded in Earth orbit by a primary power failure.

The Block L stage was a challenging design because it had to coast unpowered for

Figure 4.3 Launch vehicles for robotic spacecraft in 1961. From left, US Viking, US Jupiter-C, US Atlas-Agena, USSR 8K-72 Luna, USSR 8K-78 Molniya (from Peter Gorin in Siddiqi 2000).

almost 2 hours in Earth orbit without losing volatile propellant, orient itself to the proper firing attitude at a programmed time, and ignite its engine in a zero-G state. The engine used a more efficient 'closed-cycle' technology which US rocket makers deemed unworkable, and used gimbals for yaw and pitch control and a pair of small verniers for roll control. The stage used a cold gas attitude control system and solid rockets to provide ullage control before engine ignition in zero-G. The challenges of perfecting this planetary injection stage proved difficult. The Block L succeeded on its second opportunity on February 12, 1960, deploying Venera 1. But it failed many times thereafter, including the final planetary mission to use the Molniya launcher on March 31, 1972, when the Block L stage failed to put another spacecraft intended for Venus on an escape trajectory.

In 1962 an extended shroud was introduced to accommodate the next generation

Figure 4.4 R-7 vehicle on rail carrier.

2MV Mars and Venus spacecraft and the sustainer engines were upgraded. In 1964 a new version of the 8K78 was introduced, with improved versions of Glushko's RD-107/8 engines and an improved engine in the Block L fourth stage. This vehicle was designated 8K78M, and was known in the West variously as SL-6 and A-2-e. It was first used for the test launch of a 3MV Venera spacecraft on March 19, 1964, then used consistently for Venus missions until the introduction of the Proton-launched spacecraft in 1975. Mars launches used the 8K78 until switching to the Proton in 1969. The Ye-6 Luna probes used both the 8K78 and 8K78M vehicles until Luna 9 on January 31, 1966, when the 8K78M came into exclusive use. A variant of this vehicle was created for the Luna soft landing missions, in which the avionics were deleted from the upper stages to save mass and the Ye-6 spacecraft controlled the functioning of the third and fourth stages. This vehicle was designated with a /Ye-6 suffix. The 8K78 vehicle was completely replaced by the 8K78M after December 1965.

The 8K78M received another upgrade in 1966–67 when the core and strap-ons were replaced by those of the three-stage 'Soyuz' version used in the manned space program. In 1965, responsibility for the Block L stage was transferred from OKB-1 to NPO-Lavochkin, which introduced improvements in 1968 including upgraded avionics and a new third stage interface and fairing design. Lavochkin produced two versions of this new Block L, one for lunar and planetary missions and the other to place 'Molniya' communications satellites into highly elliptical Earth orbits. Further improvements to the 8K78M were made in 1974 and again in 1980. In its various forms the Molniya launcher was the workhorse for the lunar and planetary program in the 1960s and early 1970s, successfully deploying the Luna 4 to 14 missions from 1963–1968, Mars missions from 1960–1965 including Mars 1, and Venera 1 to 8 from 1961–1972. The versatility of the 'Semyorka' rocket is demonstrated by its continued

Figure 4.5 Molniya launch.

use up to the present day, particularly in its three-stage 'Soyuz' variant. It resumed its utility for planetary launches on June 2, 2003, with the successful launch of the Mars Express spacecraft for the European Space Agency using a Soyuz fitted with the new Fregat fourth stage.

THE PROTON LAUNCHER

The UR-500 Proton launcher was initially developed as an ICBM to carry heavier warheads over longer ranges than the R-7. By 1961 Vladimir Chelomey's OKB-52 had developed a practical storable propellant fast-response ICBM for the military and Korolev's R-7 had become a space launcher, so Khrushchev naturally enlisted Chelomey to build the larger rocket to deliver the new H-bomb. Chelomey's answer was the Universal Rocket 500, or UR-500. However, as with fission devices, the Soviets soon learned how to make much lighter thermonuclear devices and the UR-500 was canceled by the military. Chelomey convinced the government that his rocket, augmented with an upper stage, could send cosmonauts on direct flights to the Moon for circumlunar missions. At that time, Korolev was envisaging achieving this by two launches and an Earth orbital rendezvous. Chelomey's scheme would be simpler. He succeeded in wresting the circumlunar project from OKB-1 and kept the UR-500 program alive with support from Keldysh, who wisely recognized that such a booster would have many important applications. In 1965 Korolev succeeded in regaining the spacecraft and fourth stage combination for the circumlunar project, his reasoning being that Chelomey had never built a spacecraft and OKB-1 already had one in production. The mission would use Chelomey's booster but with the fifth stage from Korolev's N-1 rocket serving as its fourth stage. For use on the Proton stack, the N-1's Block D guidance package was removed and this function had to be provided by the spacecraft.

The three stages of the UR-500 were all powered by engines that burned nitrogen tetroxide and UDMH, a combination despised by Korolev. The first stage had six highly advanced and very efficient closed-cycle RD-253 engines made by Korolev's nemesis, Valentin Glushko in OKB-456. The second and third stages were powered by engines built by Kosberg's OKB-154. A feature of the first stage's design is that the UDMH is contained in six tanks arranged around the larger central oxidizer tank. This unique design was imposed by width limitations of the railway system, which precluded vehicle widths greater than 4.1 meters. The various tanks were therefore delivered separately and assembled at the launch site. A specialised rail transporter then took the completed vehicle to the pad.

The UR-500 began flying only four years after being commissioned, and showed immediate promise with a two-stage version on July 16, 1965, successfully placing into orbit a very heavy Proton satellite to study cosmic rays; hence its popular name. The fifth launch on March 10, 1967, was the first for the four-stage UR-500K, also known as the Proton-K. This had Korolev's restartable Block D upper stage, which used his preferred propellants of kerosene and LOX. The payload was the first in the

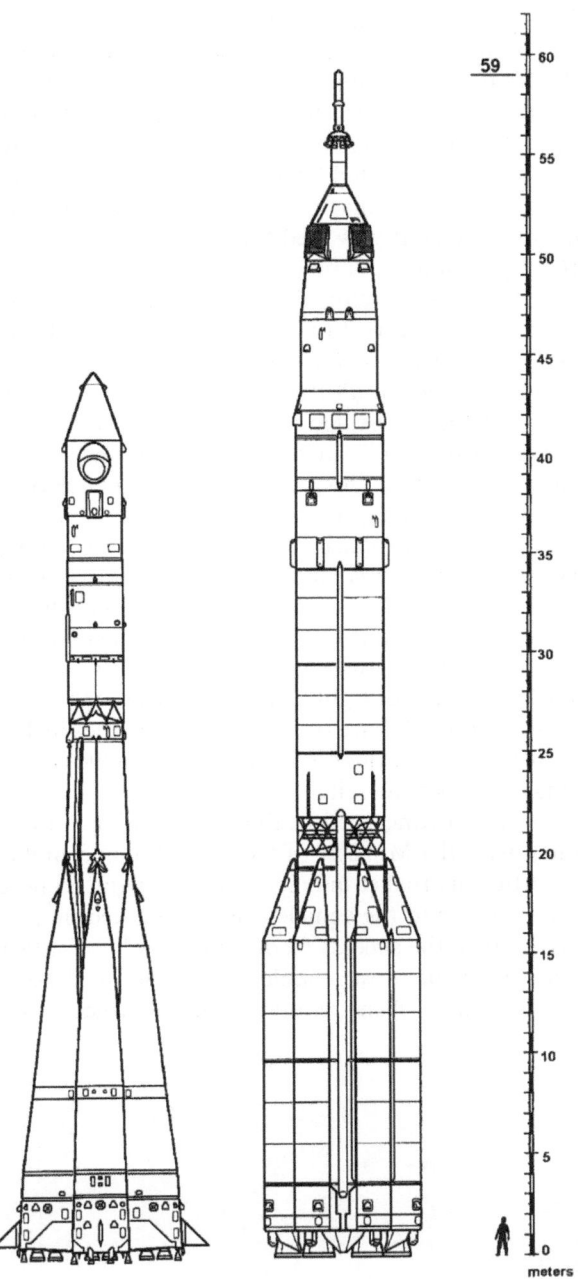

Figure 4.6 Soyuz (similar to Molniya) and Proton-K Zond launchers to the same scale (from Peter Gorin in Siddiqi 2000).

series of tests of the lunar Soyuz, disguised by the name Zond, and was considered a success. The power of the Proton-K was irresistible for lunar and planetary missions, and it replaced the Molniya for lunar and Mars missions in 1969 and Venus missions in 1975. It became the workhorse for lunar and planetary missions in the 1970s and continued into the 1990s. Being much more powerful than the Molniya it facilitated much heavier and more sophisticated lunar and planetary spacecraft. Its lift capacity made possible such missions as lunar rovers, lunar sample returns, and soft landing missions on Mars and Venus. It was used for Zond 4 to 8, Luna 15 to 24, Mars 2 to 7, Venera 9 to 16, Vega 1 and 2, Phobos 1 and 2, and Mars-96. Several versions of the Block D fourth stage were developed for the Proton-K. The original was used for Luna 15 to 23, Zond 4 to 8, Mars 2 to 7 and Venera 9 and 10; the D-1 version was used for Luna 24, Venera 11 to 16 and Vega 1 and 2; and the D-2 was employed for Phobos 1 and 2 and Mars-96. In all these missions, the spacecraft was required to supply guidance for the Block D stage.

One of the glaring reasons for so many failures in Luna, Zond, Venera and Mars missions in the late 60s and early 70s was poor performance of the Proton vehicle. Succeeding in its initial launch and in two of its next three launches in 1965–66, its initial performance appeared promising. But its record in the 3 years from March 1967 to February 1970 was abysmal. Ten of nineteen spacecraft were lost when the Proton failed to deliver the Block D to Earth orbit. Another three achieved orbit but were stranded when the second burn of the Block D failed. Only six of the nineteen launches were fully successful. Sixteen were interplanetary, and the Proton failed in eleven cases – four failures out of eight Zond launches to the Moon, five failures out of six Luna launches, and the failure of both Mars launches in 1969. Unfortunately, the failures were distributed throughout the vehicle including all stages, so it was very difficult to make the vehicle reliable.

NPO-Lavochkin was so concerned at the Proton failures that General Designer Georgi Babakin met with the Minister of General Machine Building in March 1970 to demand action. After the rocket underwent a full engineering review a number of improvements were made, and the vehicle was re-qualified in a successful test flight in August 1970. After this, the success record improved dramatically and eventually the Proton became one of the most reliable workhorses in the Soviet launcher fleet. Indeed, it today enjoys an excellent reputation and a large share of the commercial launch market.

Figure 4.7 Transport and erection of the Mars-96 Proton-K vehicle.

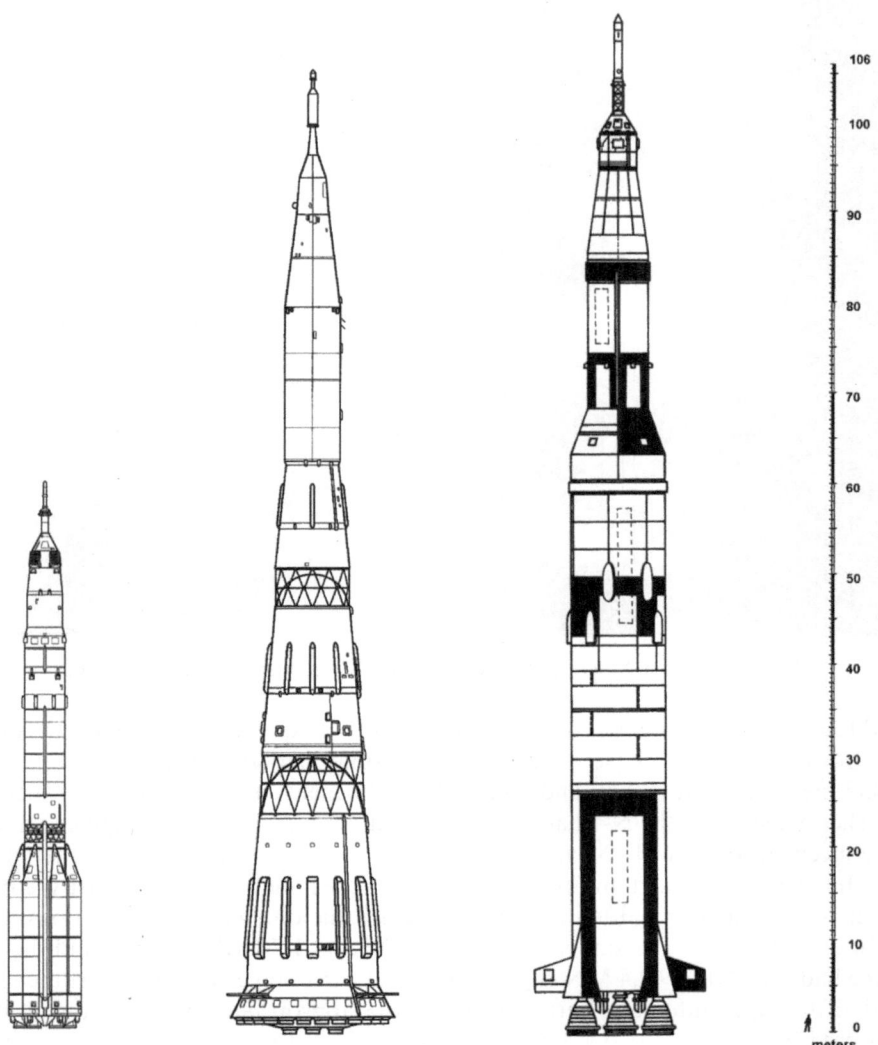

Figure 4.8 N-1 rocket compared to the Proton Zond launcher and the US Saturn V (from Peter Gorin in Siddiqi 2000).

N-1 MOON ROCKET

The N-1 launcher was the Soviet counterpart to the Saturn V, and was developed for the same role. It had five stages, stood 105 meters tall, weighed 3,025 metric tons at launch and could place 95 metric tons in low Earth orbit. In contrast, the Saturn V

Figure 4.9 First N-1 test vehicle in assembly (left) and on rollout (right).

had four stages (treating the Apollo lunar module as equivalent to N-1's fifth stage in the powered descent phase of the mission profile), stood 110 meters tall, weighed 3,039 metric tons at launch and could place 119 metric tons in low Earth orbit.

The first stage of the N-1 had 30 NK-33 non-gimbaled 1.51 MN engines arranged in concentric rings with 24 around the outer ring and 6 around the inner. Large graphite vanes mounted on four of the outer ring engines provided thrust vectoring. If one of the engines malfunctioned, both it and the one diametrically opposite had to be shut down. The second stage had eight NK-43 1.76 MN engines, and the third stage had four NK-39 0.4 MN engines. The first three stages were to insert the fourth and fifth stages into low Earth orbit. At the appropriate time the fourth stage, powered by four NK-31 0.4 MN engines, would send the fifth stage, incorporating the orbiter/lander, towards the Moon. The 'NK' engines were made by OKB-276, headed by Nikolai Dmitriyevich Kuznetsov, and used a LOX-kerosene combination. The fifth stage was the Block D which Korolev adapted to serve as the fourth stage of the Proton-K. It was powered by a single Melnikov RD-58 engine that also used LOX-kerosene, and was to perform midcourse maneuvers, lunar orbit insertion and the majority of the powered descent, being discarded in the final phase to enable the lander to use its own engine to perform the soft landing.

The N-1 failed test flights in February and July of 1969, in the latter case with a spectacular explosion at liftoff that dashed Soviet hopes of competing with the US Apollo program. The N-1 had another test flight in 1971 and a final test in 1972. The

Figure 4.10 N-1 on the pad just prior to launch.

first stage failed each time and the project was abandoned. Its Achilles' heel was the large number of engines that all had to work without adversely affecting the others. Remarkably, there were no static test firings. The launch attempts in 1969 carried an automated form of the Soyuz 7K-L1 circumlunar spacecraft and a dummy LK lunar lander. The launches in 1972 carried an automated Soyuz 7K-LOK lunar orbiter and dummy LK lunar lander. The only successful result was proof that the escape rocket could pull the crew module clear of the exploding rocket.

5

Spacecraft

LUNAR SPACECRAFT

Russian lunar spacecraft can be divided into families according to their evolution from the very first simple flyby and impactor spacecraft in 1958–1960, exemplified by Luna 1 to 3, to the first modular designs built for soft-landing culminating with Luna 9 and 13, to the final series of complex sample return and lunar rover missions beginning with Luna 15 and continuing through to Luna 24. After the first success at soft-landing, some of these spacecraft were modified to carry lunar orbital payloads, in particular to perform tasks in support of an eventual manned lunar landing. These modifications were easily and quickly accomplished because of the modular design. The final series were essentially large soft-landers with interchangeable payloads. Although they were complex, they achieved the first robotic sample return missions and first lunar rovers in addition to a pair of orbiters.

Luna Ye-1 series, 1958–1959

In the summer of 1958, the Americans and the Russians were racing to launch the first spacecraft to the Moon as a major signal of strength in rocket technology. The spacecraft were small and lightly instrumented and were flown opportunistically on what were mainly test flights of military rockets. The goals were more technological and political than scientific.

The Americans tried eight times to reach the Moon without success in 1958–1960. Only one spacecraft, Pioneer 4, was launched successfully to Earth escape velocity, but it missed the Moon by a wide margin.

To counter the American lunar campaign, the Soviet Union built the Ye-1 lunar impactor spacecraft for launch on a new three-stage Luna rocket derived from the R-7 that launched Sputnik. The Ye-1 was a very simple spherical payload similar to Sputnik, spin-stabilized, with several protruding antennas. Six such spacecraft were launched during the 12 months between September 1958 and 1959. All but two were

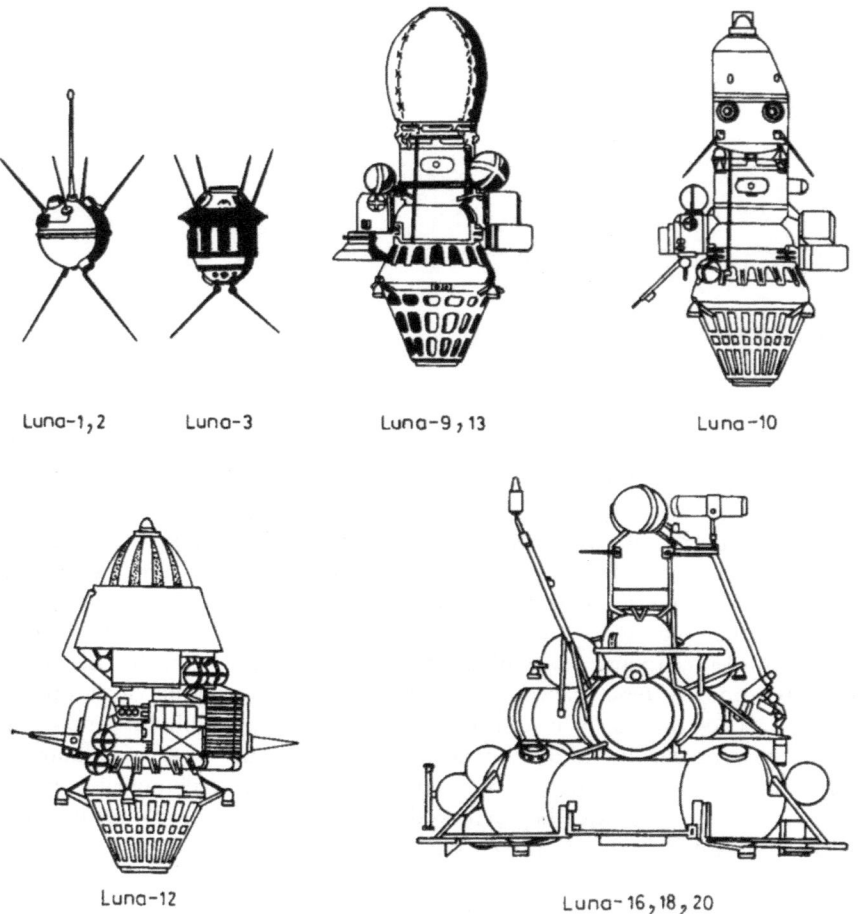

Luna-1,2   Luna-3   Luna-9,13   Luna-10

Luna-12   Luna-16,18,20

Figure 5.1 Examples to scale from the Luna series of spacecraft: Luna 1 and 2 Ye-1 impactor spacecraft; Luna 3 Ye-2 flyby spacecraft; Luna 9 and 13 Ye-6 soft-lander spacecraft; Luna 10 Ye-6S orbiter spacecraft; Luna 12 Ye-6LF orbiter spacecraft; and Luna 16, 18 and 20 Ye-8-5 sample return spacecraft in landed configuration without in-flight drop tanks (from *Space Travel Encyclopedia*).

lost to launch vehicle mishaps, but owing to the Soviet way of not naming a mission until it was successfully underway, these were Luna 1 and Luna 2. Although Luna 1 missed the Moon on January 4, 1959, it was the first spacecraft to achieve escape velocity – two months before Pioneer 4. The final spacecraft to be launched in this series, Luna 2, became the first spacecraft to impact the Moon on September 14, 1959. In effect, the Soviets had kept launching until their goal was achieved, and then they moved on.

Luna Ye-2 and Ye-3 series, 1959–1960

These series were the second generation of simple, single-module lunar spacecraft, designed for a more complex payload and flight mission. Instead of a direct flight to impact the Moon, they were placed into a highly elliptical orbit that would take them beyond the far side of the Moon, which they would photograph, and upon returning to the vicinity of Earth they would scan and transmit the pictures. The Ye-2 was the first three-axis stabilized spacecraft. It flew to the Moon in spin-stabilized mode, and then switched to three-axis stabilization and control for lunar photography. The Ye-3 had a modified attitude control system and an improved camera. One Ye-2 and two Ye-3 spacecraft were launched in the six-month period between the start of October 1959 and the end of April 1960 using the Luna launcher. Only the Ye-2 spacecraft, Luna 3, was successful. Both Ye-3 spacecraft were lost to launch vehicle failures.

Luna Ye-6 (OKB-1) series, 1963–1965

The Ye-4 impactor and Ye-5 orbiter designs were made obsolete by the Ye-6. It was a modular spacecraft with a carrier spacecraft on which could be mounted either a soft lander or an orbiter module. The first Ye-6 series were built at OKB-1 for soft landings on the Moon but managed not a single success after eleven straight launch attempts over three years between January 1963 and December 1965.

Luna Ye-6M (Lavochkin) series, 1966–1968

Responsibility for robotic lunar and planetary spacecraft design and construction was transferred from OKB-1 to NPO-Lavochkin in 1965. Lavochkin introduced its own modifications, and was immediately rewarded with Luna 9, which became the first successful lunar lander on February 3, 1966. Lavochkin also produced several orbiter versions, and over the following 14 months achieved a record of six mission successes out of nine Ye-6M launches. Both Ye-6M landers were successful, Luna 9 and Luna 13. After one failed launch the first model of the Ye-6S yielded the first lunar orbiter, Luna 10, which had instruments to measure the particles and fields in the lunar environment. It was then modified to acquire orbital photography. Both of these Ye-6LF missions, Luna 11 and Luna 12, were successful although no useful imagery was acquired in the first case. Finally, after another modification produced the Ye-6LS, two failed launches were followed by the Luna 14 orbiter.

Luna Ye-8 series, 1969–1976

The Ye-8 series were much heavier and more complex, and were launched on the powerful new Proton rocket. The design centered on four large spherical propellant

Figure 5.2 Luna 17 Ye-8 lander with Lunokhod 1 aboard.

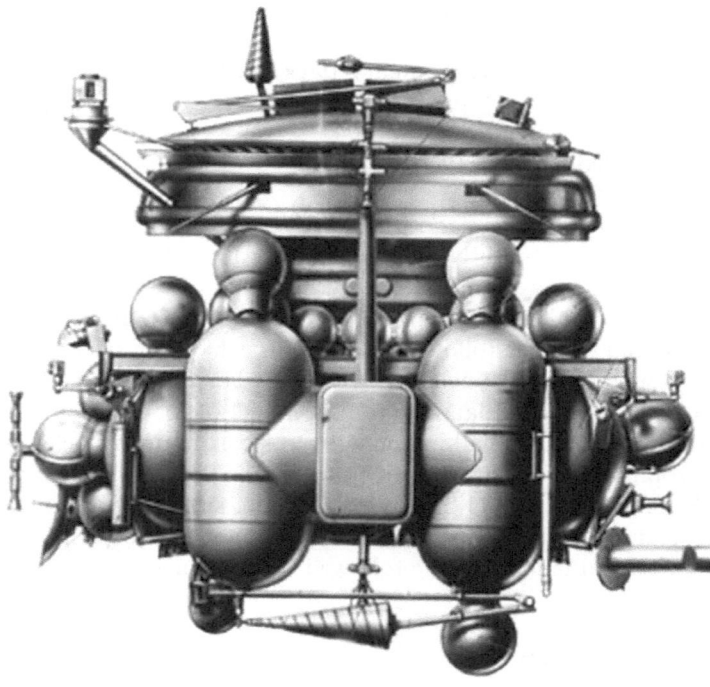

Figure 5.3 Luna 19 Ye-8LS orbiter in flight configuration including external drop tanks (courtesy NPO-Lavochkin).

tanks connected in a square using cylindrical inter-tank sections. The landing system and engine were mounted on the underside of this assembly and the lander payload on the upper side.

The principal goals of these spacecraft were first to deploy a lunar rover on the surface (the Ye-8 model) and second to return samples of the lunar surface to Earth (the Ye-8-5 model). Three Ye-8 lunar lander/rover spacecraft were launched, two of which, Luna 17 and Luna 21, were successful. Of a total of eleven Ye-8-5 sample return spacecraft launched, only three were successful, Luna 16, 20 and 24. In fact, Luna 20 and 24 were advanced Ye-8-5M models. Two additional Ye-8 models were modified as Ye-8LS lunar orbiters and both flown successfully as Luna 19 and 22. The overall record for the Ye-8 was therefore seven successes of sixteen attempts.

Lunar Soyuz (Zond), 1967–1970

As early as 1959 the Soviets had a plan for manned circumlunar flights. When the Americans decided in mid-1961 to go to the Moon, Korolev was already designing the Soyuz spacecraft for these missions. It was the same three-module arrangement with which we are all familiar, with a support module containing all the resources required for power, propulsion, communication, navigation and consumables for the cosmonauts, a descent module to carry them aloft and to return them to Earth, and a compartment to provide more room for the cosmonauts on long flights. After the Vostok and Voskhod manned capsules, this system was introduced and remains the reliable Russian system still in use today.

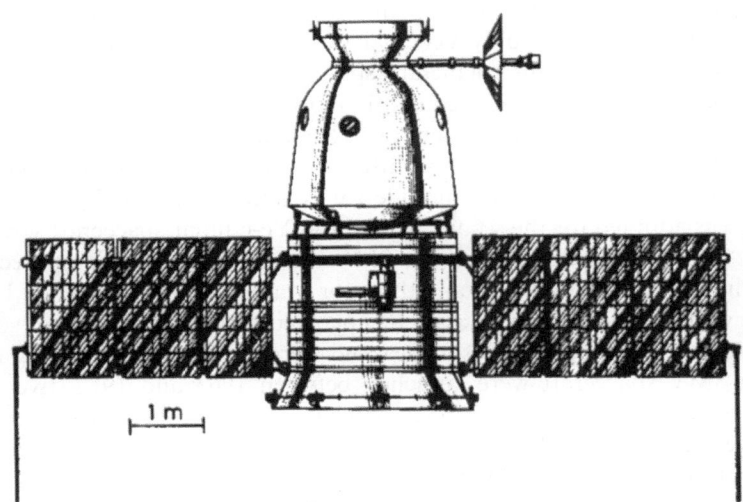

Figure 5.4 Soyuz 7K-L1 'Zond' circumlunar spacecraft (from *Space Travel Encyclopedia*).

The Soyuz 7K-L1 was a version of the 7K-LOK lunar orbital spacecraft modified to perform a circumlunar mission. Although the three-stage R-7 used for Soyuz flights in Earth orbit was replaced by the more powerful four-stage Proton, mass limitations meant that the 7K-L1 did not have the 'orbital' module and was designed to carry only two cosmonauts. The idea was to fly circumlunar missions with two astronauts using the 7K-L1 as a precursor to performing a lunar landing using the 7K-LOK version of the Soyuz (which would have an orbital module) and the LK lunar lander, all launched by the massive N-1 rocket. To prepare for the manned circumlunar missions, several automated flights of the 7K-L1 were conducted, the first two in Earth orbit and then nine others over the years 1967–1970 either to lunar distance or performing actual circumlunar flights. Zond 4 reached lunar distance before returning to Earth, but in a direction away from the Moon in order to simplify navigation, and Zond 5 to 8 each made circumlunar flights. Zond 4 self-destructed on re-entry, Zond 5 had significant but non-fatal problems with on board systems, and Zond 6 crashed on landing only a few weeks before the Apollo 8 mission. Although Zond 7 and Zond 8 were complete successes, the Soviets never used the system for a manned circumlunar flight.

PLANETARY SPACECRAFT

There were essentially three general design series of Russian planetary spacecraft. None of them resembled their American counterparts because, unlike the latter, the Russian spacecraft required pressurized containers for most of their electronics. The Venus and Mars flights in 1960–61 used the first generation spacecraft, which were simple pressurized canisters with attached solar panels and high gain antennas. Their payloads were specific to the target planet, but in general the spacecraft were the same. Of the four launched, only Venera 1 was successfully dispatched and it failed early in its cruise through interplanetary space.

The second generation introduced the first modular spacecraft, with a pressurized carrier that had the propulsion system at one end and a module for the payload at the other. They were individually outfitted for missions to Mars or Venus, with either an entry probe or a flyby module for remote sensing. (This same modular approach was adopted for the second generation Ye-6 lunar spacecraft series.) There were two sub-types of this spacecraft, 2MV and 3MV. Six 2MV spacecraft were launched in 1962, three for Venus and three for Mars, but only one, Mars 1, survived its launch vehicle. The flight of Mars 1 was plagued with problems and it succumbed half way to its target, but the lessons learned were applied in developing the 3MV. Seventeen 3MV spacecraft were launched between 1963 and 1972, five of which, Venera 4 to 8, achieved their planetary objectives. One of the Mars types, Zond 3, did achieve significant results by imaging the far side of the Moon as it departed and subsequently testing the communications system by transmitting the pictures from deep space.

The third generation planetary spacecraft were a major design change, enabled by the powerful Proton launcher with the Block D upper stage. These much larger and

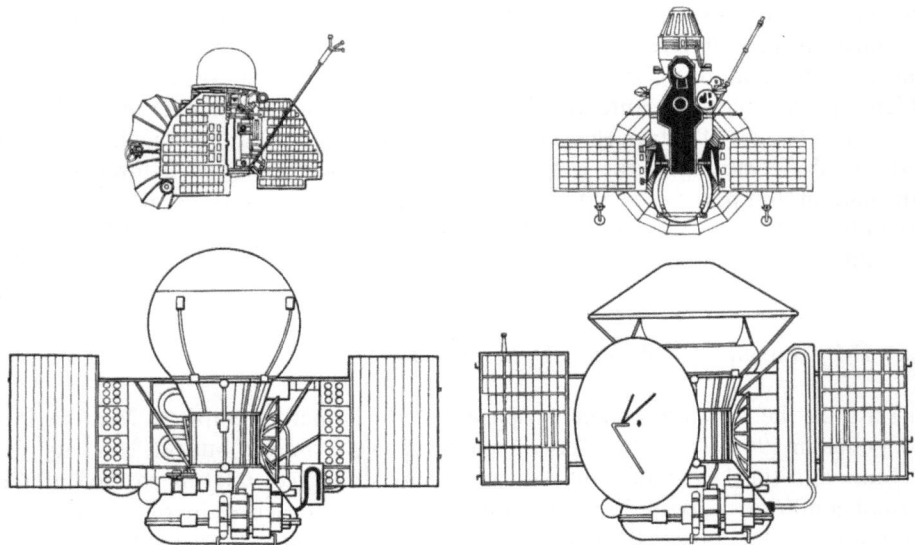

Figure 5.5 Representative Soviet planetary spacecraft to scale: first generation Venera 1 (upper left); second generation Venera 4 to 8 (upper right); and third generation Venera 9 to 14 at lower left; and Mars 2, 3, 6 and 7 at lower right (from *Space Travel Encyclopedia*).

more complex spacecraft were meant to provide planetary orbiters and soft-landers, starting with Mars in 1969 and Venus in 1975. Of twenty-two launched, Venera 9 to 16 and Vega 1 and 2 ran up a string of straight successes at Venus. The other twelve experienced a more difficult challenge at Mars, where only five can be deemed even partial successes, Mars 2 and 3, Mars 5 and 6, and Phobos 2. The Phobos missions of 1988 and the Mars-96 spacecraft were derivatives of this class, but with upgrades sufficiently significant for them perhaps to be regarded as another generation.

In normal flight, Russian spacecraft were flown in uniaxial orientation in which their static solar panels were oriented constantly towards the Sun and the craft spun at 6 revolutions per hour on the axis perpendicular to the plane of the solar panels. The command uplink was at 768.6 MHz through semi-directional conically-shaped spiral antennas which were also used for low-rate data transmission. Because these antennas generate funnel-shaped radiation patterns, several were placed around the spacecraft pointing at the Sun, and at any point in the mission the one with the best funnel angle for Earth was used. For high data rate transmissions, a parabolic high-gain antenna was affixed to the spacecraft. This had to be aimed directly at Earth by disabling the uniaxial control mode, reorienting the spacecraft appropriately, and switching to the three-axis orientation control mode. Circularly polarized decimeter ($\sim$920 MHz) and centimeter ($\sim$5.8 GHz) band transmitters shared the dish antenna.

In 2MV and 3MV missions, planetary probes and landers were designed for direct transmission to Earth by small spiral antennas with pear-shaped radiation patterns.

The heavier Proton-launched Mars and Venera landers were designed to relay their transmission through flyby or orbiter spacecraft using large meter band (186 MHz) helical antennas mounted on the rear of the solar panels. The Mars 3 class of entry vehicle carried small wire antennas on the entry stage and another set on the lander. The Venera 9 class of entry vehicle had another large helical antenna installed on top of the lander. Data from the Mars and Venus entry systems was stored for later transmission, but in the case of the Venus landers it was also relayed in real-time as a precaution. The entry system data link operated at 72,000 bits/s for Mars and at 6,144 bits/s for Venus.

Mars 1M (Marsnik-1) and Venus 1VA, 1960–1961

Russia built the first interplanetary spacecraft for launch attempts at Mars in 1960 and Venus in 1961. These two sets of spacecraft were similar, but the 1M pair built for Mars and the 1VA pair built for Venus had differences relating to the different thermal conditions expected and the communication differences involved. Each pair was identical. Only one of these spacecraft, Venera 1, survived the launch vehicle and was dispatched towards its target, but contact was lost 7 days later.

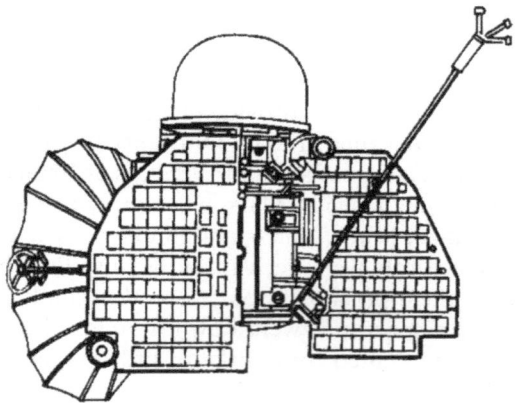

Figure 5.6 The 1VA Venera 1 spacecraft (from *Space Travel Encyclopedia*).

Mars/Venus 2MV series, 1962

After the failures of the 1M and 1VA missions, a new multi-mission spacecraft was designed for missions to Mars and Venus. The 2MV modular spacecraft had a mass of approximately 1,000 kg. The core of the spacecraft was a cylindrical pressurized 'orbital' module that had the propulsion module attached at one end and the payload at the other end. The payload could consist of either an entry probe or a pressurized module with instruments for flyby observations. Solar panels, antennas, thermal control devices, navigational sensors and several science instruments were attached to the side of the main module. The communications, attitude control and thermal control systems for the 2MV were greatly improved over the 1M and 1VA, and the same propulsion system was provided for midcourse trajectory corrections.

Four variants were designed. The 2MV-1 and 2MV-2 were Venus models, and the 2MV-3 and 2MV-4 were Mars models. The -1 and -3 versions were equipped with appropriate entry vehicles, and the -2 and -4 carried instruments for a flyby mission. Six spacecraft were launched, three to Venus (two probes and one flyby) in August and September 1962, and three to Mars (two flybys and one probe) in October and November 1962. Unfortunately all but one was lost to launch vehicle failures. The Mars 1 spacecraft launched on November 1, 1962, flew for almost 5 months before communications were lost on March 21, 1963, at what was then regarded as the vast range of 106 million kilometers from Earth. In view of this engineering success, the 2MV general design set a long-term precedent for Russian planetary spacecraft, particularly for Venus where this type of spacecraft was used until 1975.

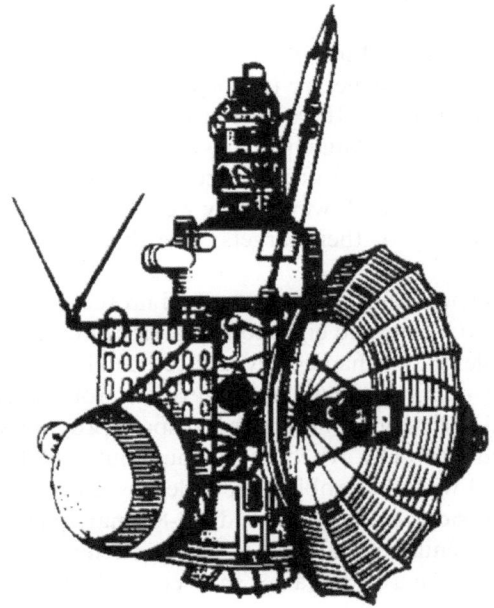

Figure 5.7 The 2MV Mars 1 flyby spacecraft (from *Space Travel Encyclopedia*).

Mars/Venus 3MV series, 1963–1972

As with the first campaign to Mars and Venus in 1960–61, the second campaign to both planets in 1962 failed. Of the ten spacecraft launched, only two survived launch and neither of those completed its cruise in a functioning state. But the long flight of Mars 1 was very encouraging, and the 2MV design was upgraded with new avionics to make the 3MV spacecraft for the 1964 launch opportunities to Venus and Mars.

Six 3MV launches were planned for the 1964 campaign, three each for Mars and Venus, but only five came off. In view of the high rate of previous failures, the first launch in each set of three was to test the 3MV/launch vehicle system. However, both the Mars vehicle test flight in November 1963 and the Venus vehicle test flight in February 1964 were lost to launch vehicle failures. Although there was little time left before the launch window to Venus opened in March 1964, it proved possible to launch the two spacecraft in late March and early April. The first mission was lost to a launch vehicle failure, and the second, designated Zond 1, failed 2 months into the cruise when pressurization was lost. A single 3MV was successfully launched to Mars in November 1964. Designated Zond 2, it failed in transit after 1 month, in this case because of avionics problems. Both of these missions were given the designation 'Zond' because it was realized shortly after launch that neither would be able reach its target in a functioning state.

The 3MV spacecraft that missed its Mars launch window in November 1964 was launched as a test spacecraft in July 1965. It conducted a successful flyby of the Moon as Zond 3, but failed its planetary test objectives when communications were lost before reaching Mars distance. It was the last 3MV launched to Mars. Later in November 1965 three more were launched to Venus. The first, Venera 2, was lost only 17 days before Venus encounter and the second, Venera 3, was lost just as the spacecraft approached the planet. However, they were both the first Soviet planetary missions to reach the vicinity of their targets. The third spacecraft was lost to a launch vehicle failure.

By March 1966 the Soviet planetary program had no success to show for nineteen launch attempts, eleven to Venus and eight to Mars, since the start of the program in October 1960. Meanwhile, the US had achieved successful flyby missions of Venus in 1962 and Mars in 1965. Also, the builder of all Soviet robotic spacecraft to date, OKB-1, was overloaded with work on the manned space program and so the robotic program was transferred to NPO-Lavochkin. Throughout 1966 Lavochkin modified Korolev's designs to deal with the problems revealed by previous flights, and began to produce their own versions of the Ye-6 and Ye-8 lunar spacecraft and the 3MV planetary spacecraft for Venus. It was decided not to attempt further 3MV missions to Mars, and instead to design a new heavier spacecraft which would enter into orbit around the planet and deliver a soft lander. This strategy was intended to upstage the US flyby missions of Venus and Mars scheduled for the 1967–1969 launch windows with entry probe and lander missions to Venus and with orbiter and lander missions to Mars.

Lavochkin prepared two new 3MV spacecraft with entry probes for the Venus 1967 opportunity. The entry probe was designed to make atmospheric measurements while descending by parachute and to survive impact on the surface for an assumed marginal atmospheric pressure. Both were launched in June 1967. The second was lost to a launch vehicle failure, but on October 18, 1967 the first, Venera 4, became

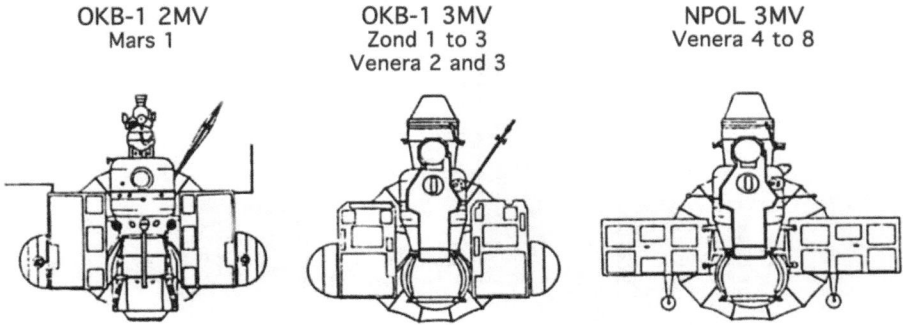

Figure 5.8 Evolution from 2MV to the 3MV models produced by OKB-1 and NPOL (by Ralph F. Gibbons).

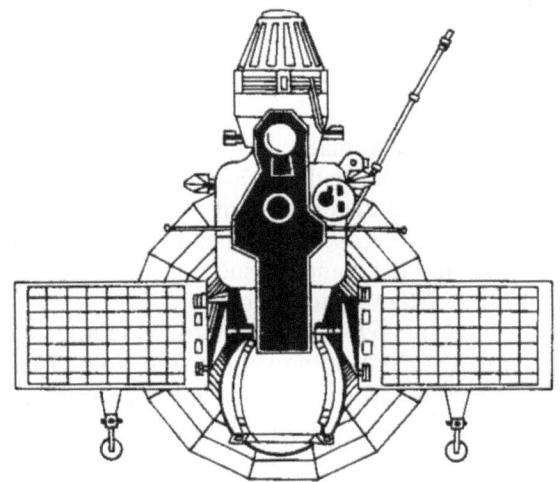

Figure 5.9 The 3MV Venera 4 spacecraft (from *Space Travel Encyclopedia*).

the first successful planetary entry probe. The Soviets initially believed that it had survived all the way to the surface, but it transpired that it had been overwhelmed by conditions while still high in the atmosphere.

This strategy of two Venus probe launches at each opportunity was repeated for the next three Venus launch opportunities in 1969, 1970 and 1972. There were four successes in six launches, Venera 5 and 6 in 1969, Venera 7 in 1970, and Venera 8 in 1972. The probes were strengthened for each opportunity until they were finally able to survive the high pressures and temperatures at the surface of Venus. The first spacecraft to land and survive on the surface of another planet was Venera 7. Venera 8 duplicated the feat in 1972 near the morning terminator on the illuminated side of Venus and with a more versatile set of measurements.

Figure 5.10 Mars-69 spacecraft with probe (not flown).

Mars-69, 1969

The Mars program was the first to make use of the powerful Proton launcher. The 1969 launch opportunity was a particularly favorable one and the large increase in available spacecraft mass offered the opportunity to attempt a soft landing mission, but it was decided on this initial campaign to send a spacecraft that would release an entry probe from orbit around the planet. If successful, it would provide the first in-situ measurements of the Martian atmosphere.

The first design for the Mars-69 spacecraft took advantage of all the work that had been done for the new lunar landing vehicle, the Ye-8. This design ultimately turned out to be impractical for Mars and a complete redesign produced an in-line modular configuration similar to, but much larger and more robust, than the preceding 3MV spacecraft.

The core of the Mars-69 spacecraft was a spherical propellant tank which had the engine beneath and a cylindrical section above, and held the solar panels, antennas and thermal control system. The navigation system and orbital instrument modules were mounted on opposite sides of the tank. The entry probe was installed above the cylinder. The avionics were a significant improvement on the 3MV spacecraft. Due to insufficient time for testing and significant growth in spacecraft mass, the entry vehicle had to be deleted from the 1969 campaign. Two identical orbiter spacecraft were launched in late March and early April, 1969, and unfortunately both were lost to launch vehicle failures.

Mars-71 and Mars-73 series, 1971–1973

The energy requirements for a Mars flight were larger in 1971 than in 1969. This, and several engineering problems with the multiple instrument modules used in the Mars-

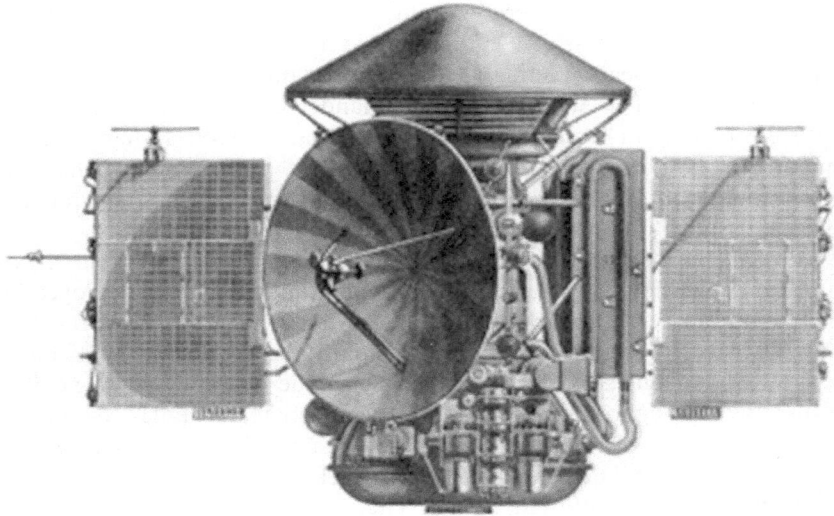

Figure 5.11 Mars 3 spacecraft (courtesy NPO-Lavochkin).

69 design, prompted yet another redesign. In the new version, the propulsion system at the base of the spacecraft formed the main structural element, and a single instrument module was mounted at the base of the cylindrical fuel and oxidizer tank system, forming a torus around the engine. As before, the solar panels, antennas, and thermal control system were attached to the side of the propellant tanks. New digital electronics were provided based on the avionics for the final stage of the N-1 rocket. Advantage was taken of this heritage to save mass by removing the control system of the Proton Block D and allowing the spacecraft to manage the upper stage engine operations.

The higher energy requirements of the 1971 launch opportunity did not allow the orbiter to carry the entry vehicle into Martian orbit, so it would have to be deployed prior to orbit insertion. The higher atmospheric entry velocities and the decision to perform a soft landing, demanded a new entry vehicle design with a larger aerobrake possessing a shallower cone angle. The parachute would have to open at supersonic velocities, which was unprecedented. The final entry vehicle design was a modular stack consisting of the aerobrake at the forward end, the egg-shaped lander nested in the aerobrake, the toroidal parachute container on top of the lander, and a propulsion assembly at the rear of the entry vehicle. For the cruise, the entry vehicle was carried on top of the orbiter.

Lacking a sufficiently precise Mars ephemeris to provide accurate targeting of the entry systems prior to launch, it was decided to send an advance spacecraft to enter orbit around the planet and provide the navigational data necessary for the following two orbiter/lander missions to target and deploy their landers inbound to the planet. Unfortunately, the launch of the orbiter failed in May 1971 due to a stored command error. This accident had two very negative effects, the first being that the American Mariner 9 spacecraft would become the first to orbit Mars, and the second being that the two orbiter/landers would have to rely on a back-up, real-time and less accurate optical targeting technique. The launches were successful, and Mars 2 and 3 were on their way. The Mars 2 lander crashed when the back-up targeting system failed. On December 2, 1971, the Mars 3 entry system succeeded and its lander became the first to touch down on Mars. Unfortunately, the lander transmitted for only 20 seconds before failing and returned no useful data. Both parent spacecraft successfully achieved orbit.

The 1973 Mars launch opportunity was even less energetically favorable, making orbiter/lander combinations impractical. The lander would have to be deployed by a flyby vehicle. Four spacecraft were launched in July and August 1973, two orbiters and two flyby/landers. The spacecraft were essentially the same as in 1971, but the 1973 spacecraft were plagued by electronics problems due to manufacturing changes in a transistor used throughout the system. The engine on Mars 4 failed to ignite and the orbiter sailed past the planet. The Mars 5 orbiter succeeded, but failed after only about one month in orbit. The Mars 6 carrier had telemetry difficulties throughout its cruise, but managed to deploy its lander. The entry vehicle performed properly and transmitted the first in-situ atmospheric data, but no signal was ever received from the lander after it was dropped in close proximity to the surface. Mars 7 failed to put its lander on a proper trajectory, causing it to miss the planet.

Venera/Vega series, 1975–1985

The Venus objectives of the 3MV series were fulfilled when Venera 7 and 8 both survived landing and provided data on surface conditions. This led to the decision to design a new spacecraft for more extensive operations on the surface of this planet. For the first time since the initial launch to Venus in 1961, a launch opportunity was skipped in 1973 to devote the time to developing a new, heavier, more complex and capable Venus orbiter/lander system based on the Proton-launched Mars spacecraft. The success of the American Mariner 9 orbiter in 1971, and the anticipated superior capability of that nation's Viking landers to be launched in 1975, led to the decision to focus the more expensive Proton-launched missions on Venus rather than Mars in the immediate future.

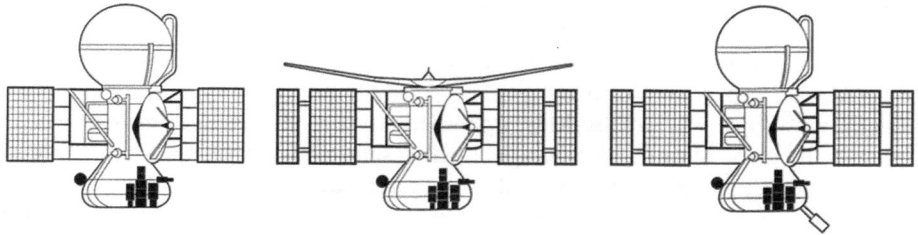

Figure 5.12 Venera 9 to 14, Venera 15 and 16, and Vega 1 and 2 spacecraft (from *Pioneering Venus*).

The new Venera orbiter was nearly identical to the Mars orbiter, with changes in solar panel size and thermal design. But the entry vehicle was significantly different. The thicker, deeper atmosphere of Venus allowed for a simpler entry and landing system consisting of a large, hollow, spherical entry vessel containing the lander and parachute system. Pairs of spacecraft were launched on three launch windows, and all were successful, the Venera 9 and 10 orbiter/landers in 1975, the Venera 11 and 12 flyby/landers in 1978, and the Venera 13 and 14 flyby/landers in 1981. Not only did the Venera 9 lander provide the first imagery and composition measurements from the surface, the parent spacecraft was the first successful Venus orbiter. Flight energy requirements in 1978 and 1981 did not allow for orbiters. In 1983 the lander module was replaced with an imaging radar, and the Venera 15 and 16 orbiters were successful in providing the first radar imagery of the surface of the planet.

Nearly simultaneously with the 1983 radar mission, another flyby/lander mission was being prepared in a French partnership to deploy a large balloon which would be equipped with a comprehensive science payload and drift around the planet in the cloud deck. But when it was recognized that the flyby spacecraft could be retargeted to Halley's comet after releasing the entry vehicle for Venus, the flyby spacecraft payload was redesigned for Halley, the balloon significantly descoped, a lander added, and the launch date adjusted to provide encounters with both Venus and Halley. Renamed Vega 1 and Vega 2, all aspects of these missions were carried out very successfully at both targets.

Phobos and Mars 96, 1988–1996

After a long run of very successful Venus missions beginning in 1967, including the highly successful Vega mission in 1985, and the clear lack of American follow-up to the Viking Mars orbiter/landers, the Soviets took the opportunity in the late 1980s to resume Mars missions. A new Universal Mars Venus Luna (UMVL) spacecraft was developed based on the highly successful Proton-launched Venera series. Two such spacecraft were built for the 1988 Mars opportunity, for a mission that would focus on the Martian moon Phobos. Once a spacecraft was in orbit around the planet, it would make a series of close encounters with Phobos, coming ever closer. When the geometry was just right, active remote sensing experiments would blast material off the surface of Phobos and two small landers would be deployed, one stationary and the other mobile. It was to be a very ambitious mission, including instruments from many international partners.

The Phobos 1 spacecraft was lost to a command error during the interplanetary cruise. Its partner achieved Mars orbit and returned very useful remote sensing data on Mars as it trimmed its orbit to approach Phobos. Unfortunately, communications with Phobos 2 were lost just days before its first planned rendezvous, and only very limited remote sensing data on this target were transmitted.

Encouraged by the Phobos effort, a Mars orbiter and ambitious surface mission was planned. This was originally scheduled for launch in 1992 using a new version of the UMVL spacecraft but budget constraints led to it being descoped and slipped to

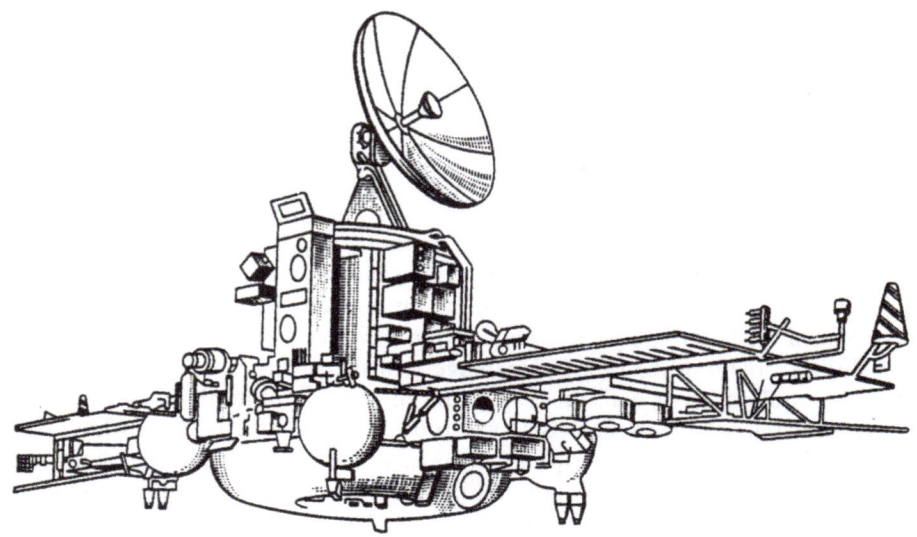

Figure 5.13 The UMVL Phobos spacecraft.

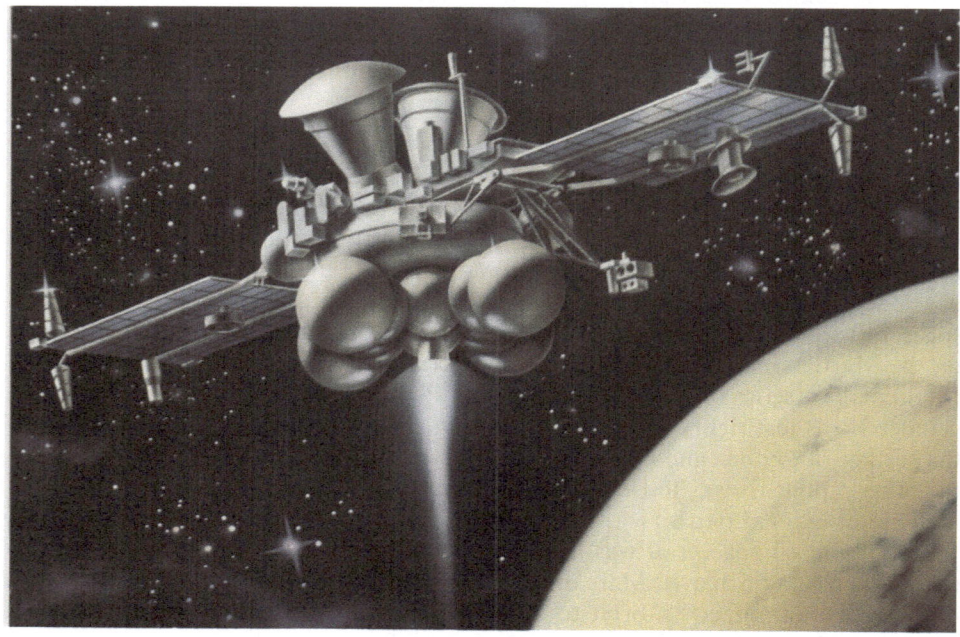

Figure 5.14 The Mars-96 spacecraft (courtesy NPO-Lavochkin).

1994, and then further delayed to 1996. In addition to a large orbital science payload, the orbiter had two small soft-landers similar to the previous Mars landers and two penetrators. This project involved even more international cooperation than the Phobos effort. However, this time only a single spacecraft was built, and when it was launched on November 16, 1996, failures in the control system between the spacecraft and the Block D upper stage resulted in the escape burn causing re-entry. Having lost Mars-96 so embarrassingly, the Russian planetary exploration program entered a hiatus which continued through the end of the 20th Century. It is scheduled for renewal with the planned launch of the Phobos-Grunt sample return spacecraft in late 2011.

Part II

Putting the pieces together: flying to the Moon, Venus and Mars

6

Breaking free of Earth

TIMELINE: AUG 1958–SEP 1960

The space age began on October 4, 1957, with the launch of Sputnik during a test flight of Korolev's new R-7 launcher. The ignition of the rocket's engines on the pad in Baikonur on that day was the explosion that opened the floodgates of space exploration. A little over ten months later, on August 17, 1958, the first attempt was made to send a spacecraft to the Moon, a tiny orbiter, and this time launched by the US, but the rocket exploded. On September 23, the USSR attempted to send a lunar impactor to the Moon using a new variant of the R-7 augmented with a small third stage to reach escape velocity. The booster failed and was destroyed. The race to the Moon and planets was on.

In the three years 1958–1960, the US attempted nine times to send a small Pioneer class spacecraft to the Moon. All failed in one way or another. In the two years 1958–1959, the USSR also attempted nine times to send a spacecraft to the Moon. Of these, six were lost to launch vehicle failures, the Luna 1 impactor missed the Moon by 6,000 km, Luna 2 succeeded in impacting the Moon, and Luna 3 traveled beyond the Moon and sent back grainy pictures of its hitherto mysterious far side.

THE YE-1 LUNAR IMPACTOR SERIES: 1958–1959

Campaign objectives:

After the launch of Sputnik, Korolev took advantage of the world's reaction to push the Soviet government into approving plans for non-military applications of his R-7 rocket, including lunar exploration. An earlier attempt in 1955 had been of no avail, but now the time was ripe. He established three new design groups at OKB-1, one for

Launch date

1958

17 Aug	Pioneer lunar orbiter	Booster exploded
23 Sep	Luna impactor	First stage destroyed
11 Oct	Pioneer 1 lunar orbiter	Reached 115,000 km altitude
11 Oct	Luna impactor	First stage destroyed
8 Nov	Pioneer 2 lunar orbiter	Third stage failure
4 Dec	Luna impactor	Second stage premature shutdown
6 Dec	Pioneer 3 lunar flyby	Reached 107,500 km altitude

1959

2 Jan	Luna 1 impactor	Missed Moon by 5,965 km
3 Mar	Pioneer 4 lunar flyby	Missed Moon by 60,030 km
18 Jun	Luna impactor	Second stage guidance failure
12 Sep	Luna 2 impactor	Successful lunar impact on Sep 14
24 Sep	Pioneer lunar orbiter	Pad explosion during test
4 Oct	Luna 3 circumlunar flyby	Success, returned lunar far side images
26 Nov	Pioneer lunar orbiter	Shroud collapsed during launch

1960

15 Apr	Luna circumlunar flyby	Third stage malfunction
19 Apr	Luna circumlunar flyby	First stage disintegrated
25 Sep	Pioneer lunar orbiter	Second stage malfunction

communications satellites, one for manned space flight, and the third for robotic lunar spacecraft. Mikhail Tikhonravov and Gleb Maksimov were in charge of the latter. Mstislav Keldysh provided specific scientific goals. After a few months work, Korolev and Tikhonravov sent a letter to Moscow on January 28, 1958, proposing a lunar impactor and a lunar flyby mission to photograph the far side. Korolev and Keldysh jointly convinced the government, and on March 20 approval was granted. In fact, lunar spacecraft designs were underway and in February Korolev had begun to develop the third stage required for his R-7 launcher. He was very aware of well-publicized plans in the US for a lunar orbiter to be launched in the summer of 1958, and he wanted to be first.

While preparing for the launch of the first spacecraft to the Moon in the summer, Korolev and Tikhonravov expanded the scope of their plan for the conquest of space by the Soviet Union. This plan was finished in early July 1958, but was held secret outside of a few people in the closed Soviet space circle. It called for upgrading the R-7 to three stages to launch robotic lunar landers and photographic flyby missions, then upgrading the R-7 again to four stages to launch spacecraft to Mars and Venus, and developing orbital rendezvous and other techniques and technologies to enable men to fly around and land on the Moon, as a precursor to creating a colony on the Moon and visiting Mars and Venus.

Maksimov and Tikhonravov also prepared detailed designs for five types of lunar spacecraft in the spring of 1958:

Ye-1   Lunar impact spacecraft, 170 kg
Ye-2   Lunar far-side photographic flyby, 280 kg
Ye-3   Same as Ye-2 with improved photographic equipment
Ye-4   Lunar impact spacecraft with explosives, possibly nuclear, 400 kg
Ye-5   Lunar orbiter

The Soviets were concerned about proving that their spacecraft had hit the Moon, not recognizing at first that they could be tracked by any other country with the right equipment. Telemetry cessation was not definitive and so the notion was considered of exploding a device on the Moon for all to see, hence the Ye-4 with sufficient payload capacity to carry a nuclear or a large conventional explosive. Korolev was reluctant to use a nuclear explosive, and after consultation with recognized nuclear physicist Yakov Zeldovich and several other physicists this idea was dropped. The Ye-4 itself was eventually abandoned as the technical and political problems became clear, and tracking was recognized as the solution. Even so, the Ye-1 third stage was outfitted with a device to release a sodium cloud for optical tracking and for general observation around the world. The Ye-5 lunar orbiter project was canceled when the new 8K73 rocket which was to launch it suffered engine development problems and was itself canceled.

As the summer of 1958 arrived, Korolev rushed to launch his first lunar impactor before the date set by the Americans for their tiny orbiter. Although he was having technical problems with his first three-stage vehicle, he decided to take the risk and readied his R-7 on the same day as the Americans, but stood down when he learned that the US rocket had exploded. But the extra time was of no avail, and one month later his launch vehicle also exploded after a short flight. No matter, the space race was on and its public characteristics were now well defined.

In the Soviet program, planning and launch information were held secret and only successful launches reported. Spacecraft that were launched successfully but failed in their objectives had their missions redefined in the Soviet press to make all appear successful. By contrast American plans were announced well in advance, and open to press and public scrutiny. It made for a dramatic contest on each side; one well aware of the other's plans and proceeding to blind-side its competitor, and the other mostly unaware of the other's plans and groping almost blindly to seize and retain a leading position.

Six Ye-1/1A spacecraft were launched in the 12-month period between September 1958 and 1959. Only two, Luna 1 and 2, escaped launch vehicle mishaps. Luna 1 and its flight past the Moon was a sensation to match that of Sputnik almost a year earlier. The Soviet press referred to the whole system, launcher and all, as the 'First Cosmic Rocket', and when it passed the Moon the spacecraft was renamed 'Mechta' (Dream). Years later, it was retroactively named Luna 1. It was intended to hit the Moon but on January 4, 1959, it missed its target by about 6,000 km. Nonetheless, it was the first spacecraft to attain Earth escape velocity. On September 14, 1959, the second spacecraft to be launched successfully, Luna 2, became the first spacecraft to impact the Moon, thereby fulfilling Korolev's program goals for this series.

Spacecraft Launched

First spacecraft: Ye-1 No.1
Mission Type: Lunar Impactor
Country/Builder: USSR/OKB-1
Launch Vehicle: Luna
Launch Date/Time: September 23, 1958 at 09:03:23 UT (Baikonur)
Outcome: Booster failure.

Second spacecraft: Ye-1 No.2
Mission Type: Lunar Impactor
Country/Builder: USSR/OKB-1
Launch Vehicle: Luna
Launch Date/Time: October 11, 1958 at 23:41:58 UT (Baikonur)
Outcome: Booster failure.

Third spacecraft: Ye-1 No.3
Mission Type: Lunar Impactor
Country/Builder: USSR/OKB-1
Launch Vehicle: Luna
Launch Date/Time: December 4, 1958 at 18:18:44 UT (Baikonur)
Outcome: Second stage failure.

Fourth spacecraft: Luna 1 (Ye-1 No.4)
Mission Type: Lunar Impactor
Country/Builder: USSR/OKB-1
Launch Vehicle: Luna
Launch Date/Time: January 2, 1959 at 16:41:21 UT (Baikonur)
Encounter Date/Time: January 4, 1959
Outcome: Missed the Moon, entered solar orbit.

Fifth spacecraft: Ye-1A No.5
Mission Type: Lunar Impactor
Country/Builder: USSR/OKB-1
Launch Vehicle: Luna
Launch Date/Time: June 18, 1959 at 08:08:00 UT (Baikonur)
Outcome: Second stage failure.

Sixth spacecraft: Luna 2 (Ye-1A No.7)
Mission Type: Lunar Impactor
Country/Builder: USSR/OKB-1
Launch Vehicle: Luna
Launch Date/Time: September 12, 1959 at 06:39:42 UT (Baikonur)
Encounter Date/Time: September 14, 1959 at 23:02:23 UT
Outcome: Success, impacted Moon.

The scientific goals of these first interplanetary spacecraft were to study cosmic radiation, ionized plasma, magnetic fields and the micrometeoroid flux in the region between the Earth and Moon known as cislunar space. In addition to assisting with optical tracking, the sodium release experiment would also allow for visualization of the magnetosphere and diffusion in the upper atmosphere of the Earth in transit.

Spacecraft:

The Ye-1 spacecraft were spherical in shape, similar to but slightly larger than the first satellite. Sputnik 1 was 56 cm in diameter and the Ye-1 and -1A were 80 cm in diameter, made from aluminum-magnesium alloy, and four times heavier. Antennas and instrument ports protruded from the surface of the sphere. The spacecraft was highly reflective and spin stabilized at about one revolution every 14 minutes. It had no propulsion system. The interior held 1.3 bar of nitrogen that was circulated by a fan between the cold outer shell and the warm electronics for thermal control in the range 20 to 25°C, and contained a meter band radio telemetry system, silver-zinc and mercury-oxide batteries, the science payload, and commemorative medallions. The receiver operated at 102 MHz and the transmitter at 183.6 MHz and 1 kbits/s. There was a backup telemetry system operating in the short-wave at 19.993 MHz.

Launch mass: 361.3 kg (Luna 1) 390.2 kg (Luna 2)

Payload:

The payload consisted of five scientific instruments for studying interplanetary space and two spheres covered by pentagonal medallions that were to break up and scatter across the surface on impact.

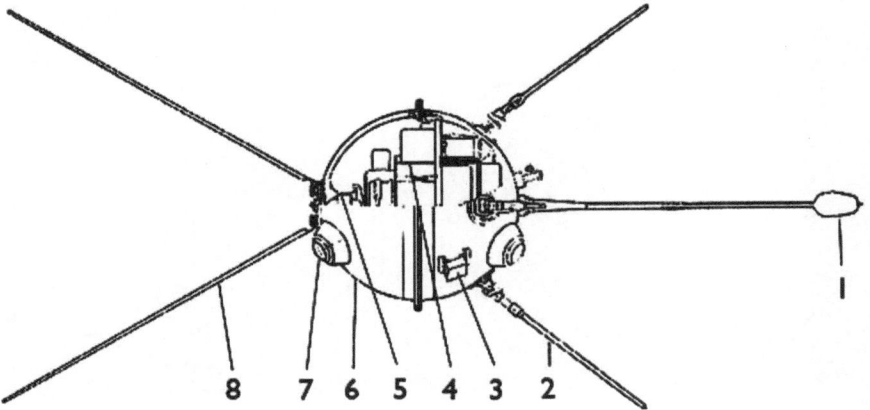

Figure 6.1 Luna 1 Ye-1 spacecraft (from Don Mitchell): 1. Magnetometer; 2. 183.6 MHz antenna; 3. Micrometeorite counter; 4. Batteries and electronics; 5. Ventilator, fan; 6. Spacecraft shell; 7. Ion traps; 8. Ribbon antenna for 19.993 MHz.

1. Triaxial fluxgate magnetometer for magnetic fields
2. Cherenkov detector for cosmic radiation
3. Scintillation and gas discharge Geiger counters for cosmic radiation
4. Piezoelectric micrometeoroid detector
5. Ion trap detectors for interplanetary plasma

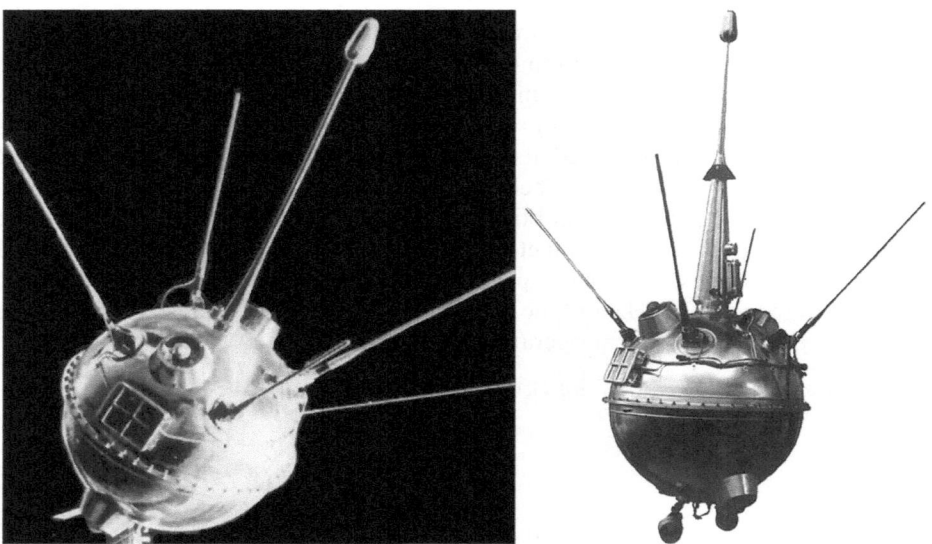

Figure 6.2 Luna 1 (left) and Luna 2 (right).

Figure 6.3 Spherical medallions carried by Luna 1 and Luna 2.

The two cosmic radiation instruments were mounted just inside the sphere, which provided for aluminum shielding, and the other instruments were mounted outside. Luna 2 had additional Geiger counters both inside and outside the sphere. In each case the third stage carried additional cosmic radiation experiments, a capsule with aluminum medallion strips, and a sodium release experiment to be activated while in the Earth's magnetosphere on the escape trajectory.

Mission description:

Of the six Ye-1 spacecraft launched, only two survived the process. A seventh was returned to the barn after its launcher failed to lift off. Aware from press reports that the Americans were to try for the Moon on August 17, Korolev managed with great effort to prepare a vehicle for the same day. There were a number of malfunctions during pre-launch preparations, but he knew that his flight path to the Moon was shorter than the Americans so he waited before risking a launch to see if the Florida launch succeeded. When the US rocket blew up after only 77 seconds of flight, he stood down in order to perform more careful preparations and additional testing. However, the launch on September 23 failed when the strap-on boosters of the first stage developed resonant longitudinal vibrations in the second minute of flight. The various stages separated at 93 seconds, fell back and exploded. Reacting to pressure to beat the US to the Moon, Korolev is said to have lost his temper and replied "Do you think only American rockets explode?" Indeed not, and he could not know that the Soviet lunar program would be plagued with rocket failures for years to come.

The second American attempt at the Moon was scheduled for October 11, amid a flurry of press coverage. Korolev was again ready that same day. The whole world was aware of events in Florida, but only a few in the USSR were aware of Korolev with his finger on the launch button at Baikonur, ready to beat the Americans to the Moon using a faster trajectory. News of the US launch was relayed to Korolev. But the third stage failed, preventing Pioneer 1 from reaching the Moon. Sitting now in the catbird seat, Korolev proceeded with his launch. Later the same day, the second Luna 8K72 launcher blew up 104 seconds into its flight due to the same vibrations which had destroyed the first vehicle.

The two failures in row were demoralizing. It was discovered from analyzing the wreckage that the additional mass of the third stage was creating resonant vibration in the basic R-7 booster which had not been present before. The problem was solved with minor design changes, but it would take two months and Korolev had to watch over his shoulders as America made a third attempt on November 8, but this too fell short.

The third launch on December 4 failed yet again, this time caused by a different problem. The rocket sailed through the period when vibrations broke up the previous two rockets, but after 4 minutes of flight the thrust of the second stage engine began to diminish and then the engine shut down due to a gear box failure in the hydrogen peroxide turbine pump. Frustrated, but relieved by the fourth American failure two days later on December 6, Korolev prepared for another

Figure 6.4 Luna launch preparations.

Figure 6.5 Luna launch.

Figure 6.6 Luna 2 mounted on the Block L fourth stage prior to launch.

attempt. The fourth launch was a success and on January 2, 1959, put Ye-1 No.4 on a trajectory to the Moon. The spent third stage released a 1 kg cloud of sodium gas on January 3, some 113,000 km from Earth, producing a glowing orange trail visible over the Indian Ocean with the brightness of a sixth-magnitude star. The experiment provided data on the behavior of ionized gas in near-Earth space, and was used for tracking. Luna 1 (as the probe was later named) missed its target and passed within 5,965 km of the lunar surface on January 4 after 34 hours of flight. The miss was caused by a late second-stage shutdown command from the ground radio guidance. Nevertheless, Luna 1 holds three cosmic 'firsts', being the first spacecraft to achieve escape velocity, the first spacecraft to fly close by the Moon, and the first spacecraft to enter an independent heliocentric orbit. Contact was lost on January 5, after 62 hours of flight, possibly when its battery drained.

Luna 1 was a major success and feather in the cap of Soviet space exploration, but it failed to impact the Moon as planned and the program goal was not yet fulfilled. After further problems with the R-7 in the beginning of 1959, another spacecraft was readied that incorporated modifications to the magnetometer, Geiger counters and micrometeoroid detectors resulting from the successful in-flight measurements of both Luna 1 in January 1959 and the American Pioneer 4 lunar flyby in March. The modifications earned it a new Ye-1A designation. The first attempt to launch a fifth spacecraft, Ye-1A No.5, was aborted on June 16, 1959, when it was discovered that the third stage tank had been filled with the standard kerosene instead of the higher density type required for this mission. The tanks were emptied, refilled with the proper fuel, and a second launch attempt made two days later. All went awry when the launch vehicle deviated from the planned trajectory after 153 seconds. One of the gyroscopes in the inertial guidance system had failed, and the launcher was destroyed by ground command.

An aborted launch of a sixth spacecraft occurred on September 9, 1959, when the core sustainer engine failed to reach full thrust upon ignition. The launcher remained on its mount, and all the engines were shut down after 20 seconds. The rocket was replaced by a backup. The spacecraft on the aborted rocket was probably the Ye-1A No.6 model. Three days later Luna 2 (Ye-1A No.7) was successfully launched on a lunar trajectory. On 13 September, at a distance of 156,000 km, the spent third stage released its sodium cloud. Luna 2 impacted the Moon at 23:02:23 UT September 14, after 33.5 hours of flight, near the Autolycus crater in the Marsh of Decay region at about 29.1°N 0.0°E. Some 30 minutes later, the third stage of the Luna launcher also impacted the Moon.

Luna 2 was the first spacecraft to impact another celestial body. The Soviets had announced their transmission frequencies and Jodrell Bank in England tracked the spacecraft through its the final plunge to silence. There had been some claims in the West that Luna 1 was a fraud, but Sir Bernard Lovell's tracking and radio recordings provided all the proof needed that Luna 2 had hit the Moon. Nikita Khrushchev celebrated the achievement by presenting President Eisenhower with duplicates of the Soviet emblems that had been carried to the Moon at a United Nations meeting in New York on September 15, 1959.

Results:

Luna 1 was the first spacecraft to reach the vicinity of the Moon. The measurements obtained provided new data on the Earth's radiation belt, and discovered the solar wind – a thin, energetic ionized plasma flowing outward from the Sun past the Earth and Moon. It established that the micrometeoroid flux between Earth and Moon was small, and placed an upper limit on the strength of any magnetic field that the Moon may possess at no more than 1/10,000th that of Earth.

Luna 2 was the first spacecraft to impact on the Moon. It verified at much closer distance that the Moon had no appreciable magnetic field, and found no evidence of radiation belts around the Moon.

THE YE-2 AND YE-3 LUNAR FLYBY SERIES: 1959–1960

Campaign objectives:

Korolev's second step after demonstrating the ability to hit the Moon was to obtain photographs of its far side, which can never be observed from Earth. The Television

Scientific Research Institute developed a camera for the mission; a facsimile system using film developed on board and scanned by a photometer for transmission. The camera was fixed and had to be pointed at the Moon by appropriately orienting and stabilizing the spacecraft, which required a 3-axis pointing and control system rather than the spin stabilization used by the Ye-1 spacecraft. The Ye-2 would be the first to accomplish this vital mode of attitude control. In addition, the spacecraft had to be placed onto a trajectory that would enable it to view the far side of the Moon from a close range and under suitable illumination, and then return to the vicinity of Earth in order to transmit its pictures.

Spacecraft launched

First spacecraft: Luna 3 (Ye-2A No.1)
Mission Type: Lunar Circumlunar Flyby
Country/Builder: USSR/OKB-1
Launch Vehicle: Luna
Launch Date/Time: October 4, 1959 at 00:43:40 UT (Baikonur)
Encounter Date/Time: October 7, 1959
Mission End: October 22, 1959
Outcome: Success, photographed the lunar farside.

Second spacecraft: Ye-3 No.1
Mission Type: Lunar Circumlunar Flyby
Country/Builder: USSR/OKB-1
Launch Vehicle: Luna
Launch Date/Time: April 15, 1960 at 15:06:44 UT (Baikonur)
Outcome: Upper stage failure.

Third spacecraft: Ye-3 No.2
Mission Type: Lunar Circumlunar Flyby
Country/Builder: USSR/OKB-1
Launch Vehicle: Luna
Launch Date/Time: April 19, 1960 at 16:07:43 UT (Baikonur)
Outcome: Booster failure.

Keldysh's Applied Mathematics Institute designed special orbits that would allow the spacecraft to photograph the far side of the Moon and then return to the vicinity of Earth over the USSR to transmit the pictures back at close range. There were only two launch opportunities for these restricted types of orbits, one in October 1959 for photography on approaching the Moon, and another in April 1960 for photography on receding from it. One Ye-2A spacecraft was launched in October and two of the Ye-3 type were assigned to the follow up. The first, Luna 3, was successful, but both of the more advanced spacecraft were lost to launch vehicle failures.

The Luna 3 mission was a momentous achievement for that time, and its pictures excited the world. But no one outside the USSR knew of the failures in the program. Of nine launches, six were total losses. Luna 1 failed to achieve its prime mission, but

Luna 2 was successful. And although Luna 3 took and transmitted pictures, they were of poor quality. Nevertheless, to the outside world it appeared that the Soviets had successfully launched three lunar missions of progressively greater complexity, and could do almost anything at will. In stark contrast, at the end of 1960 America appeared to be incompetent with nine embarrassing public failures yielding just one wide miss of the Moon.

After the success of Luna 3, the Soviet lunar program experienced a 3-year hiatus as the focus shifted to the more challenging planetary targets, Venus and Mars, and a new robotic spacecraft was developed for soft landing on the Moon.

Spacecraft:

Two competing telecommunication systems were started for the lunar photography, one by Bogomolov labeled Ye-2 and the other by Ryazansky labeled Ye-2A. It was decided to use the Ye-2A system. The Ye-2A spacecraft designed by Gleb Maximov was a cylindrically shaped canister 130 cm in length with hemispherical ends and a 120 cm wide flange near the top. The cylindrical section was approximately 95 cm in diameter. The canister was hermetically sealed at 0.23 bar and held the cameras and film processing apparatus, communications equipment, thermal control fans, gyroscopes, and rechargeable silver-zinc batteries. Uplink was at 102 MHz and downlink at 183.6 MHz. A backup telemetry system operated at 39.986 MHz. The spacecraft had six omnidirectional antennas, four protruding from the top and two from the bottom. The thermal control system was to prevent the internal temperature from exceeding 25°C by using passive flaps mounted along the cylinder. There were micrometeoroid detectors, cosmic ray detectors, and solar cells for recharging the batteries on the exterior. The upper hemisphere of the probe housed the camera port, and the lower hemisphere housed the cold gas 3-axis attitude control jets. There was no propulsion system with which to perform midcourse maneuvers. The spacecraft was to be spin stabilized under cruise, switch to 3-axis stabilization for photography, and then resume spin stabilization. Photoelectric cells were used to maintain orientation with respect to the Sun and Moon.

The follow up spacecraft, originally designated Ye-2F, were intended to acquire more and improved images of the lunar far side. As these were being prepared, there was a parallel rush to get the new four-stage R-7 and the Mars and Venus spacecraft ready for launch starting in the fall of 1960. The Ye-3 project was canceled when its camera system was judged too complicated and unreliable, and the Ye-2F spacecraft was re-designated Ye-3 shortly before launch. These two spacecraft were essentially the same as the Ye-2A but with improved imaging and radio systems.

Ye-2A launch mass: 278.5 kg

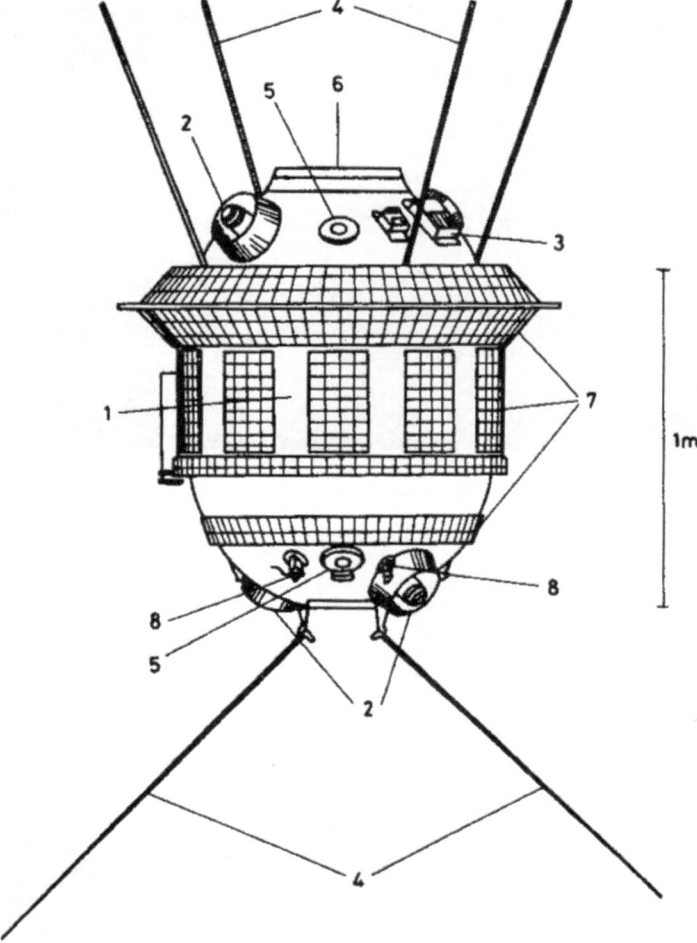

Figure 6.7 Luna 3 diagram (from *Space Travel Encyclopedia*): 1. Thermal control louvres; 2. Ion traps; 3. Micrometeorite detector; 4. Antennas; 5. Sun sensors; 6. Camera port; 7. Solar panels; 8. Attitude control microjets.

Payload:

1. Yenisey-2 photo-television facsimile camera system
2. Micrometeoroid detector
3. Ion traps (3)
4. Cherenkov radiation detector
5. Scintillation and gas discharge Geiger radiation counters
6. Mass spectrometer (not flown)

Figure 6.8 Luna 3 spacecraft.

Figure 6.9 Yenisey-2 photo-facsimile imaging system.

Several instruments from Luna 1 and 2 were flown in addition to the new camera system. A mass spectrometer based on an instrument that was flown successfully on Sputnik 3 was planned but canceled owing to mass and time constraints.

Unlike the Americans who chose to use television vidicon tube cameras for their early deep space photography missions (except the Lunar orbiter series), the Soviets used a film camera system. This was mechanically complex and heavy but provided higher resolution, greater sensitivity, better quality, and was distortion free. The Yenisey-2 facsimile imaging system on the Ye-2 and -3 spacecraft consisted of a 35 mm film camera equipped with 200 mm f/5.6 and 500 mm f/9.5 lenses, an automatic film processing unit, and a photomultiplier film scanner with a resolution of 1,000 pixels/line. The 200 mm objective was sized to image the full disk of the Moon. The camera cycled through four exposure times from 1/200th to 1/800th second. It exposed adjacent frame pairs simultaneously, one through each lens, and was capable of taking 40 frames at 1,000 × 1,000 pixel resolution using temperature and radiation resistant isochromatic film. The developed film could be scanned and rewound at ground command, and could be transmitted at either 1.25 lines/second or 50 lines/second depending on the range from Earth. The video signal was sent using the 3-W 183.6 MHz transmitter. After the Cold War, it was revealed that the Soviets did not have radiation-resistant film and used US radiation-resistant film acquired by scavenging downed American spy balloons flown over the USSR from Western Europe.

Mission description:

Only the first of these missions survived its launch vehicle. The Luna 3 spacecraft (Ye-2A No.1) was successfully launched on October 4, 1959, into an elliptical Earth orbit that took it close to the south pole of the Moon, whereupon lunar gravitation redirected the trajectory back to the vicinity of the Earth, forming a figure-of-eight loop. The spacecraft, which the Soviet press dubbed the 'Automatic Interplanetary Station', experienced severe overheating with consequent ragged telemetry shortly after launch. This was alleviated somewhat by reorienting the spin axis and shutting off some equipment. To prepare for photography, the spin was stopped and the gyro-controlled 3-axis orientation system activated. It flew within 6,200 km of the south pole of the Moon when at closest approach at 14:16 UT on October 6, and then crossed through the plane of the Moon's orbit out over the sunlit far side. Early on October 7 the photocell on the top end of the spacecraft detected the sunlit Moon at a distance of 65,200 km and initiated the 40 minute photography sequence. Twenty-nine frames were exposed before the mechanical shutter jammed. The final image was taken at a distance of 66,700 km.

After photography was complete, the spacecraft resumed spinning and the first attempt was made to retrieve images. The signal strength was low and intermittent, and only one image with almost no detail was received. A second attempt was made near apogee at 470,000 km from Earth, but again the transmission quality was poor. The antenna patterns on the spacecraft may not have been optimal. It was decided to wait for the most ideal situation when the spacecraft returned to the vicinity of the Earth ten days later. As the spacecraft approached Earth, several attempts to retrieve the images at fast playback did not yield good results. The signals were weak, with a lot of static and radio noise. To reduce the latter, Soviet engineers enforced radio silence in the Black Sea in the vicinity of the Yevpatoria receiving antenna. Finally, on October 18 the signals improved abruptly and 17 resolvable but noisy pictures were successfully received. By design, the mission was undertaken when a portion of the near side was illuminated to provide a point of reference, so only 70% of the far side was sunlit. Contact with Luna 3 was lost on October 22 and it burned up in the Earth's atmosphere in April 1960.

Both Ye-3 spacecraft fell victim to their launchers. The third stage of the rocket carrying Ye-3 No.1 cutoff prematurely. The kerosene tank had not been completely filled! At a range of only about 200,000 km from the Earth the spacecraft fell back and burned up in the atmosphere. The Ye-3 No.2 launch failed spectacularly when at the moment of liftoff one of the strap-on boosters failed to reach full thrust, placing abnormal loads on the vehicle. Three of the strap-ons separated at only a few meters altitude, resulting in violent maneuvers of the four separated pieces of the rocket and powerful explosions. There was considerable damage to the pad and buildings at the launch site. This brought a fiery end to the first series of Soviet lunar spacecraft and the final use of the 8K72 R-7E Luna launcher for lunar missions.

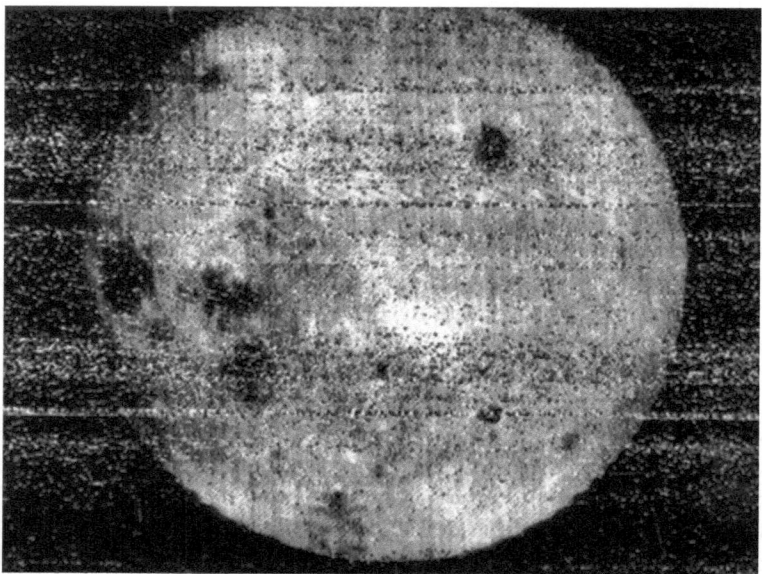

Figure 6.10 First image of the far side of the Moon returned by Luna 3. The dark area at lower left is Mare Smythii on the near side. The right-most three-quarters of the image shows part of the far side. The dark spot at upper right is Mare Moscoviense and the small dark circle at lower right is the crater Tsiolkovsky with its central peak.

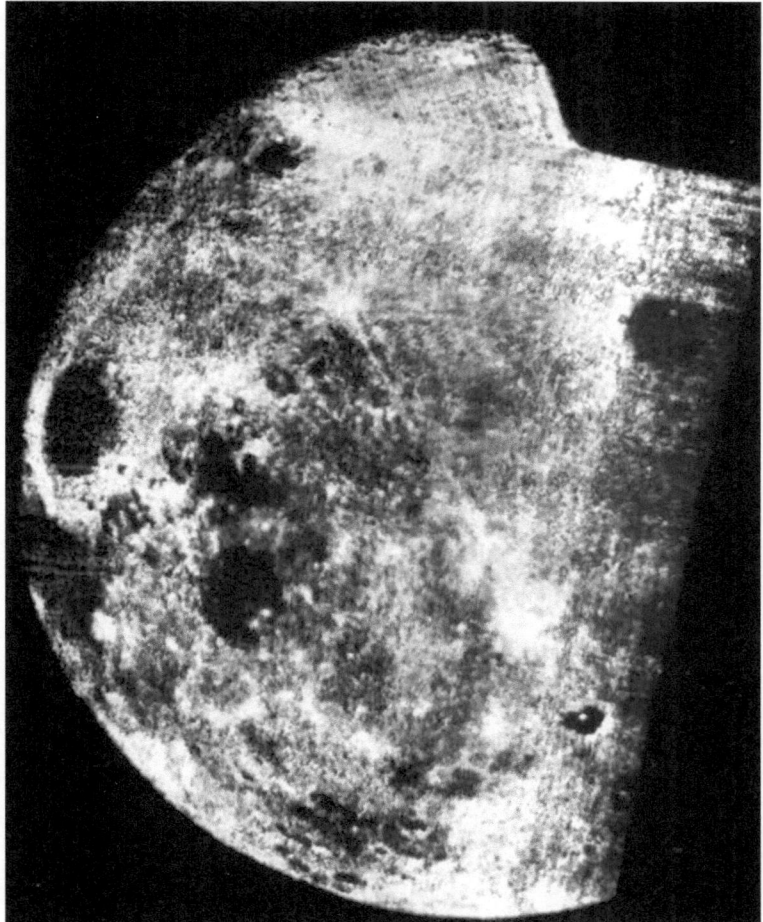

Figure 6.11 Mosaic of Luna 3 images showing the far side of the Moon.

Results:

Luna 3 was the first spacecraft to photograph the lunar far side, but the 17 pictures successfully received were very noisy and of low resolution. Only six of these were published. A tentative atlas was compiled showing the far side to be very different to the near side, being predominantly bright highland terrain, without extensive mare. Two small dark regions were named Sea of Moscow and Sea of Dreams, the latter in honor of the Mechta first flyby mission.

7

Launching to Mars and Venus

TIMELINE: OCT 1960–FEB 1961

The Moon had been the principal target for the USSR in 1958 and 1959, and for the US as well through 1960, but both had ambitions to send spacecraft to the planets. Venus was the closest and most accessible, but Mars, slightly farther away, was the most fascinating. Having been successful at the Moon, the Soviet Union was ready to start launching planetary missions in 1960. The US, struggling to achieve success at the Moon, decided to put off attempting planetary missions.

Korolev developed a four-stage version of the R-7 rocket to launch missions to the planets, and a spacecraft quite different from the initial Luna series to meet the challenges of interplanetary flight. The new rocket and spacecraft were ready for the launch opportunities for Mars in late 1960 and for Venus in early 1961. The first attempts to send a spacecraft to Mars were on October 10 and 14, 1960, and in both cases the third stage failed, giving the new fourth stage and spacecraft no chance to perform. Later on February 4, 1961, the first attempt to send a spacecraft to Venus was foiled when the engine of the new fourth stage failed to ignite. Finally, on its fourth launch on February 12, 1961, the new rocket succeeded in sending its payload on a trajectory towards the planet Venus. Unfortunately, the Venera 1 spacecraft had a number of problems and failed early in its flight.

Launch date		
1960		
10 Oct	Mars flyby	Third stage failure
14 Oct	Mars flyby	Third stage failure
15 Dec	Pioneer lunar orbiter	Booster exploded
1961		
4 Feb	Venera impactor	Fourth stage failure
12 Feb	Venera 1 impactor	Communications lost in transit

THE FIRST LAUNCH TO MARS: 1960

Campaign objectives:

Mankind's first venture to the planets began in 1960, known only to those in the Soviet Union who performed the task and to the elite in the American spy services. The first Mars space flight campaign consisted of two identical spacecraft built for flyby exploration of the planet. They were launched in October 1960 and preceded the first US attempt at Mars by four years. Two similar spacecraft for Venus were launched in February 1961. These four spacecraft were the first designed to directly investigate our neighboring planets.

Chief Designer and Academician S.P. Korolev began work on Mars and Venus missions at OKB-1 in late 1958, during a hectic time in which not only was his R-7 ICBM being developed and tested but also the second silo-based R-9 ICBM. Despite flight tests often occurring at a rate of several per month, he worked on adapting the R-7 for the non-military space exploration role that had always been his dream. He created a small third stage to attain Earth escape velocity, and was ready for the first lunar launch attempts in late 1958. With the successful impact of Luna 2 and far side photography of Luna 3 in 1959, the objectives of the initial lunar spacecraft series were satisfied and Korolev was able to move on to Mars and Venus. He had planned to use an upgraded form of the third stage of the R-7E Luna launcher for planetary spacecraft later in 1959 and 1960, but the Institute of Applied Mathematics of the Soviet Academy of Sciences convinced him that a four-stage rocket would be much more efficient and robust. When American plans for a Venus launch were postponed to 1962, Korolev decided to skip the 1959–60 planetary windows in order to gain the time to develop a four-stage R-7 that would have a third stage derived from the R-9 second stage and a wholly new fourth stage with a restartable engine. The first three stages and the initial burn of the fourth stage would insert the fourth stage into a low orbit around Earth. At the appropriate time, the fourth stage would restart to gain the desired interplanetary trajectory and release its payload. This 8K78 vehicle became known as the 'Molniya' launcher. It was capable of sending 1.5 tons to the Moon or just over 1 ton to either Mars or Venus.

In early January 1960, Khrushchev aired his concern about the growing US space program at a meeting with Korolev and other space leaders. Pushed by his political master to send a spacecraft to Mars, by the end of February Korolev had a schedule for launching to Mars in that fall. His team balked at the 8-month timescale because the four-stage R-7 was still a 'paper' vehicle and the spacecraft was not yet designed – there were not even any working drawings. By today's standards the schedule was ludicrous, but Korolev and his team had, as they put it "a fervent desire to beat the Americans...and were in a desperate hurry".

The 1M (Mars) and 1V (Venus) spacecraft originally envisioned in 1958 were far more complex than the Luna missions. They were fully 3-axis stabilized spacecraft with attitude control and propulsion systems, solar arrays and battery power, thermal control, and long range communications. Korolev's plan called for three launches in the Mars window to dispatch two flyby spacecraft and one lander, with the optimum date being September 27. A lander had to undertake the most difficult of planetary missions – to pass through the atmosphere, survive impact with the surface, and take photographs. But at that time information on the atmospheric properties of the two planets was unreliable, particularly for Mars. Korolev assumed a surface pressure for Mars between 60 and 120 millibars, which was thin but feasible, and that Venus had an atmosphere more like Earth's. Experiments in the summer of 1960 using the R-11A scientific suborbital rocket (a version of the 'Scud' military rocket) took test entry vehicles up an altitude of 50 km for drop tests using parachutes. But designing to the uncertainties, in so short a time, forced the engineers to give up on a lander for Mars and to settle for the simpler flyby task, and for Venus it was decided to design a probe to report on conditions in the atmosphere without the need to survive impact with the surface. Nevertheless, even these simpler missions were very challenging, not least owing to the large uncertainties in the ephemerides of the two planets, which for Mars exceeded the diameter of the planet itself.

The scientific objectives for the Mars flyby spacecraft were specified by Mstislav Keldysh in a document dated March 15, 1960:

1. Photograph the planet from a range of 5,000 to 30,000 km at a surface resolution of 3 to 6 km with the coverage including of one of the polar regions
2. Coverage of the infrared C-H band in the reflection spectrum, to search for plant or other organic material on the surface
3. Research into the ultraviolet band of the Martian spectrum.

The instruments and spacecraft were rushed through development in order to meet the deadline. There were many problems at the factories. The spacecraft delivered to the launch site at the end of August were a shambles. Korolev's team worked around the clock to solve the numerous technical problems, constantly taking subsystems apart for repair and retesting. The communications system gave the most trouble. In fact, full scale integrated testing did not begin until September 27, which was the optimum launch date!

Korolev also raced to assemble his first four-stage R-7. The new third stage was a conversion from another vehicle, so the real task was to rapidly develop the entirely new fourth stage with its restartable engine. The pressure on the rocket team was not eased by knowing that the first test launch would be a full-fledged attempt at Mars.

Ultimately, the payload had to be slashed in order to make mass available for test instrumentation on the new upper stage combination. The heaviest instruments – the camera, infrared spectrophotometer and ultraviolet spectrometer – were deleted. The two spacecraft did not reach the launch pad until well after the optimum launch date, which meant they would not be able to approach as close to Mars as intended. But in the event this was of no consequence because they were both victims of their launch vehicles.

Had these Mars spacecraft and the Venus spacecraft in February 1961 succeeded, then the world would have been treated to a spectacular planetary exploration coup in May 1961. The Venus probes would have arrived at their destination on May 11

Spacecraft launched

First spacecraft:	1M No.1 [Mars 1960A, Marsnik 1]
Mission Type:	Mars Flyby
Country/Builder:	USSR/OKB-1
Launch Vehicle:	Molniya
Launch Date/Time:	October 10, 1960 at 14:27:49 UT (Baikonur)
Outcome:	Launch failure, third stage malfunction.
Second spacecraft:	1M No.2 [Mars 1960B, Marsnik 2]
Mission Type:	Mars Flyby
Country/Builder:	USSR/OKB-1
Launch Vehicle:	Molniya
Launch Date/Time:	October 14, 1960 at 13:51:03 UT (Baikonur)
Outcome:	Launch failure, third stage did not ignite.

and 19, and the Mars flybys would have occurred on May 13 and 15. Following only one month after Gagarin's orbital flight in April, the effect of these triumphs on the West would have exceeded even Sputnik.

The fact that the new four-stage R-7 and Mars spacecraft were even launched is a testament to the can-do attitude of Korolev's team against almost impossible odds. Nevertheless, a typical Russian sense of resignation ran underneath the optimism. In early September, during the scramble to make the launch date, a spacecraft engineer remarked, "Forget about that radio unit and all the Mars problems. The first time we won't fly any farther than Siberia!" He was right.

Spacecraft:

The spacecraft was essentially a cylindrical container, 2.035 meters long and with a diameter of 1.05 meters, pressurized to 1.2 bar with nitrogen for the avionics and instruments, with a dome on top housing the propulsion system, which was a fixed 1.96 kN KDU-414 restartable liquid hypergolic rocket engine that burned nitric acid and dimethylhydrazine. The engine was capable of making one or more trajectory correction maneuvers with a total firing time of 40 seconds.

The power system consisted of two fixed 1.6 × 1.0 meter solar panels populated by a total of 2 square meters of solar cells, and had silver-zinc batteries for storage. Thermal control was achieved using internal circulation fans in association with shutters on the exterior to stabilize the internal temperature to 30°C.

The avionics included a telemetry tape recorder and a program timer (actually a clockwork event sequencer) that had to be preset for a specific time of launch. The communication system consisted of three units. A directional system used a high-gain 2.33 meter diameter fine copper net parabolic dish for 8 cm (3.7 GHz) and 32 cm (922 MHz) band transmitters. This antenna was to open automatically when the spacecraft separated from the fourth stage of the launcher. Two cross-shaped semi-

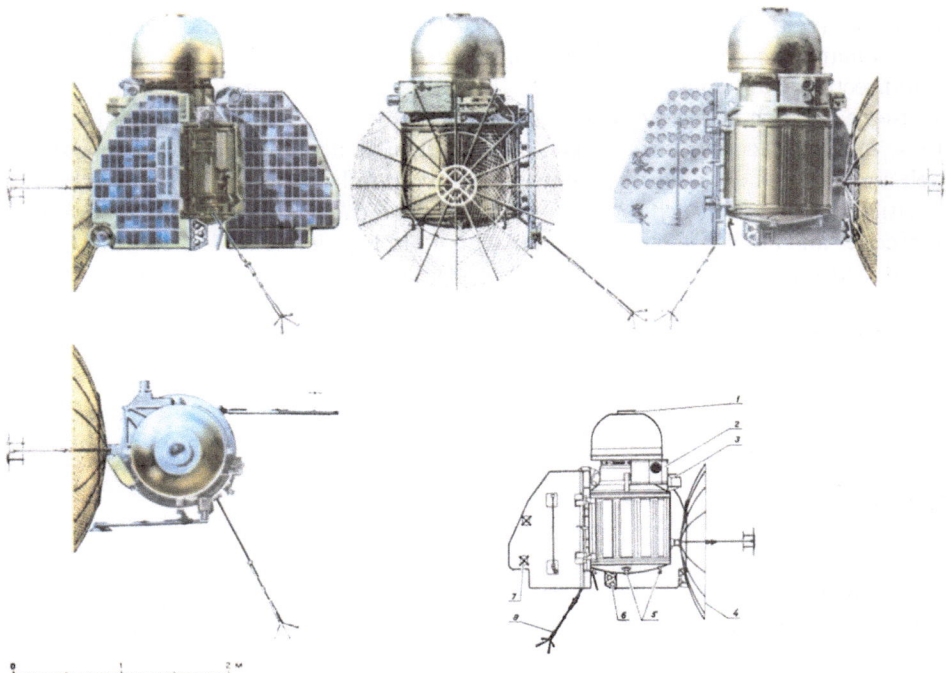

Figure 7.1 Diagram of the 1M spacecraft: 1. Propulsion system nozzle; 2. Sun and star sensors; 3. Earth sensor (1VA only); 4. Parabolic high gain antenna; 5. Attitude control jets; 6. Thermal sensors; 7. Medium gain antenna; 8. Boom omni antenna.

Figure 7.2 The 1M Mars spacecraft.

directional medium-gain antennas were mounted on the back of the solar panels for the command receiver and for low bandwidth telemetry at 922.8 MHz. A low-gain omnidirectional antenna was affixed to the end of the 2.2 meter magnetometer boom for use near Earth in the 1.6 meter band. Commands were sent at 768.6 MHz at 1.6 bits/s. Before executing an uplinked command sequence, the spacecraft repeated it back and awaited acknowledgement from the ground.

Attitude sensing was achieved using fixed Sun and star sensors in combination with gyroscopes and accelerometers, and a system of nitrogen gas jets adopted from Luna 3 provided attitude control and 3-axis stabilization. In cruise mode the solar panels were maintained within 10 degrees of perpendicular to the Sun. For telemetry sessions, the spacecraft used radio bearings to turn and lock onto Earth. For the 1VA Venus spacecraft, this mode was improved by using a separate Earth optical sensor.

Launch mass: 650 kg (*dry mass* ~480 kg)

Payload:

1. Boom mounted triaxial fluxgate magnetometer to search for a Martian magnetic field
2. Ion trap charged particle detectors to investigate the interplanetary plasma medium
3. Micrometeoroid detector to investigate interplanetary spacecraft hazards
4. Cosmic ray detectors to measure radiation hazards in space
5. Infrared radiometer to measure the Martian surface temperature
6. Facsimile film camera system to image the surface (not flown)
7. Infrared 3 to 4 micron C-H band spectrometer to search for organic compounds (not flown)
8. Ultraviolet spectrometer to determine atmospheric composition (not flown)

Most of the instruments were externally mounted. The camera and spectrometer were inside the pressurized module with their optics observing through a port. The camera was the same facsimile film system as that flown on Luna 3 to photograph the Moon and used the 3.7 GHz channel for transmission. It would be triggered by a Mars sensor.

The interplanetary cruise instruments were derived from balloon and sounding-rocket experiments. Shmaia Dolginov supplied the magnetometer and Konstantin Gringauz the two ion traps. The cosmic ray detectors consisted of two Geiger counters and one sodium iodide scintillator inside the pressurized container, and one cesium iodide scintillator mounted externally, all of which were supplied by Sergey Vernov. Tatiana Nazarova provided the micrometeoroid sensor.

All three key planetary instruments outlined in Keldysh's memo in March were ultimately deleted. The schedule was set in February and the launch window opened on September 20, leaving very little time to build the instruments. The spacecraft itself had a large number of problems in development and testing. On September 20 the radio was still at the factory and the electrical system was not working. The radio had further problems after it arrived for integration with the spacecraft. By now the minimum energy launch date on September 27 had passed, and every day thereafter the mass that could be launched diminished. To save mass as the days passed, the camera system was deleted. It had suffered test and integration problems of its own. Finally, as the end of the launch window approached, the infrared spectrometer, which had failed to detect life during a field test in Kazakhstan, and the ultraviolet spectrometer were deleted. Pressure integrity tests of the avionics compartment were never done. After the launch of the first spacecraft failed and the mass constraint tightened, the entire science payload and midcourse engine were removed. As it was too late in the launch window to attempt the desired close flyby of Mars, the goal of the mission was reduced to simply gaining flight experience with the spacecraft.

Payload mass: 10 kg

Mission description:

Spacecraft 1M No.1 arrived at the pad on October 8 and was launched on October 10, towards the end of the launch window. It did not achieve Earth orbit. Resonant vibrations in the launcher during the second stage burn caused a gyroscope in the avionics to malfunction. At 309 seconds into the flight, after third stage ignition, the vehicle pitched over beyond permissible limits and the engine was shut down. The stack crashed in eastern Siberia.

Figure 7.3 Preparing for the first test of the new four-stage R-7 on October 10, 1960, carrying the 1M Mars flyby spacecraft.

1M No.2 did not achieve Earth orbit either. Its launcher failed after 290 seconds, when the third stage engine failed to ignite. An oxidizer leak on the pad had frozen the kerosene in the feed pipes. The launch window closed before the planned third spacecraft could be launched. The failure of the third stage on both launches robbed the new fourth stage and the spacecraft of any chance to perform.

The Soviets made no announcement of either launch, since they had not reached orbit. But the US was cognizant, having tracked them from its surveillance station in Turkey and from a reconnaissance aircraft flying between Turkey and Iran. Tracking stations along the southern borders of the USSR readily picked up radio traffic prior to launches, and the telemetry from early Soviet launch vehicles was unencrypted. The deployment of tracking ships was a further indication that a launch attempt was imminent, and in any case planetary launch windows were well known. Only a few lunar and planetary launches escaped detection by the Americans.

Nikita Khrushchev was in New York for a United Nations meeting during the launch period. He had a model of the 1M Mars probe with him as a bragging piece. After the first failure, instead of boasting with his model, he made his famous speech with accompanying shoe-banging on October 12. When the second launch failed, he was on his way back to the USSR with his model.

Results:

None.

THE FIRST VENUS SPACECRAFT: 1961

Campaign objectives:

The first ever Venus campaign consisted of two spacecraft, each almost identical to the two lost in the failed Mars launches four months earlier. As had been the case for the Mars spacecraft, the Venus spacecraft were also built in a great hurry. Although there was additional time, the schedule did not provide the iterative design process and extensive ground testing employed by later flight programs. Korolev's engineers had to spend considerable time and effort debugging the systems. There were many disassemble, reassemble and test cycles to fix failed items, and after one fix another failure would occur. Once again the communications system was a major problem. Design issues emerged and workarounds had to be devised.

The 1V Venus spacecraft had originally been intended to be a lander with camera, but by the time the Mars spacecraft were launched in October 1960 it had become clear that the lander would not be ready for the Venus launch window that opened in January, and the payload mass had to be significantly reduced to accommodate the instrumentation for the new launcher. The lander was abandoned and the mission

Spacecraft launched

First spacecraft:	1VA No.1 [Sputnik 7]
Mission Type:	Venus Impactor
Country/Builder:	USSR/OKB-1
Launch Vehicle:	Molniya
Launch Date/Time:	February 4, 1961 at 01:18:04 UT (Baikonur)
Outcome:	Failed to depart Earth orbit, fourth stage failure.
Second spacecraft:	Venera 1 (1VA No.2)
Mission Type:	Venus Impactor
Country/Builder:	USSR/OKB-1
Launch Vehicle:	Molniya
Launch Date/Time:	February 12, 1961 at 00:34:37 UT (Baikonur)
Mission End:	February 17, 1961
Encounter Date/Time:	May 20, 1961
Outcome:	Failed in transit, communications lost.

descoped to a simple impactor. The goal was changed to undertaking science during the interplanetary cruise and in the environment of Venus prior to impact. A small passive entry capsule was carried containing medallions. The 1VA redesign used as much of the 1M spacecraft as possible. Launched in February 1961, these were the second set of spacecraft to be launched to a planet, and the first to Venus, preceding the first US attempt at Venus by 18 months.

Only one 1VA spacecraft, Venera 1, was successfully launched on a trajectory to Venus. It was the first spacecraft ever successfully sent on a trajectory to another planet. Unfortunately it suffered from severe attitude control and thermal problems, and was lost after less than a week's flight time.

The truncated flight of Venera 1 was offset by the triumph of the orbital flight by Yuri Gagarin on April 12, 1961. These achievements, together with the capability to launch heavy satellites and three successful Luna missions in 1959, established the USSR as pre-eminent in space flight in mid-1961. All that America could claim was eight lunar mission launch failures and one launch which, due to insufficient boost, resulted in a distant lunar flyby, all involving tiny spacecraft that had been created more as an afterthought to rocket development than as deliberate designs for space exploration.

Spacecraft:

The spacecraft were 2.035 meter long canisters, 1.05 meters in diameter, that were pressurized to 1.2 bar. They had 1 square meter solar arrays, medium-gain antennas on each panel, a boom omnidirectional antenna, and a dome-shaped propulsion unit on top. The sequencer, communications, attitude control, navigation, and propulsion systems were the same as the 1M spacecraft. The attitude control system had three modes of operations: a 3-axis cruise mode for continuous solar pointing, a

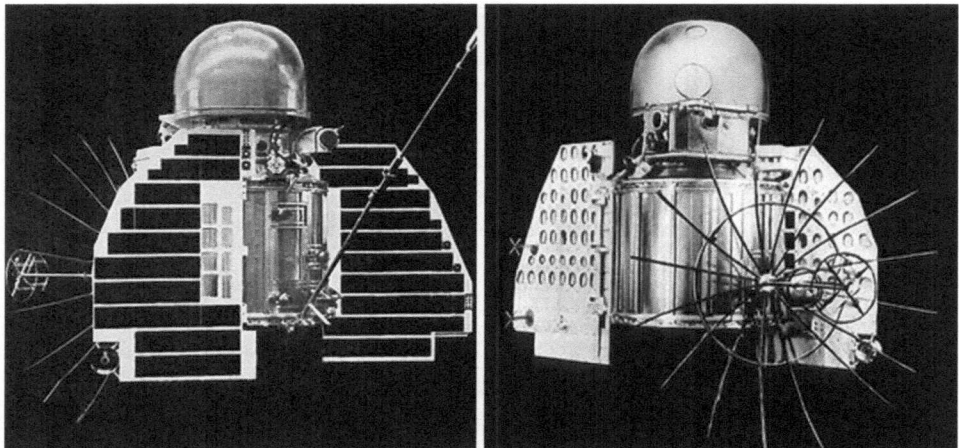

Figure 7.4 Venera 1 spacecraft, front and back.

back-up system for spinning about the solar axis in the event of some failure in the primary system, and a 3-axis Earth pointing system for communication using the 2.33 meter high-gain parabolic mesh antenna. Thermal control was by passive louvers activated by internal temperature. A key difference with the Mars spacecraft was the addition of an Earth sensor, instead of a radio beacon, for more precise orientation during a high gain telemetry session.

The spherical entry device was mounted inside the pressurized canister, and not separable. Having been encased with thermal shielding, it was expected to survive as the rest of the spacecraft burned up on atmospheric entry. It was to free-fall through the atmosphere and impact the surface. Although it was not designed to survive an impact with a solid surface, it was expected to be able to float if it happened to come down on an ocean.

At that time the Venus ephemeris was more poorly known than Mars, the errors being about 15 times its radius, so achieving an impact was not an easy task. Radar ranging of Venus was obtained for the first time in early April, while the planet was at inferior conjunction, enabling the ephemeris error to be reduced to 500 km. It is possible that if Venera 1 was still functioning, the Soviets would have used this new data to program a trajectory correction maneuver a few weeks prior to its arrival at the planet in May.

Launch mass: 643.5 kg

Payload:

Main spacecraft:

1.   Boom-mounted triaxial fluxgate magnetometer to search for a Venus magnetic field

2.   Ion trap charged particle detectors to investigate the interplanetary medium
3.   Micrometeoroid detector to investigate interplanetary spacecraft hazards
4.   Cosmic ray detectors to measure radiation hazards in space
5.   Infrared radiometer for Venus temperature

These instruments were identical to those on the 1M Mars spacecraft. It is also reported to have had a pair of parallel magnetometers to measure the interplanetary magnetic field.

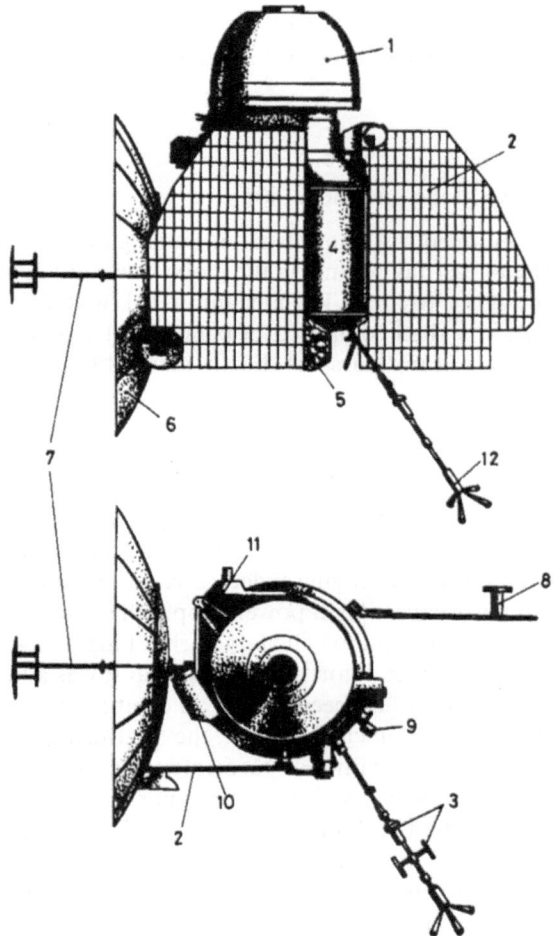

Figure 7.5 1VA Venera 1 diagram (from *Space Travel Encyclopedia*): 1. Propulsion module; 2. Solar panels; 3. Magnetometer; 4. Thermal control shutters; 5. Thermal sensors; 6. High gain antenna; 7. Dipole emitters; 8. Medium gain antenna; 9. Ion trap; 10. Earth sensor; 11. Sun and star sensor; 12. Boom omni antenna.

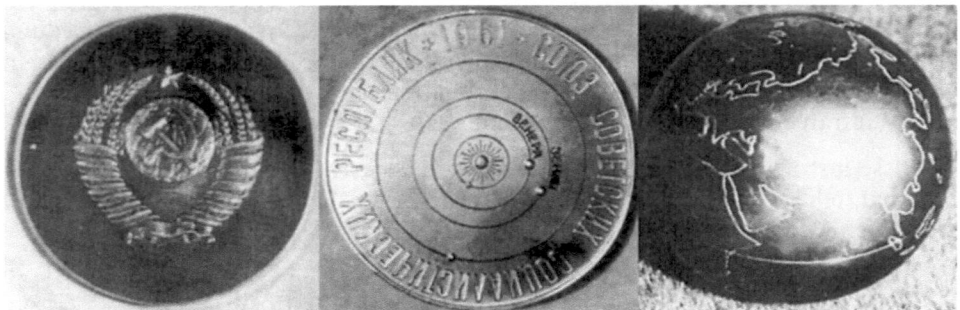

Figure 7.6 Medallions contained in the Venera 1 entry probe.

Entry probe:

1.   Commemorative globe and medallion

The entry probe contained a 70 mm diameter metal globe with a commemorative medallion inside. The terrestrial oceans on the globe were blue-tinted and continents gold-tinted. It was designed to float. The medallion disk was inside the globe, which in turn was contained in a shell composed of pentagonal stainless steel elements on each of which was inscribed (in Russian) 'Earth-Venus 1961'.

Mission description:

On February 4, 1961, the new Molniya planetary rocket managed for the first time to deliver its fourth stage with attached spacecraft into a low 'parking' orbit. After 60 minutes of unpowered coast, the engine failed to reignite, stranding the 1VA No.1 spacecraft. The failure was caused by a power supply that used a transformer which not been designed to work in vacuum! The large orbital mass, 6,483 kg including the propulsive stage, prompted speculation in the West that it was a failed manned craft. The Soviets later said that they had been testing an orbiting platform from which an interplanetary probe could be launched. In fact, the 'platform' was no more than the new fourth stage of the launcher with the spacecraft attached. The Soviet description was doubtless highlighting the introduction of the parking orbit technique for deep space missions. It was designated Sputnik 7 in the US. On February 26 it re-entered the atmosphere over Siberia. Interestingly, the wreckage was discovered by a young boy and the heat-damaged pennant handed over to the KGB. The recovered articles were returned to the Academy of Sciences, which later sold them at auction in New York in 1996 to raise money for Russia's impoverished science programs.

The power supply problem in the first launch was traced to improperly mounting a transformer outside where it would be exposed to vacuum. A quick fix was rigged in time for the second launch by sealing the apparatus inside a vacuum-tight battery box.

On February 12, 1961, an 'Automatic Interplanetary Station' that was later named Venera 1 was successfully boosted out of parking orbit. Communications sessions 2 hours and 9 hours after launch confirmed that the spacecraft was on a 96-day type I trajectory that would take it to the vicinity of Venus. Subsequent tracking indicated that a large midcourse correction would be required, but the target was in the cross-hairs! Analysis of the telemetry showed that operation in the Sun-pointing mode was unstable. The spacecraft automatically switched to the backup spin-stabilized mode in which most electrical systems except the sequencer and thermal control were shut down. This was a serious design error, since the command receiver was also turned off and denied the ground control of the spacecraft. In this safe mode, the spacecraft would re-activate the communication system every 5 days for a session with Earth. The high gain antenna could not be used because the spacecraft could not point at Earth. After an agonizing 5 days, the spacecraft contacted Earth on February 17 at a distance of 1.9 million km. The session was used to check the primary Sun-pointing operation, which failed again. On February 22 the spacecraft failed to respond, and no signal was received. The Soviets asked Jodrell Bank to listen for telemetry, and sent a team to England to assist, but nothing was heard. Attempts by Yevpatoria on March 4 and 5 also failed to receive any signal. Due to the inability to conduct a midcourse maneuver, Venera 1 flew silently past the planet at 100,000 km distance. In case it was silently continuing its mission, commands were sent on May 20, 1961, the day of the encounter, without result.

It was later determined that the attitude control system failure was due to the Sun sensor overheating. The thermal control design had only considered the average temperature for the instrument, and not the localized temperature at an unpressurized sensitive element. The lack of response after February 17 was attributed to a failure of the sequencer for the communications system. There was also evidence that the motorized thermal control shutters were not operating properly.

The flight of Venera 1 was followed worldwide as the first mission to another planet – another coup for the Soviet Union. But failure followed quickly. Radio Moscow announced the loss on March 2, noting that an investigation was underway and that sabotage was not excluded. The window closed on February 15, before the third 1VA could be launched.

Results:

None for Venus. Results were obtained from the Venera 1 instruments during its short cruise period. A faint interplanetary magnetic field on the order of 3.5 nT was reported and the solar wind plasma flow discovered by Luna 1 to 3 was found to be present beyond the Earth's magnetopause in deep space. Venera 1 marked the first flight of a true interplanetary spacecraft with all the capabilities necessary for such a mission, including flexible attitude stabilization modes and midcourse maneuvering.

8

New spacecraft, new failures

TIMELINE: AUG 1961–NOV 1962

The short flight of Venera 1 revealed much about the requirements for planetary spacecraft, and in the time remaining to the next launch windows Sergey Korelev's engineers developed the 2MV design that was to become the basis for many Venera and Mars spacecraft in years to come.

Meanwhile, the Americans worked on their first true lunar spacecraft, which was much heavier than its precursors owing to the use of the new Atlas-Agena launcher. This Ranger lunar spacecraft was also the basis for the successful Mariner series of planetary spacecraft, both built by the Jet Propulsion Laboratory. The US skilled up by a process of trial and error, with the Ranger series suffering six failures before the launcher and spacecraft were perfected. The Mariner series was more successful at the outset and, ironically, the US had a successful mission to Venus in 1962 before it had one to the Moon! The Soviets were quite chagrined that the US had beat them to Venus despite their own early and extensive lead in rockets and spacecraft.

The Soviet Union launched six 2MV spacecraft in 1962, three for Venus and three for Mars. Only one survived its launch vehicle, Mars 1. Launch vehicle failures were to continue to be a major cause of lunar and planetary mission losses throughout the 1960s, with the dominant cause being fourth-stage problems. Mars 1 flew for almost 5 months and exposed numerous problems with the new spacecraft design before it finally failed in transit.

A BETTER SPACECRAFT: A SECOND TRY AT VENUS: 1962

Campaign objectives:

After the failures of the 1M Mars missions in October 1960 and the 1VA Venus missions in February 1961, Korolev resolved to develop an improved, second

Launch date

1961

23 Aug	Ranger 1 lunar mission test	Upper stage failed second burn
18 Nov	Ranger 2 lunar mission test	Upper stage failed second burn

1962

26 Jan	Ranger 3 lunar hard lander	Missed Moon by 37,745 km
23 Apr	Ranger 4 lunar hard lander	Failed, impacted lunar far side
22 Jul	Mariner 1 Venus flyby	Launch vehicle failure
25 Aug	Venera entry probe	Fourth stage failure
27 Aug	Mariner 2 Venus flyby	Successful Venus flyby on Dec 14
1 Sep	Venera entry probe	Fourth stage failure
12 Sep	Venera flyby	Third and fourth stage failures
18 Oct	Ranger 5 lunar hard lander	Spacecraft failed, flew past Moon
24 Oct	Mars flyby	Fourth stage exploded
1 Nov	Mars 1 flyby	Failed in transit Mar 21, 1963
4 Nov	Mars entry probe	Fourth stage failure

generation planetary spacecraft. In the spring of 1961 he directed that a new multi-mission spacecraft be designed that could be configured for either flyby or entry probe missions at either Mars or Venus. This new series was the first modular interplanetary spacecraft, with a standardized multipurpose 'orbital' module (in the US vernacular this was a carrier vehicle) to guide the spacecraft to either planet, and a separate module to carry a science payload tailored to the planet and mission type. Two standard types of science module were provided, the first a pressurized vessel to accommodate instruments for studying the planet during a flyby, and the second an entry vehicle for atmospheric probe or lander missions. For the latter, the entry vehicle was detached at arrival and the carrier vehicle discarded and left to burn up in the atmosphere. This was a major improvement over the 1VA design, where the probe was retained and simply expected to survive the destruction of the spacecraft on entry.

The communications, attitude control, thermal control, entry, and propulsion systems were much improved over the 1M and 1VA spacecraft. This new generation set the design precedent for all Molniya-launched planetary missions until the more capable Proton launcher was introduced. The initial design, designated 2MV, lasted only for the 1962 Mars and Venus opportunities. Six were built and launched, three for Venus and three for Mars. Korolev also upgraded the 8K78 launcher to lift these heavier spacecraft by improving the strap-on booster engines and lengthening both the interstage between the third and fourth stages and the aerodynamic shroud. Only one 2MV survived its launch vehicle, Mars 1. After the 1962 campaign, the design was upgraded to produce the 3MV.

For the 1962 launch opportunity for Venus the Soviets prepared two 2MV-1 entry probe spacecraft and one 2MV-2 flyby spacecraft. The mission of the entry probes was to penetrate below the veil of clouds, survive landing, and return data profiling

Spacecraft launched

First spacecraft: 2MV-1 No.3 [Sputnik 19]
Mission Type: Venus Atmosphere/Surface Probe
Country/Builder: USSR/OKB-1
Launch Vehicle: Molniya
Launch Date/Time: August 25, 1962 at 02:18.45 UT (Baikonur)
Outcome: Failed to leave Earth orbit, fourth stage failure.

Second spacecraft: 2MV-1 No.4 [Sputnik 20]
Mission Type: Venus Atmosphere/Surface Probe
Country/Builder: USSR/OKB-1
Launch Vehicle: Molniya
Launch Date/Time: September 1, 1962 at 02:12:30 UT (Baikonur)
Outcome: Failed to leave Earth orbit, fourth stage failure.

Third spacecraft: 2MV-2 No.1 [Sputnik 21]
Mission Type: Venus Flyby
Country/Builder: USSR/OKB-1
Launch Vehicle: Molniya
Launch Date/Time: September 12, 1962 at 00:59:13 UT (Baikonur)
Outcome: Failed to leave Earth orbit, third and fourth stage failures.

the temperature, pressure, density, and composition of the atmosphere, and then the composition of the surface. The flyby spacecraft would photograph the planet using an upgraded version of the camera originally intended to fly on the 1M spacecraft. Frustratingly, all three spacecraft would be lost to failures of the launcher's fourth stage.

Spacecraft:

The 2MV spacecraft was 1.1 meters in diameter and 3.3 meters long in total, and measured 4 meters across the solar panels with the thermal radiators deployed. It was divided into two attached parts. The main spacecraft, known as the 'orbital' (or carrier) module, was 2.7 meters long including the 60 cm long propulsion system at one end. The carrier's pressurized compartment contained the flight system avionics and scientific instrumentation. The propulsion system used cold gas jets for attitude control and the KDU-414 gimbaled engine that delivered a thrust of 2 kN and was capable of more than one midcourse correction, for a total firing time of 40 seconds. Attached to the other end of the carrier module was either a 60 cm long pressurized flyby instrument module or a detachable 90 cm diameter spherical entry probe.

In addition, significant changes were made to the communication system. A 1.7 meter parabolic antenna and radio system transmitting at either 5, 8 or 32 cm was provided for high rate communications during the interplanetary cruise and for data transmission on arrival at the target. A separate omnidirectional antenna and meter

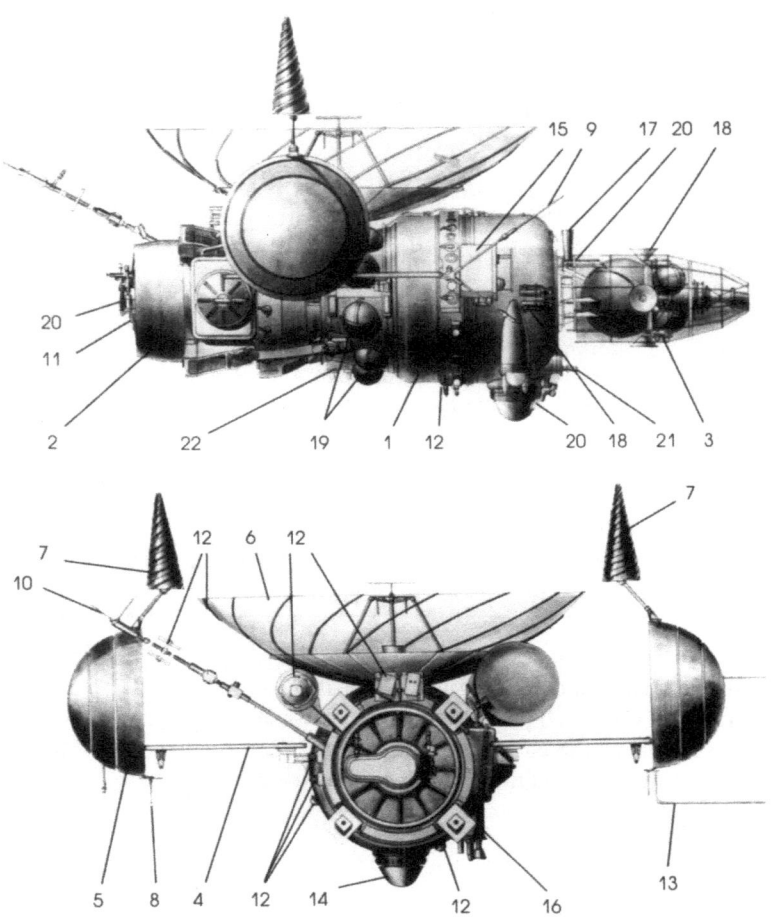

Figure 8.1 The 2MV flyby spacecraft (courtesy Energiya Corp): 1. Pressurized orbital module; 2. Pressurized imaging module; 3. Propulsion system; 4. Solar panels; 5. Thermal control radiators; 6. High gain parabolic antenna; 7. Low gain omnidirectional antennas; 8. Low gain omnidirectional antenna; 9. Meter-band antenna; 10. Emergency omni antenna; 11. Camera and planet sensor port; 12. Science instruments; 13. Meter band antenna; 14. Sun and star tracker; 15. Emergency radio system; 16. Continuous sun sensor; 17. Earth tracking antenna; 18. Attitude control nozzles; 19. Attitude control nitrogen tanks; 20. Attitude sensor light baffle; 21. Coarse sun tracker; 22. Sun tracker.

band transmitter was added to supplement the decimeter high gain directional unit. Commands were received at 39 cm (768.96 MHz) using semi-directional antennas attached to the thermal radiators, which were also used for transmission at 32 cm (922.776 MHz) in the vicinity of Earth and at a reduced rate at longer range in the event of an emergency. A backup 1.6 meter band radio was provided for operations

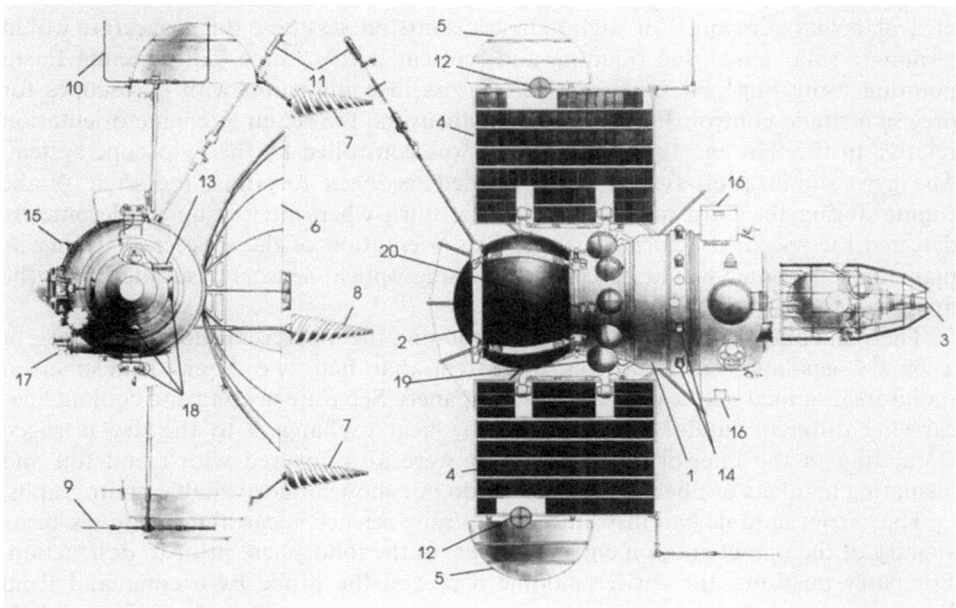

Figure 8.2 The 2MV probe spacecraft (courtesy Energiya Corp): 1. Orbital module; 2. Entry capsule; 3. Propulsion system; 4. Solar panels; 5. Thermal control radiators; 6. High gain antenna; 7. Medium gain antennas; 8. Entry capsule test antenna; 9. Meter band transmit antenna; 10. Meter band receiver antenna; 11. Magnetometer and boom antenna; 12. Low gain antennas; 13. Earth sensor; 14. Science instruments; 15. Sun/Star sensor; 16. Emergency radio system; 17. Sun sensor; 18. Attitude control nozzles; 19. Nitrogen tanks; 20. Sun sensor.

near Earth at 115 and 183.6 MHz using whip antennas mounted on top of the solar panels. For flyby missions, the camera in the instrument module had its own 5 cm band impulse transmission system. The high gain antenna was fixed, pointing in the opposite direction to the solar arrays. To use it, the spacecraft was required to adopt Earth-pointing attitude. An onboard tape recorder was provided to store data while Sun pointing and to replay it when the high gain antenna had locked on. The power supply system consisted of 2.6 square meters of solar cells that supplied 2.6 kW to a 42 amp-hour NiCd battery array.

The attitude control system was upgraded by providing an Earth sensor for high gain antenna pointing, instead of using a radio bearing. And based on the Venera 1 experience, the Sun, Earth, and star sensors were repositioned inside the controlled environment of the carrier module, looking out through a quartz window dome. New and more reliable sequencers were used, with element-by-element redundancy. As with Venera 1, several orientation modes were provided. While cruising, the spacecraft was to maintain a low-precision 3-axis Sun pointing mode in order to keep the solar panels illuminated. To avoid losing control of the spacecraft as with Venera 1, it was decided never again to turn off the receivers during the cruise phase of a planetary mission. For high gain transmission sessions, the spacecraft would terminate solar panel Sun pointing and reorient itself to high gain antenna Earth pointing using Sun and Earth optical sensors in conjunction with gyroscopes for precise attitude control. For midcourse maneuvers, the required engine orientation relative to the Sun and the star Canopus was controlled by the gyroscope system. The gyro stabilization system also provided feedback to adjust the angle of the engine during the burn and terminated the burn when integrating accelerometers detected the specified velocity change. The orientation of the spacecraft during its planetary encounter would be controlled using optical sensors associated with the imaging system in the instrument module.

Thermal control was improved by abandoning the motorized shutters in favor of a binary gas-liquid thermal control system that had two liquid hemispherical radiators mounted on the ends of the solar panels. Separate heating and cooling lines carrying different liquids were coupled by heat exchangers to the dry nitrogen circulating in the interior. The spacecraft were also covered with metal foil and insulating blankets of fiberglass cloth that do not show in the available photographs.

The carrier module had instruments for cruise science, measurements in the near-vicinity of the planet and, on entry missions, in the ionosphere prior to destruction. For entry missions, the carrier module deployed the probe by a command from Earth that triggered a timer just prior to entering the atmosphere. Pyrotechnic charges were fired to release the restraining straps and the entry probe was ejected by a spring-like mechanism. The entry system was a 90 cm diameter sphere protected by an ablative aeroshell material. In addition to the science instruments, the entry probe contained a three-stage parachute system, silver-zinc batteries, and a decimeter band radio with a semi-directional antenna for direct transmission to Earth. Based on the best guess of surface conditions at the time, the probes were designed to survive pressures up to 5.0 bar and temperatures up to 77°C. The Venus and Mars probes were almost identical, but those for Venus had thicker shells and smaller parachutes and whereas the Mars probes were cooled by air circulation the Venus probes were cooled using a passive ammonia-based system. Unlike Venera 1, the new probes were chemically sterilized by being soaked in an atmosphere of 60% ethylene oxide and 40% methyl bromide in order to prevent biological contamination of the surface of their target on landing.

Launch mass:	1,097 kg (probe version)
	~890 kg (flyby version)
Probe mass:	~305 kg

Payload:

Carrier spacecraft:

1.   Magnetometer to measure the magnetic field
2.   Scintillation counters to detect radiation belts and cosmic rays
3.   Gas discharge Geiger counters

4. Cherenkov detector
5. Ion traps for electrons, ions and low energy protons
6. Radio to detect cosmic waves in the 150 to 1,500 meter band
7. Micrometeoroid detector

This list is for the Mars 1 payload, and it is assumed here that the carrier modules of all the 2MV series were similarly instrumented. The magnetometer was mounted on a 2.4 meter boom, and ribbon antennas were extended for the cosmic wave radio detector. Starting with these 2MV spacecraft, piezoelectric micrometeoroid detectors with a total area of 1.5 square meters were attached to the rear of the solar panels.

Descent/landing capsule:

1. Temperature, pressure and density sensors
2. Chemical gas analyzer
3. Gamma-ray detector system to measure radiation from the surface
4. Mercury level wave motion detector

The chemical gas analyzer consisted of simple chemical test cells, precursors for the proper chemical test instruments that would be flown on later missions. Platinum wire resistance thermometers were utilized, and the density gauge was an ionization chamber for measurements in the upper atmosphere where the pressure was less than 10 millibars.

Flyby instrument module:

Instrumentation was probably the same as the Mars flyby module with the exception of the infrared spectrometer, which for Venus was designed to study the atmosphere instead of the surface.

1. Facsimile imaging system to photograph the surface
2. Ultraviolet spectrometer in the camera system for ozone detection
3. Infrared spectrometer to study the thermal balance of the atmosphere

The imaging system was complex and heavy. It weighed 32 kg and was mounted inside the pressurized instrument module, peering out through portholes on the end. It focused 35 mm and 750 mm lenses on 70 mm film with a capacity of 112 images, alternately shot with square frames and 3×1 rectangular frames. Individual frames could be scanned or rescanned at 1,440, 720, or 68 lines and stored on wire tape for later transmission. An ultraviolet spectrograph projected its spectrum onto the film alongside the images. The imaging system had a dedicated 5 cm (6 GHz) impulse transmitter housed inside the instrument module. This transmitter would issue short 25 kW pulses with an average power output of 50 W. The transmission rate was 90 pixels/second, requiring about 6 hours to transmit a high resolution image of 1,440 $\times$ 1,440 pixels. The pixels were probably encoded as analog pulse position rather than as binary values. The infrared spectrometer was on the exterior of the instrument module and bore-sighted with the camera.

Mission description:

All three missions were lost to fourth-stage failures after successful insertion into parking orbit. On the 2MV-1 No.3 mission, only three of four ullage control solid rocket motors on this stage fired, causing it to somersault after 3 seconds. The main engine did ignite, but because of the tumbling motion it burned for only 45 of the planned 240 seconds. Several pieces were left in orbit. On the 2MV-1 No.4 mission, a stuck valve blocked the fuel line and the fourth stage failed to reignite.

The 2MV-2 No.1 Venus flyby spacecraft was lost due to a violent shutdown of the third stage. An engine in the third stage exploded at shutdown because the LOX valve did not close, continuing to feed LOX into the combustion chamber. The third stage broke up into seven pieces. The fourth stage continued into parking orbit, but the tumbling imparted to it by the destruction of the third stage induced cavitation in the oxidizer pump which caused the engine to shut down less than a second after it was reignited for the escape burn.

Results:

None.

THE FIRST MARS SPACECRAFT: 1962

Campaign objectives:

After the three Venus launches failed in late August and early September, Korolev's team scrambled to prepare for three more launches to Mars in late October and early November. Many measures were taken to enhance the reliability of the fourth stage. There was some pressure to abandon the Mars attempts until the problems with this stage were solved, but Korolev blazed ahead.

The 1962 Mars campaign consisted of two flyby missions and one entry probe. The objectives of the entry probe were to obtain in-situ data on the composition and structure of the atmosphere, and data on surface composition. The objectives of the flyby missions were to examine the interplanetary environment between Earth and Mars, to photograph that planet in several colors, to search for a planetary magnetic field and radiation belt, to search for ozone in the atmosphere, and to search for organic compounds on the surface. A comprehensive payload was prepared for each spacecraft, but apart from the camera and a magnetometer most of the payload was deleted when it was decided instead to install instrumentation to monitor the fourth stage to find out why it was suffering so many failures. These missions then became primarily engineering test flights of the 8K78 fourth stage, with Mars as a secondary objective.

Spacecraft launched

First spacecraft:	2MV-4 No.3 [Sputnik 22]
Mission Type:	Mars Flyby
Country/Builder:	USSR/OKB-1
Launch Vehicle:	Molniya
Launch Date/Time:	October 24, 1962 at 17:55:04 UT (Baikonur)
Outcome:	Failed in Earth orbit, fourth stage explosion.

Second spacecraft:	Mars 1 (2MV-4 No.4) [Sputnik 23]
Mission Type:	Mars Flyby
Country/Builder:	USSR/OKB-1
Launch Vehicle:	Molniya
Launch Date/Time:	November 1, 1962 at 16:14:16 UT (Baikonur)
Mission End:	March 21, 1963
Encounter Date/Time:	June 19, 1963
Outcome:	Failure in transit, communications lost.

Third spacecraft:	2MV-3 No.1 [Sputnik 24]
Mission Type:	Mars Atmosphere/Surface Probe
Country/Builder:	USSR/OKB-1
Launch Vehicle:	Molniya
Launch Date/Time:	November 4, 1962 at 15:35:15 UT (Baikonur)
Outcome:	Failed in Earth orbit, fourth stage disintegrated.

Although the fourth stage failed again on two of the launches, the second of three worked and provided the Soviets with their first spacecraft to Mars. Unfortunately, as in the case of Venera 1 it was immediately clear that Mars 1 had attitude control problems. The inability to perform a midcourse maneuver ruled out the desired close flyby of Mars. On the other hand, communications with Mars 1 were maintained for almost 5 months before it fell silent about half way to its target.

Spacecraft:

The 2MV Mars spacecraft were virtually identical to the versions described in detail above for the 1962 Venus missions. Although we have no description of the 300 kg entry probe of the 2M-3 No.1 spacecraft we know it was not designed as a lander but as a simple spherical entry system containing a parachute, radio, and instruments intended for measurements during descent. Surviving impact must have been more a hope than a goal. In fact, since the designers had no idea just how thin the Martian atmosphere is, the entry probe would have crashed into the surface before any useful data could have been returned.

The Mars 1 spacecraft is depicted in Figure 8.4 in a stand. Above the stand is the pressurized compartment containing the scientific instruments for the flyby. Next is the 'orbital' compartment. The large port in the front is the star sensor, and to the right of that is the Sun sensor. The gas bottles for the attitude control system are on

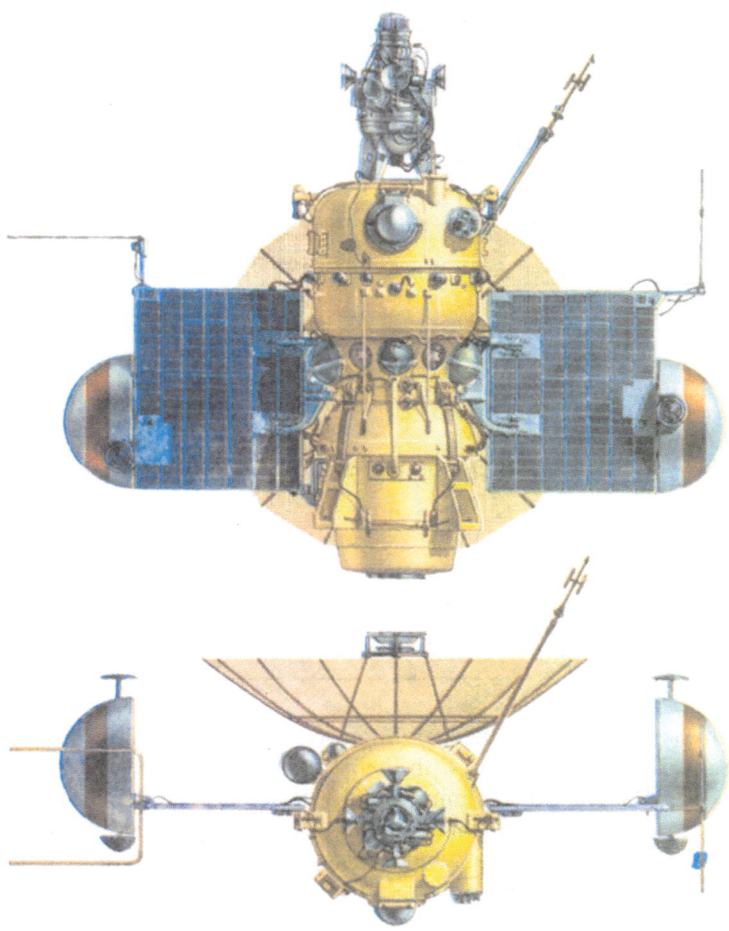

Figure 8.3 Diagram of the Mars 1 spacecraft.

the waist separating the two compartments. Topping the spacecraft is the propulsion system. The parabolic high gain antenna is fixed pointing in the opposite direction to the solar panels, and there are hemispherical radiators mounted on the ends of the panels.

Launch mass: 893.5 kg (Mars 1)
                        1,097 kg (probe version)
Probe mass: 305 kg

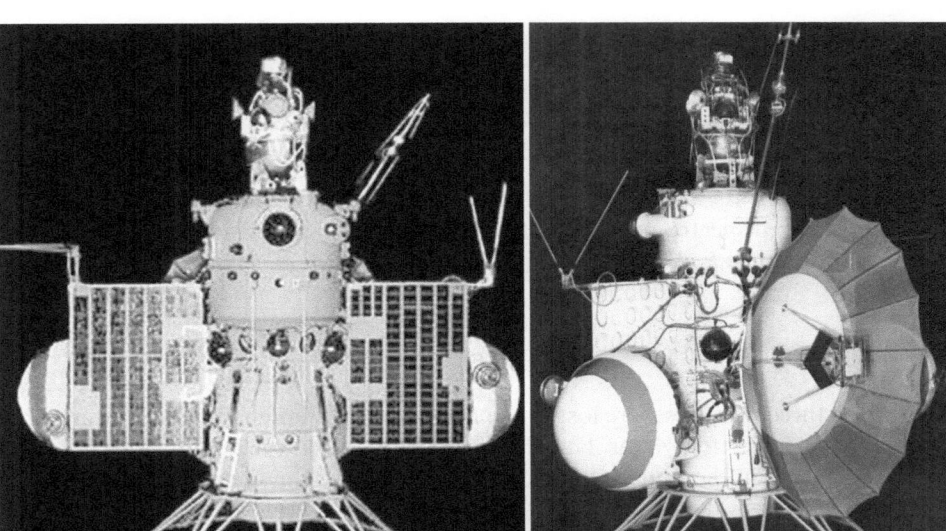

Figure 8.4 Mars 1 spacecraft, front (left) and back (right) views.

Payload:

Many of the instruments developed for the 2MV Mars spacecraft were removed in order to accommodate systems to monitor the fourth stage of the launcher. There is no information on how many were actually removed, but the magnetometer and the flyby imaging system are known to have been carried by Mars 1.

The original set of instruments is given in this list.

Carrier spacecraft:

1. Magnetometer to measure the magnetic field
2. Scintillation counters to detect radiation belts and cosmic rays
3. Gas discharge Geiger counters
4. Cherenkov detector
5. Ion traps for electrons, ions and low-energy protons.
6. Radio to detect cosmic waves in the 150 to 1,500 meter band
7. Micrometeoroid detector

Descent/landing capsule:

1. Temperature, pressure and density sensors
2. Chemical gas analyzer
3. Gamma-ray detector system to measure radiation from the surface
4. Mercury level movement detector

Flyby instrument module:

1.   Facsimile imaging system to photograph the surface
2.   Ultraviolet spectrometer in the camera system for ozone detection
3.   Infrared spectrometer to search for organic compounds

These instruments were identical to those built for the Venus mission, except that the Mars infrared spectrometer operated in the 3 to 4 micron C-H band to search for organic compounds and vegetation on the surface of Mars.

Mission description:

Two of the three missions were lost to the new and as yet unreliable fourth stage. The 2MV-4 No.3 Mars flyby was launched on October 24, 1962, but failed to leave parking orbit when the fourth stage turbo pump failed after 17 seconds due either to a foreign particle in the assembly or to the pump overheating after a lubricant leak. The fourth stage and spacecraft broke into five large pieces that re-entered over the course of the next few days. The US Ballistic Missile Early Warning System radar in

Figure 8.5 Mars 1 shortly prior to liftoff.

Alaska, which was at a state of high alert in the midst of the Cuban missile crisis, detected the debris after launch and was initially concerned that it might represent a Soviet nuclear ICBM attack, but rapid analysis of the debris pattern put this fear to rest.

The rocket carrying the second spacecraft was rolled out to the pad the next day, October 25, at the peak of the missile crisis. Shortly thereafter the firing range was ordered to battle readiness, which required the preparation for launch of the two R-7 combat missiles. One of these was stationed at the launch site where the Mars rocket stood. Stored in a corner of the Assembly and Test Building, it was uncovered and the launch team switched from supporting the Mars launch to preparing the missile. Fortunately, when the order to stand down came on October 27 the Mars rocket had not yet been removed from the launch pad. The 2MV-4 No.4 flyby spacecraft was successfully launched on the optimum date of the window, November 1, and became the first spacecraft to be sent towards Mars. The mission was named Mars 1. Just as in the case of Venera 1, a serious problem was discovered immediately after launch. The pressure in one of the two nitrogen gas containers was dropping rapidly because of a leaking valve. Later analysis showed that manufacturing had allowed debris to foul one of the valves. The outgassing caused the spacecraft to tumble out of control. When the tank drained after several days, ground controllers managed to use the gas in the remaining tank to halt the tumbling, restore the spacecraft to the desired Sun pointing attitude and spin it at 6 revolutions per hour so that the batteries would be continuously recharged from the solar panels. But by then most of the dry nitrogen for the cold gas jets of the attitude control system and for pressurizing the engine was expended. The backup gyro system used for attitude control was not designed for continuous use. Stuck in the backup Sun pointing spin mode, the spacecraft was unable to point its high gain antenna at the Earth or to make a midcourse correction. The Earth link was maintained through the UHF system and the medium-gain semi-directional antennas. Contact was established every 2 days for the first 6 weeks, and then every 5 days thereafter. On March 2, 1963, the signal strength began to decline and communications were lost on March 21, probably due to a final breakdown of the attitude control system at the unprecedented range of 106,760,000 km. The silent spacecraft would have passed Mars at a distance of about 193,000 km on June 19, 1963; the intended flyby distance was between 1,000 and 10,000 km.

The third spacecraft to be launched, 2MV-3 No.1, was stranded when the fourth stage failed to reignite properly. Vibrations in the core stage caused by cavitation in its oxidizer lines had dislodged a fuse and igniter in the fourth stage. Its engine was commanded to shut down after 33 seconds. The Americans detected five pieces of debris whose origins were unclear. The spacecraft is believed to have re-entered on January 19, 1963.

Of the six 2MV spacecraft launched between August and November 1962, four were lost to failures of the fourth stage, one was lost to a failure of both the third and fourth stages. The other one was launched successfully and named Mars 1, but failed in transit. No more 2MV spacecraft were built. The design was improved to produce the 3MV spacecraft for the next series of Mars and Venus missions in 1964–1965.

In the US, the orbital remains of the 1962 Venus and Mars spacecraft, including

Mars 1, were designated as Sputniks 19 to 24 in order of launch. All the spacecraft stranded in parking orbit re-entered within days.

Results:

No information was obtained on Mars. However, Mars 1 did acquire data during its cruise before it fell silent. The radiation zones around Earth were detected, and the distribution and flux of particles were measured. A third zone at 80,000 km was detected. The solar wind and magnetic fields were measured in interplanetary space to a farther distance than Venera 1. A solar wind storm was measured on November 30, 1962. The intensity of cosmic rays had almost doubled since 1959 due to a less active Sun. The micrometeoroid collision rate decreased with distance from Earth and showed intermittent increases as meteoroid showers were traversed. The Taurid meteor shower was encountered twice at ranges from 6,000 to 40,000 km, and again at distances from 20 to 40 million km, with a strike rate of one every 2 minutes on average.

9

Three more years of frustration

Between their success with Luna 3 in October 1959 and the opening of 1963, the Soviets suffered two failed lunar missions, five failed Mars missions, and five failed Venus missions. During those long 39 months, Sergey Korolev's engineers had been working on a new lunar soft-lander to take advantage of the four-stage version of the R-7. Unfortunately, this new Luna would suffer an even longer and more frustrating series of failures than the Ranger crash-lander being developed by the Americans. In fact, there would be eleven failures over the three years 1963–65 before managing a soft landing, with four of the six launch failures being caused by malfunctions of the fourth stage. Of those that were successfully deployed, Luna 4 to 8, two missed the Moon and the other three crashed onto it. The real objectives of these missions were not revealed at the time. Meanwhile, in 1964–65 the US finally had a string of successes with Ranger 7 to 9.

In parallel, the Soviets were applying the lessons from Venera 1 and Mars 1 in the development of the 3MV planetary spacecraft, basically an improved version of the 2MV, for the planetary launch windows of 1964 and 1965. To gain some experience with the new spacecraft, it was decided in late 1963 to make two interim flight tests, but these spacecraft were lost to launch vehicle failures on November 11, 1963 and February 19, 1964. Undeterred, when the Venus launch window opened the Soviets proceeded to launch spacecraft to Venus on March 27 and April 2, 1964. Only the second spacecraft was launched successfully, but because it was immediately clear that on board failures would prevent it from reaching its target it was named Zond 1 instead of being given a Venera designation. A launch to Mars was accomplished successfully on November 30, but once again the spacecraft was sufficiently crippled that it would not make its target and so it was named Zond 2.

The loss of so much hard work must have been doubly galling with the success of the Mariner 4 flyby of Mars on July 15, 1965. As with Venus, the US had somehow reached Mars first! Troubled by the significant problems suffered by Zond 1 and 2, the Soviets decided to conduct another 3MV test flight. Launched on July 18, 1965, outside a launch window, Zond 3 was unable to reach Mars but it could fly the kind of trajectory that would be used by a real mission. The timing of the departure was arranged so that it could test its camera system by photographing the far side of the Moon. It exercised its navigation systems, propulsion system, flight and instrument operations, and returned its pictures. Unfortunately communications were lost before it reached the equivalent of Mars distance, preventing it from demonstrating its deep space capabilities.

Launch date

1963

4 Jan	Luna lander	Fourth stage failure
3 Feb	Luna lander	Launcher veered off course
2 Apr	Luna 4 lander	Navigation failed, missed Moon
11 Nov	Mars test flight	Fourth stage failure

1964

30 Jan	Ranger 6 impactor	Impacted Moon, but cameras failed
19 Feb	Venera test flight	Third stage engine exploded
21 Mar	Luna lander	Third stage engine failure
27 Mar	Venera entry probe	Fourth stage engine failed
2 Apr	Zond 1 Venera entry probe	Communications failed in transit
20 Apr	Luna lander	Upper stage failures
28 Jul	Ranger 7 impactor	Success, images returned on Jul 31
5 Nov	Mariner 3 Mars flyby	Shroud failure
28 Nov	Mariner 4 Mars flyby	Successful Mars flyby Jul 15, 1965
30 Nov	Zond 2 Mars flyby	Communications failed after one month

1965

17 Feb	Ranger 8 impactor	Success
12 Mar	Luna lander	Fourth stage failure
21 Mar	Ranger 9 impactor	Success
10 Apr	Luna lander	Fourth stage failure
9 May	Luna 5 lander	Crashed on the Moon on May 12
8 Jun	Luna 6 lander	Midcourse failed, missed Moon
18 Jul	Zond 3 lunar flyby/Mars test	Successful lunar flyby, later lost
4 Oct	Luna 7 lander	Crashed on the Moon on Oct 7
12 Nov	Venera 2 flyby	Failed at flyby Feb 27, 1966
16 Nov	Venera 3 entry probe	Failed 17 days before arrival
23 Nov	Venera flyby	Upper stage failures
3 Dec	Luna 8 lander	Crashed on the Moon on Dec 6

The Zond 3 experience paid off on the flights of Venera 2 and Venera 3, both of which were launched later in November 1965, the first equipped to make a flyby and the second with an entry probe. They flew without serious problems and became the first Soviet planetary spacecraft to reach their target. But they added to their maker's woes by both losing communication, Venera 2 just as it arrived and Venera 3 when seventeen days from its target on a collision trajectory. Neither mission returned any data from Venus.

The years 1964–65 were crucial turning points in the Soviet space program. Since 1959 OKB-1 had been working on a plan to send cosmonauts on circumlunar flights, and had designed the Soyuz system for this objective. It had no plans to land people on the Moon. The Soviets initially saw the US stated intention to land a man on the Moon as mere hyperbole, but by 1964 it had become clear that major resources had been assigned to the program and that the development of the necessary rockets and spacecraft was progressing. Confident that their Soyuz could beat the Americans to a circumlunar mission but reluctant to be upstaged by a lunar landing, the Politburo directed Korolev in August 1964 to proceed with landing cosmonauts on the Moon in addition to performing the circumlunar program. However, in what would prove to be a huge management mistake, the government divided the development work for these programs amongst competing design bureaus without central leadership or responsibility. The space program was reorganized and additional resources applied. To manage all the work, new design bureaus were established and responsibilities distributed. At this point Soviet space policy underwent a fundamental change, and instead of pursuing long standing plans for the conquest of space it began to directly compete with the American program.

The other issue forcing a shake up in the Soviet space program was the long string of robotic lunar and planetary program failures between 1960 and 1965. Thus far, Korolev had been responsible for virtually all types of Soviet spacecraft and now he had been given the further task of overtaking the Americans in the race to the Moon. By his own admission, OKB-1 was overburdened and unable to devote enough time and resources to the robotic missions. In March 1965, on the advice of Keldysh, he asked his friend Georgi Babakin, who headed NPO-Lavochkin, to take over after the current production run of lunar and planetary spacecraft at OKB-1 ran its course. In April, Korolev handed over OKB-1's plans and knowledge base to Lavochkin. The remaining robotic spacecraft were launched during the rest of the year, ending with Luna 8 and Venera 3, while Lavochkin designed modifications and set up to produce new spacecraft.

THE YE-6 LUNAR LANDER SERIES: 1963–1965

Campaign objectives:

After having focused for several years on Mars and Venus, the Moon reasserted itself as a priority in concert with the progress of the manned space flight program. The 1959 Soviet plan to send cosmonauts on circumlunar flights had also envisaged robotic orbiters and landers. An early proposal for a Ye-5 lunar orbiter to respond to the first American attempts at small lunar orbiters was canceled along with its three-stage 8K73 launcher in favor of launching Ye-6 landers and Ye-7 orbiters using the four-stage 8K78 'Molniya' developed for planetary missions. These new spacecraft were also to exploit the design and flight experience of the second generation Mars and Venus spacecraft launched in 1962. The 2MV was a modular spacecraft with a common flight module and a mission-specific flyby or entry probe payload module. Unlike the earlier Luna spacecraft, which were launched directly towards the Moon, the Ye-6 series and all subsequent lunar missions were placed in Earth orbit for later injection onto a lunar trajectory by the restartable fourth stage.

Spacecraft launched

First spacecraft:	Ye-6 No.2 [Sputnik 25]
Mission Type:	Lunar Lander
Country/Builder:	USSR/OKB-1
Launch Vehicle:	Molniya
Launch Date/Time:	January 4, 1963 at 08:49:00 UT (Baikonur)
Outcome:	Failed to leave Earth orbit.

Second spacecraft:	Ye-6 No.3
Mission Type:	Lunar Lander
Country/Builder:	USSR/OKB-1
Launch Vehicle:	Molniya
Launch Date/Time:	February 3, 1963 at 09:29:14 UT (Baikonur)
Outcome:	Launch vehicle veered off course.

Third spacecraft:	Luna 4 (Ye-6 No.4)
Mission Type:	Lunar Lander
Country/Builder:	USSR/OKB-1
Launch Vehicle:	Molniya
Launch Date/Time:	April 2, 1963 at 08:16:37 UT (Baikonur)
Encounter Date/Time:	April 5, 1963
Mission End:	April 6, 1963
Outcome:	Navigation failed in transit, missed Moon.

Fourth spacecraft:	Ye-6 No.6
Mission Type:	Lunar Lander
Country/Builder:	USSR/OKB-1
Launch Vehicle:	Molniya-M
Launch Date/Time:	March 21, 1964 at 08:15:35 UT (Baikonur)
Outcome:	Upper stage failure. Did not reach orbit.

Fifth spacecraft:	Ye-6 No.5
Mission Type:	Lunar Lander
Country/Builder:	USSR/OKB-1
Launch Vehicle:	Molniya-M
Launch Date/Time:	April 20, 1964 at 08:08:28 UT (Baikonur)
Outcome:	Upper stage failure. Fourth stage failed to fire.

Sixth spacecraft:	Ye-6 No.9 (Cosmos 60)
Mission Type:	Lunar Lander
Country/Builder:	USSR/OKB-1
Launch Vehicle:	Molniya

Launch Date/Time: March 12, 1965 at 09:30:00 UT (Baikonur)
Outcome: Failed to leave Earth orbit.

<u>Seventh spacecraft:</u> Ye-6 No.8
Mission Type: Lunar Lander
Country/Builder: USSR/OKB-1
Launch Vehicle: Molniya
Launch Date/Time: April 10, 1965 (Baikonur)
Outcome: Upper stage failure. Did not reach orbit.

<u>Eighth spacecraft:</u> Luna 5 (Ye-6 No.10)
Mission Type: Lunar Lander
Country/Builder: USSR/OKB-1
Launch Vehicle: Molniya-M
Launch Date/Time: May 9, 1965 at 07:49:37 UT (Baikonur)
Encounter Date/Time: May 12, 1965 at 19:10 UT
Outcome: Crashed.

<u>Ninth spacecraft:</u> Luna 6 (Ye-6 No.7)
Mission Type: Lunar Lander
Country/Builder: USSR/OKB-1
Launch Vehicle: Molniya-M
Launch Date/Time: June 8, 1965 at 07:40:00 UT (Baikonur)
Encounter Date/Time: June 11, 1965
Outcome: Midcourse maneuver failed, missed moon.

<u>Tenth spacecraft:</u> Luna 7 (Ye-6 No.11)
Mission Type: Lunar Lander
Country/Builder: USSR/OKB-1
Launch Vehicle: Molniya
Launch Date/Time: October 4, 1965 at 07:56:40 UT (Baikonur)
Encounter Date/Time: October 7, 1965 at 22:08:24 UT
Outcome: Crashed.

<u>Eleventh spacecraft:</u> Luna 8 (Ye-6 No.12)
Mission Type: Lunar Lander
Country/Builder: USSR/OKB-1
Launch Vehicle: Molniya
Launch Date/Time: December 3, 1965 at 10:46:14 UT (Baikonur)
Encounter Date/Time: December 6, 1965 at 21:51:30 UT
Outcome: Crashed.

The early Ye-6 series, built at OKB-1, was designed to accomplish the first lunar soft landing. Unfortunately, it suffered eleven straight failures between January 1963 and December 1965. Four spacecraft were lost to booster failures, two were stranded in Earth orbit by fourth-stage failures, two failed in transit and missed the Moon, and three failed at the target by crashing.

The years 1962–65 were dismal for Soviet robotic lunar and planetary exploration. The early successes of Luna 1, 2, and 3, and the encouraging but ultimately fruitless flights of Venera 1 and Mars 1, had built expectations for more success. But by the end of 1962 the Molniya launcher had failed in all but one of ten launches and the truncated flight of Mars 1 had revealed the shortcomings of the 2MV series. These problems were addressed with the 3MV series, essentially the same spacecraft with advanced engines and avionics, and these advances were incorporated into the first Ye-6 series. Nevertheless by the end of 1965 three 2MV Mars missions, three 2MV Venus missions, two test and one 3MV Mars missions, one test and five 3MV Venus missions, and eleven Ye-6 lunar missions – a total of twenty-six missions – had been lost without a single success at the assigned targets. Ironically, in the midst of this awful record, one of the test 3MV Mars spacecraft did achieve a measure of success at the Moon, when Zond 3 provided far-side photography of better quality than that from Luna 3. It was the only lunar accomplishment in this period. Such a long string of failures could well have shut down an American program, so vulnerable to public criticism, but in the Soviet Union it led to the determination to succeed, although not without a great deal of internal criticism by the government and outright threats of punishment.

Spacecraft:

The Ye-6 spacecraft consisted of three sections totaling 2.7 meters in height. The first section consisted of the Isayev midcourse correction and descent engine, which produced a thrust of 4.64 tons using hypergolic nitric acid/amine propellants. Four smaller 245 N thrusters mounted on outriggers were used for attitude control during the descent. The main pressurized cylindrical compartment containing avionics and communication equipment was mounted above the engine. A pair of cruise modules were attached to the central cylinder. One held both attitude control thrusters for the translunar flight and a radar altimeter to trigger the landing sequence, and the other contained avionics sensors for attitude reference and control during the cruise. Both were discarded after the altimeter triggered the landing sequence. The lander capsule was strapped to the top of this stack. Unlike their planetary cousins, these spacecraft carried no solar panels because the flight time for the carrier module and the time on the surface for the lander were sufficiently short that the batteries would not require a recharge.

A new autonomous control system, the I-100, was made for the Ye-6 which not only controlled the spacecraft but also the attitude and firings of both the third and fourth stages of the launcher. This approach deviated from usual practice but saved a great deal of weight by eliminating the third and fourth stage controllers with their associated cabling and connectors. However, this had never been tried before, and would be the cause of further problems for a launcher that had already failed in nine out of ten attempts.

The lander capsule comprised a 105 kg hermetically sealed 58 cm sphere encased in two hemispherical airbags sewn together. It carried communications equipment, a program timer, heat control systems, batteries, and scientific instruments including a television system. Once the lander was on the surface, it would deploy four petals to

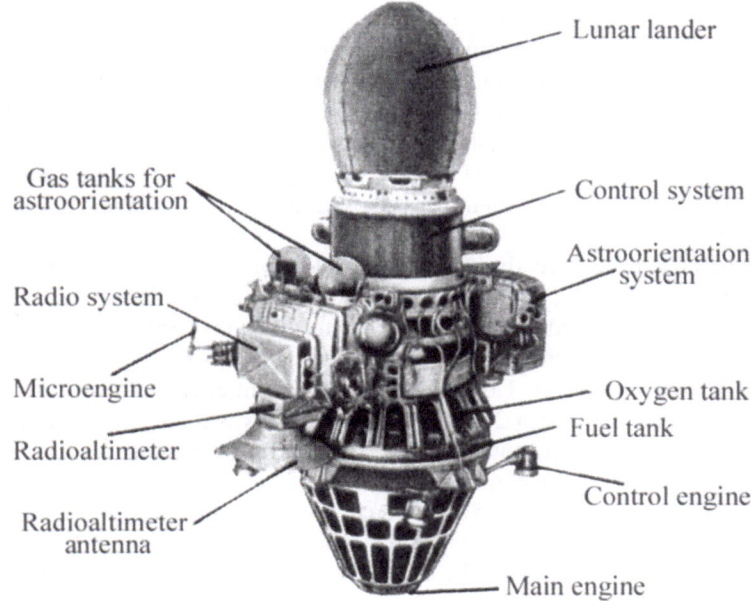

Figure 9.1 Ye-6 lunar soft-lander spacecraft.

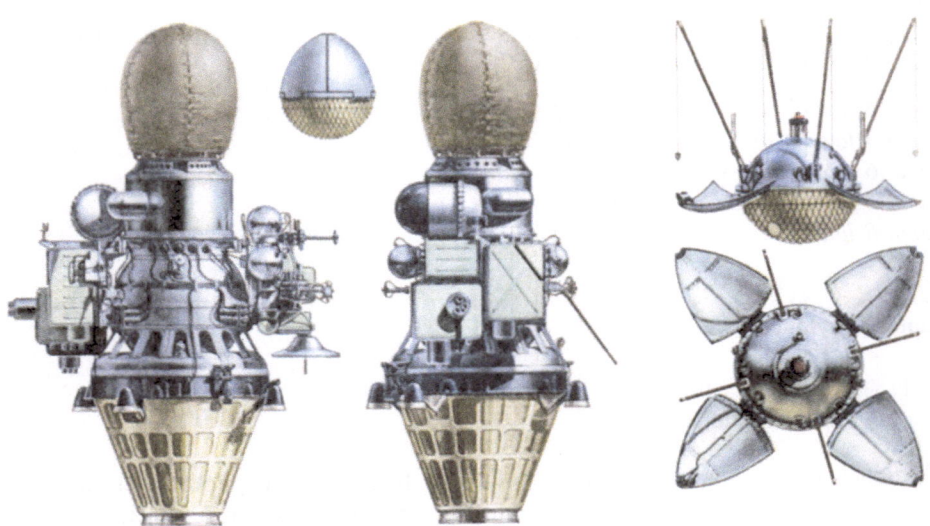

Figure 9.2 Drawings of the Ye-6 spacecraft and lander.

expose its upper hemisphere and raise four 75 cm antennas. The batteries were to supply power for a total of 5 hours over a period of 4 days, with its activities being driven either by timer or by command from Earth. The mass distribution was biased towards the bottom to assist the lander in turning upright on the surface when the petals were opened. The ideas of using air bags for impact and articulating petals to ensure a final upright stance on the surface were both quite clever, but not patented, and so the Americans adapted them for the pyramidal lander of the Mars Pathfinder mission in 1996.

After a direct approach to the target site on the Moon, the landing sequence was initiated at an altitude of 8,300 km. The attitude thrusters stabilized any roll that the spacecraft might possess and aligned the vehicle to the lunar vertical. At about 70 to 75 km altitude the radar altimeter was triggered, sending a signal to jettison the two cruise modules, inflate the airbags to 1 bar and ignite the main engine. At this time its speed relative to the Moon was about 2,630 m/s. The engine was to be shut off at an altitude of 250 to 265 meters and the four outriggers ignited for terminal descent. When a 5 meter long boom made first contact with the surface, the capsule would be ejected vertically to reduce its velocity to 15 m/s. The impact would be absorbed by the airbags. Four minutes after landing, the airbag cover would be severed along the joining seam and discarded. One minute later the lander would right itself by opening the four spring-loaded petals that formed its upper hemisphere, then raise its antennas.

The sites that could be reached by this type of mission were severely constrained, because the final approach of the translunar trajectory had to be perpendicular to the surface to direct the entire thrust of the retro-rocket straight downward. The control system of the vehicle was incapable of dealing with lateral velocity components. In practice, this limited the targets to western longitudes at latitudes that varied with the time of the year.

Luna 4 launch mass:	1,422 kg
Luna 5 launch mass:	1,476 kg
Luna 6 launch mass:	1,442 kg
Luna 7 launch mass:	1,506 kg
Luna 8 launch mass:	1,552 kg

Lander payload:

1.   Panoramic camera
2.   Radiation detector

The camera weighed 3.6 kg and drew 15 W. It was a single photometer directed at the zenith inside a pressurized glass cylinder and used a nodding and rotating mirror to scan the scene both horizontally and vertically. It could expose a full 360 degree panorama in an hour with a resolution of 5.5 mm at a distance of 1.5 meters. Three small dihedral mirrors on deployable poles facilitated 3-dimensional views of small strips of the surface. Calibration targets were dangled from the four whip antennas, which also provided a measurement of the lander's tilt on the surface. The radiation detector was a miniature gas discharge Geiger counter.

Mission description:

Six of the first eleven Ye-6 spacecraft were lost to launch vehicle failures, and none of those that flew to the Moon achieved a soft landing.

The first spacecraft to launch, Ye-6 No.2, was stranded in Earth orbit on January 4, 1963, when the failure of the PT-500 transformer in the power supply of the new I-100 controller prevented the fourth stage from reigniting. This was the sixth failure for the fourth stage out of eight attempts to use it. The object was designated Sputnik 25 by the Americans but was not acknowledged by the Soviets, and it re-entered the following day. The second attempt with Ye-6 No.3 on February 3, 1963, failed even to reach orbit because the I-100 provided an improper pitch angle to the trajectory control system after the separation of the core stage. The third stage did not fire and the remaining stack fell into the Pacific near Hawaii – although this was commented upon by the American press no explanation was forthcoming from the USSR.

With the I-100 control unit fixed, Ye-6 No.4 was successfully sent towards the Moon on April 2, 1963, as Luna 4. The Soviet press announced the launch, saying that scientists were working on the task of landing on the Moon, and pontificated on the possibility of human flights. But the mood soon changed. By the next day it was clear that the navigation system had malfunctioned and that it would not be possible to make the planned midcourse correction. Luna 4 missed the Moon by 8,336 km at 13:25 UT on April 5, and a miffed Soviet press claimed that a flyby was all that had been intended. The spacecraft ceased to transmit on April 6. The Soviet Academy of Sciences undertook a review of the program, but could not determine precisely why Luna 4's navigation system had failed. However, some issues were identified, and it was apparent that the rushed program was suffering quality control problems. It was decided to add a backup radio direction finding system, but this took time and it was a year before the next launch.

Unfortunately Ye-6 No.6 failed to reach orbit on March 21, 1964, when the third stage had an oxygen valve problem, failed to deliver full thrust, and cut off early. An upper stage failure also caused the loss of Ye-6 No.5 on April 20, 1964, when the command to fire the fourth stage failed. Suspicion fell on either the PT-500 current converter or the I-100 controller, and extensive new testing began on these devices. It took almost a year to complete testing and modifications. The sixth attempt with Ye-6 No.9 on March 12, 1965, was lost when the fourth stage did not ignite due to a failed transformer in the power system. Unlike the case for the first Ye-6 launch, the spacecraft was acknowledged by the Soviets and designated Cosmos 60, but it was obviously a failed lunar mission. After so many problems, the entire guidance and control system for the upper stages was reworked using a new three-phase converter, and separate guidance systems installed on the third and fourth stages. This change did not even get a test when the seventh attempt on April 10, 1965, failed because a

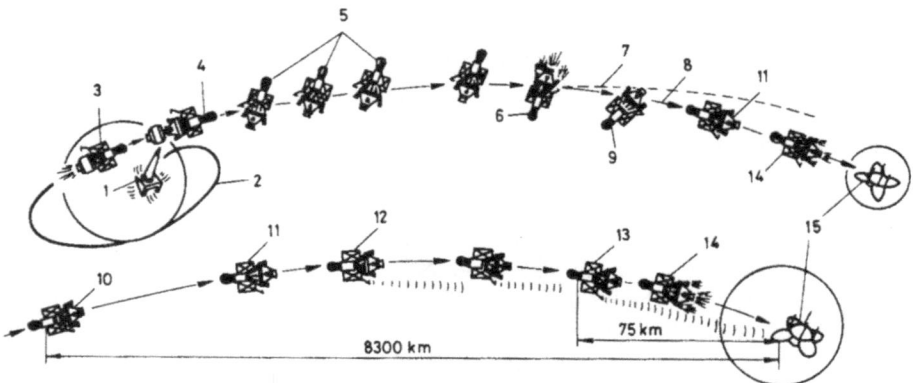

Figure 9.3 Ye-6 flight profile (from *Space Travel Encyclopedia*): 1. Launch; 2. Parking orbit; 3. Translunar injection; 4. Fourth stage separation; 5. Telemetry for trajectory determination; 6. Trajectory correction; 7. Original trajectory; 8. Corrected trajectory; 9. Landing sequence initiation; 10. Determine lunar vertical; 11. Orient to lunar vertical; 12. Radar altimeter activated; 13. Altimeter fires retrorocket system; 14. Retrorocket burn;15. Landing.

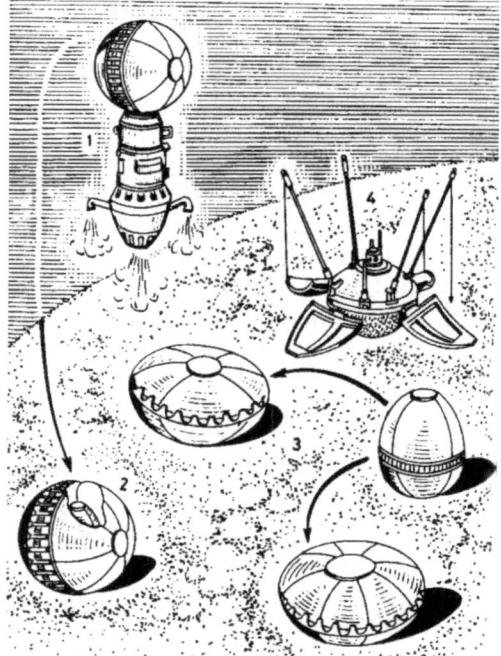

Figure 9.4 Ye-6 soft landing profile (from *Space Travel Encyclopedia*): 1. Balloons inflated, encapsulated lander ejected at 14 m/s; 2. Impact with several bounces to final complete stop; 3. Balloon hemispheres separated by firing stitches around circumference 4. Petals are deployed from upper hemisphere to insure lander rests upright.

failed oxidizer pressurization system prevented the third stage engine from igniting, and the spacecraft, Ye-6 No.8, never reached orbit.

But Ye-6 No.10 was successfully dispatched towards the Moon on May 9, 1965, and announced as Luna 5. During the midcourse maneuver attempt on May 10, the gyroscopes in the I-100 guidance system were not given sufficient time to warm up and the spacecraft began to spin around its longitudinal axis. Engineers brought the spacecraft back under control and attempted the maneuver a second time, but sent it an incorrect command. By the time this was diagnosed it was too late to perform the maneuver. With the spacecraft on course to hit the Moon, albeit obliquely, it was decided to attempt to initiate the terminal maneuvers to exercise the system, but the guidance failed again and the engine did not fire. On May 12, the spacecraft hit the Moon at 1.6°S 25°W instead of the planned site at 31°S 8°W, becoming the second Soviet spacecraft to do so. Moscow, without portraying the mission as a failure, said a lot of information had been obtained "for the further development of a system for a soft landing on the Moon's surface".

Ye-6 No.7 was launched on June 8, 1965, and successfully sent toward the Moon as Luna 6. The midcourse correction on June 9 began well, but a command error prevented the engine from cutting off, and it fired until the fuel was exhausted. This deflected the trajectory to such an extent that the spacecraft missed the Moon by 160,935 km on June 11, 1965. However, the engineers successfully put it through all of its landing sequence events.

An attempt to launch Ye-6 No.11 was canceled on September 4, 1965, when the core stage avionics failed in pre-flight testing. The vehicle was returned to the barn for major repairs to its control system. A month later, on October 4, this same rocket successfully dispatched Luna 7. This time the midcourse maneuver was performed successfully, making this the first Ye-6 in ten launches to be given the opportunity actually to attempt a lunar landing. However, in making its approach it lost attitude control, which prevented the retro-rocket from firing, and it crashed in the Ocean of Storms at 22:08:24 UT on October 7, at 9.8°N 47.8°W, west of the crater Kepler. An optical sensor had been set at the wrong angle and had lost sight of Earth during the attitude control maneuver immediately prior to starting the retro-rocket. As Moscow reported in its first admission of a failure, "Certain operations were not performed in accordance with the program and require additional optimization."

Leonid Brezhnev, who had ousted Khrushchev the previous year, called Korolev to Moscow to account for the long string of failures. Korolev's political charm stood him in good stead as he explained the difficulties and promised success with the next mission, due to launch in December. Although he did not deliver on this promise, he never had to face the new leadership again because he died during colon surgery on January 14, 1966. After the Moscow summons, Boris Chertok, a deputy at OKB-1, investigated the reliability and testing of spacecraft subsystems, and identified a lack of integrated testing of some subsystems during spacecraft assembly as a particular problem. Although corrective action was taken for the next launch, this was not in itself sufficient.

On December 4, 1965, Ye-6 No.12 was launched into a lower inclination parking orbit than its predecessors, at 51.6 degrees instead of 65 degrees. This allowed for a mass increase beyond 1,500 kg. The fourth stage then sent the spacecraft towards to Moon as Luna 8. The midcourse maneuver went well the following day, but alas the second Ye-8 to be presented with an opportunity to make a lunar landing failed. Just prior to retro-rocket ignition, the two airbags were inflated, as planned, but one was pierced by an improperly manufactured mounting bracket on a lander petal and the thrust of the escaping gas caused the spacecraft to spin. As a result, the retro-rocket cut off after just 9 of the required 42 seconds. The spacecraft crashed in the Ocean of Storms at 21:51:30 UT on December 6, at 9.1°N 63.3°W, to the west of the crater Kepler. The bracket problem was fixed, and on future missions the airbags would be inflated only after the retro-rocket had completed its burn.

Luna 8 was the eleventh straight failure in the Ye-8 program and the last before NPO-Lavochkin took over management of the Soviet lunar and planetary programs.

Results:

None.

A NEW SPACECRAFT AND ANOTHER TRY FOR MARS: 1963–1965

Campaign objectives:

Korolev's team had failed in their first two campaigns at both Mars and Venus. In 1960–61 only one mission of four, Venera 1, succeeded in reaching interplanetary space but it failed soon thereafter. In 1962 they had a new multipurpose spacecraft ready, the 2MV series, and launched three each to Mars and Venus. This time, only one of six, Mars 1, was successfully dispatched and it fell silent before reaching its target. Meanwhile, the Americans had frustrations of their own, suffering fourteen failed lunar missions through 1962. Their only success, Mariner 2 at Venus in 1962, served to further frustrate the Soviets who had worked hard to beat the Americans to our neighboring planets.

By now it was evident that there were serious problems with the reliability of the 8K78 launcher, in particular its fourth stage, and with the spacecraft. The troubled but lengthy flight of Mars 1 revealed problems serious enough to merit a redesign of the 2MV, and Korolev directed that these lessons be applied to building a new 3MV series for the Venus and Mars windows in 1964, and that test flights be conducted in between planetary opportunities. And, of course, he continued to instrument the fourth stage to diagnose its problems. The test flights were intended to validate the whole system from launcher to spacecraft.

Spacecraft launched

First spacecraft:	Cosmos 21 (3MV-1A No.2)
Mission Type:	Mars Spacecraft Test Flight
Country/Builder:	USSR/OKB-1
Launch Vehicle:	Molniya
Launch Date/Time:	November 11, 1963 at 06:23:35 UT (Baikonur)
Outcome:	Stranded in Earth orbit, fourth stage failure.

Second spacecraft:	Zond 2 (3MV-4 No.2)
Mission Type:	Mars Flyby
Country/Builder:	USSR/OKB-1
Launch Vehicle:	Molniya
Launch Date/Time:	November 30, 1964 at 13:12 UT (Baikonur)
Mission End:	May 5, 1965
Encounter Date/Time:	August 6, 1965
Outcome:	Lost in transit, communications failure.

Third spacecraft:	Zond 3 (3MV-4 No.3)
Mission Type:	Mars Spacecraft Test
Country/Builder:	USSR/OKB-1
Launch Vehicle:	Molniya
Launch Date/Time:	July 18, 1965 at 14:38:00 UT (Baikonur)
Encounter Date/Time:	July 20, 1965 (Moon)
Mission End:	March 3, 1966
Outcome:	Succeeded at Moon, failed to reach Mars distance.

The 3MV spacecraft was similar to the 2MV but with improved avionics. Special versions, designated 3MV-1A and 3MV-4A, were built for test missions simulating flights to Venus and Mars. These were lighter test models and did not carry a full set of science instruments. The first 3MV was launched in November 1963. The intent was to test planetary flyby operations and the camera system at the Moon, and then perform operations to Mars distance before the Mars launch window opened a year later. The launch failed. It was followed in February 1964 with a launch of a test flight to Venus distance just prior to the opening of the Venus window in late March. This launch also failed. Despite these two losses, the Soviets had little option but to proceed with the 1964 program. Two of 3MV spacecraft were launched in the Venus window in March and April, the first succumbing to its launch vehicle and the second, Zond 1, failing in transit.

Undaunted, the Soviets continued with preparations for Mars. Although two flyby spacecraft and at least one entry probe were built, there were technical problems and only one spacecraft made it to the launch pad. The 3MV-4 No.2 flyby spacecraft was successfully dispatched on November 30, 1964. When it became clear that the spacecraft would not be able to meet its objectives, it was named Zond 2. The other Mars spacecraft prepared for this launch window were scrubbed and stored while the problems with the 3MV were investigated. They would subsequently be used for the Zond 3 mission and for the Venus campaign in 1965.

Following the string of five 3MV mission failures in 1963–1964, it was decided to conduct another test. The 3MV-4 No.3 Mars flyby spacecraft that missed its window in 1964 was launched 8 months later. Its task was the same as the spacecraft lost in November 1963, to test the spacecraft and science instruments in a lunar flyby and then test the deep space capabilities of the spacecraft by flying to Mars distance even though the planet would not be present upon arrival. After a successful launch, the spacecraft was designated Zond 3. (The Zond designation had initially been assigned to spacecraft that were clearly not going to be able to meet their objectives, as in the cases of Zond 1 and 2, and would henceforth be used for spacecraft launched either for deliberate testing purposes or to conduct science.) The lunar flyby was timed to photograph the far side of the Moon using the Mars camera, and Zond 3 successfully achieved its test objectives at the Moon. It failed to reach Mars distance but was able to maintain communications for almost 8 months, finally falling silent at a range of over 150 million km.

The Zond 3 spacecraft was the last of the 3MV Mars series to be launched before the robotic lunar and planetary programs were transferred to NPO-Lavochkin, where it was decided to abandon the troublesome 3MV design for Mars and instead design a new, heavier and much more capable spacecraft for launch by the Proton. None of the two 1M, three 2MV and three 3MV spacecraft launched to Mars, a total of eight including the two 3MV tests, had reached their targets, although Zond 3 did succeed at the Moon.

Spacecraft:

The 3MV spacecraft was similar in appearance and general function to the 2MV. It was slightly longer at 3.6 meters and had the same inline modular design consisting of a pressurized avionics or 'orbital' module, a propulsion system, and a pressurized flyby instrument module or entry probe. Minor changes were made to the shape in order to modify the moment of inertia and to account for solar wind torques, but the other dimensions were the same as the 2MV. A black shield was added in order to prevent scattered light from interfering with the optical sensors.

A thermal protection cowl was added to the Isayev KDU-414 propulsion system. This system was used for the 1M, 1VA, and all 2MV and 3MV spacecraft through to Venera 8, with variously sized tanks for its unsymmetrical dimethylhydrazine and nitric acid propellants. It was capable of multiple firings, and on the 2MV and 3MV was gimbaled for thrust vectoring under gyroscopic control. The propulsion system assembly, including its tanks, was about 1 meter in length. For 3MV Venus missions the compressed nitrogen gas bottles used to pressurize the engine propellants and for cold gas attitude control jets were mounted on the engine cowling. For 3MV Mars missions these bottles were on the collar between the avionics module and the flyby or entry module. Major improvements were made to the avionics, and redundancy was added to the attitude control system jets. The high gain antenna was increased to a diameter of 2.3 meters. Low gain omnidirectional antennas were installed on the

Figure 9.5 Zond 2 and 3 spacecraft, front and back.

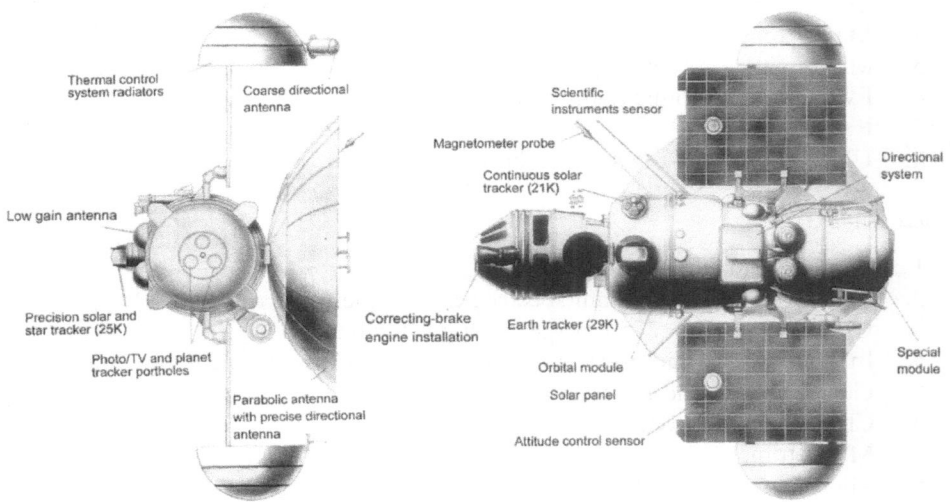

Figure 9.6 The 3MV-4 Mars flyby spacecraft (courtesy Energiya Corp).

hemispherical radiators that circulated liquid. In addition to the attitude, navigation, thermal and operational control systems, the avionics module held 32 cm and meter band transmitters, 39 cm and meter band receivers, and two tape recorders. The solar panels charged a 112 amp-hour NiCd battery array that supplied the spacecraft with DC power at 14 volts.

In addition to the science instruments and the 5 cm impulse image transmitter, the flyby module contained an 8 cm continuous wave transmitter for backup image or spacecraft data transmission, and backup forms of the command receiver and other avionics capable of operating the spacecraft in the event of a failure in the avionics module.

Each of these spacecraft, both Mars and Venus versions, had experimental plasma pulse engines on the engine cowl for attitude control in addition to the standard cold gas jets. They were tested successfully on Zond 2, and were later perfected and used regularly on Soviet spacecraft.

Launch mass: 800 kg (Cosmos 21)
                        950 kg (Zond 2)
                        960 kg (Zond 3)

Payload:

3MV-1A No.2:

1.   Facsimile imaging system
2.   Radiation detector
3.   Charged particle detector
4.   Magnetometer
5.   Micrometeoroid detector
6.   Lyman-alpha atomic hydrogen detector
7.   Radio telescope
8.   Ultraviolet and x-ray solar radiation experiment

Zond 2 and 3:

1.   Facsimile imaging system
2.   Ultraviolet 285 to 355 nm spectrograph in the camera system
3.   Ultraviolet 190 to 275 nm spectrograph for ozone
4.   Infrared 3 to 4 micron spectrometer to search for organic compounds
5.   Gas discharge and scintillation counters to detect Martian radiation belts
6.   Charged particle detector
7.   Magnetometer
8.   Micrometeoroid detector

After the 1962 campaign, a major improvement was made to the facsimile film imaging system for the flyby missions. The imager mass was reduced from 32 to 6.5 kg while using 25.4 mm film capable of storing 40 images. Zond 2 carried two of these cameras equipped with 35 and 750 mm lenses. Zond 3 carried one camera with a 106.4 mm lens. Alternative exposures at 1/100th and 1/300th of a second were used, and an image could be taken and developed every 2.25 minutes. The 25 mm film could be repeatedly rewound for scanning at 550 or 1,100 lines per frame. Imaging data were stored on the tape recorder that was included in the infrared spectrometer electronics. The 5 cm impulse transmitter and modulation scheme were improved for a factor of four decrease in image transmission times. In the high quality mode, camera images were transmitted at 550 pixels/second, which was 2 seconds per scan line, requiring a total of 34 minutes to send a 1,100 × 1,100 image. If necessary, the images could be sent at much slower rate by the 8 cm continuous wave transmitter. An ultraviolet spectrometer operating in the 285 to 355 nm range was built into the camera and recorded its data on three frames of the film. These instruments were carried inside the flyby module and observed through three portholes – one for each lens and the spectrometer. A second ultraviolet spectrometer operating in the 190 to 275 nm range was carried externally and produced digital data. The optical system for the infrared spectrometer was also mounted externally, and was equipped with a small visible wavelength photometer to provide a reference signal. All of the optical instruments were bore-sighted.

Mission description:

The 3MV-1A No.2 mission ended in failure when the spacecraft was stranded in low Earth orbit. The third and fourth stages apparently separated abnormally. The fourth stage diverged in attitude during coast, and was incorrectly aligned when the engine ignited. Telemetry was lost at 1,330 seconds into the flight and the fourth stage with its payload remained in Earth orbit. With this mission, the Soviets initiated a policy of designating lunar and planetary missions stranded in parking orbit as 'Kosmos', a designation that was previously used for scientific satellites, in an effort to obscure their intended purpose. Today, Cosmos is used to designate military missions. The failed 3MV-1A became Cosmos 21 and it re-entered 3 days later.

One year later, after another failed test launch of a 3MV Venus spacecraft and two launches to Venus, including Zond 1, the 3MV-4 No.2 spacecraft was launched on November 30, 1964, for what was intended to be a flyby of Mars at a distance of 1,500 km. However, one of the solar panels did not open due to a broken pull cord. The second panel was finally deployed on December 15 after several engine firings shook it loose, but by then it was too late to perform the first trajectory correction. It suffered other problems, including a timer that failed to activate the thermal control system properly. Unlike Zond 1 earlier in the year, the Soviets revealed Mars as an objective but, knowing that the flyby would not occur in the planned manner, they named the spacecraft Zond 2 and said its objectives were to carry out experiments "in the vicinity of Mars".

During the last authenticated communications session on December 18, 1964, the plasma engines were successfully tested. After that, communications became erratic. Jodrell Bank monitored transmissions from Zond 2 in January and on February 3, 10 and 17, but it is unclear if any further operations were conducted. The Soviets finally announced on May 5 that contact had been lost. The USSR lost the opportunity to be the first to fly past Mars. This honor went to Mariner 4 on July 15, 1965, which the US had launched 2 days before Zond 2. On August 6 the silent Zond 2 flew by Mars at a range of 650,000 km.

Zond 3 was launched successfully on July 18, 1965. Approximately 33 hours later the imaging sequence began at a range of 11,570 km from the lunar near side, and continued through the lunar far side passage over a period of 68 minutes as the range closed to 9,960 km. The closest point of approach had been 9,219 km on the far side. A total of 28 images were developed on board and transmitted on July 29, by which time the spacecraft was 1.25 million km from Earth. The spacecraft continued on its deep space test flight. A midcourse correction of 50 m/s was made on September 16 at a range of 12.5 million km. The images were rebroadcast from 2.2 million km and again at 31.5 million km to test the capabilities of the communications system. The last communication was on March 3, 1966, at a range of 153.5 million km, well on the way to the orbital distance of Mars.

Results:

There were no results for Mars. Zond 2 made a successful technology demonstration that was important for later deep space missions by operating its six plasma engines prior to the loss of communications, but they were found insufficient to control the attitude of the spacecraft. Zond 3 photographed 19 million square kilometers of the lunar surface including the 30% of the lunar far side that had been in darkness for Luna 3. The twenty-five visible-band images and three ultraviolet-band images were of much better quality than the Luna 3 pictures. The Soviets achieved an engineering success with their first course correction to be performed using both solar and stellar references.

Figure 9.7 Lunar far side image from Zond 3.

THE SECOND VENUS SPACECRAFT: 1964

Campaign objectives:

By the end of 1962, the Soviets had made five attempts at Venus. Only one mission, Venera 1, survived its launch vehicle and succeeded in reaching interplanetary space but the spacecraft failed early in transit. All six spacecraft in the second-generation 2MV Mars/Venus series, including the three Venus missions, fell victim to launch failures. Adding to the frustrations, the Americans had a successful flyby of Venus in 1962 with their Mariner 2. Undaunted, the Soviets improved the 2MV as the 3MV for the 1964 opportunities to Venus and Mars.

The initial flight tests of the new spacecraft were lost to launch vehicle failures, the first in November 1963 on a test to Mars distance and the second in February 1964 on a test to Venus distance. Despite of these losses, the Soviets proceeded with their 1964 program for Venus and Mars.

Spacecraft launched

First spacecraft: 3MV-1A No.4A
Mission Type: Venus Spacecraft Test
Country/Builder: USSR/OKB-1
Launch Vehicle: Molniya-M
Launch Date/Time: February 19, 1964 at 05:47:40 UT (Baikonur)
Outcome: Launch failure, third stage exploded.

Second spacecraft: 3MV-1 No.5 (Cosmos 27)
Mission Type: Venus Atmosphere/Surface Probe
Country/Builder: USSR/OKB-1
Launch Vehicle: Molniya-M
Launch Date/Time: March 27, 1964 at 03:24:42 UT (Baikonur)
Outcome: Failed to leave Earth orbit, fourth stage failure.

Third spacecraft: Zond 1 (3MV-1 No.4)
Mission Type: Venus Atmosphere/Surface Probe
Country/Builder: USSR/OKB-1
Launch Vehicle: Molniya-M
Launch Date/Time: April 2, 1964 at 02:42:40 UT (Baikonur)
Mission End: May 25, 1964
Encounter Date/Time: July 19, 1964
Outcome: Failed in transit, pressurization lost.

Launches to Venus were attempted on March 27 and April 2, straddling the ideal date. The first spacecraft was stranded in parking orbit but the second was deployed on a trajectory for Venus. Just as the launch vehicle was plagued by a troublesome fourth stage, the spacecraft had troublesome avionics systems. Immediately named Zond 1 when it became evident 3MV-1 No.4 would not be able to fulfill its mission at Venus, the spacecraft fell silent after less than 2 months in flight. Even if it had succeeded in deploying its entry probe, the decent capsule would not have survived to the surface. It was designed to survive only to 77°C and withstand pressures up to 5 bar. At the time there were two opposing theories for conditions on the surface of Venus. The high brightness temperatures measured by terrestrial radio telescopes, and confirmed by Mariner 2 during its flyby in 1962, could be interpreted either as a surface as hot as 400°C or as a hot ionosphere and cool surface. The easier design path was for the popular vision of a planet like Earth with a cool surface and maybe even an ocean. When the 1964 Venus launch window opened and Zond 1 set off, the controversy was not yet settled, although the weight of evidence was leaning to a hot surface. Radio observations of the planet from Earth later in 1964 would discredit the cool surface theory, but by then it was too late to redesign the Venus probes for the 1965 launches. The hot surface theory was firmly established after the flight of Venera 4 in 1967, which coincided with the highly successful Mariner 5 flyby.

Spacecraft:

Carrier spacecraft:

The 3MV Venus spacecraft were almost identical to their 3MV Mars counterparts, although the solar panels were less densely populated with solar cells. On the probe versions of the 2MV and 3MV, the entry vehicle was to be deployed just prior to the spacecraft entering the atmosphere and burning up. Instruments for interplanetary science and measurements in the near vicinity of the planet prior to destruction were carried on the main spacecraft. Communications from the probe would be direct to the Earth.

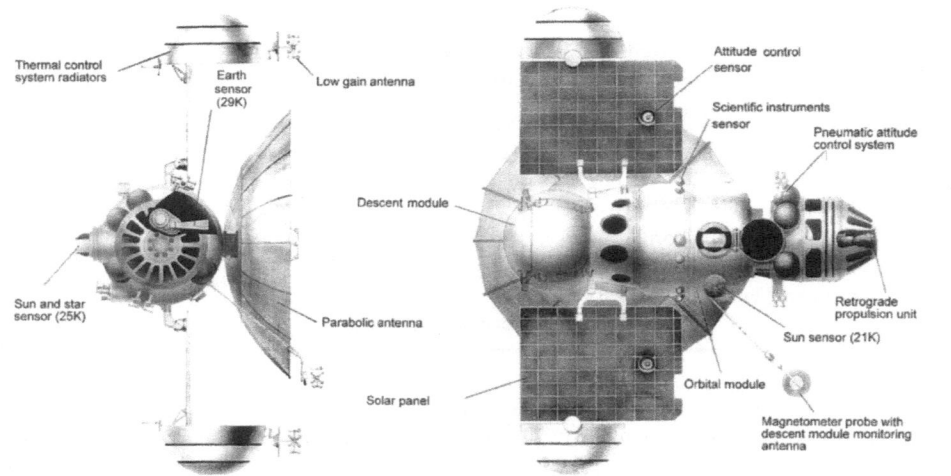

Figure 9.8 The 3MV-1 Venus probe spacecraft (courtesy Energiya Corp).

Launch mass:	800 kg (3MV-1A No.4A)
Launch mass:	948 kg (Cosmos 27 and Zond 1)
Entry system mass:	290 kg

Entry vehicle:

The 3MV entry probes were intended to obtain data during the descent through the atmosphere, survive impact, and return data from the surface. In the case of Venus the dense atmosphere that meant the probes would impact slowly enough to have a fair chance of surviving and operating for a short period of time on the surface. The 3MV probes were similar to the 2MV ones, being 90 cm in diameter and containing parachutes, batteries, sequencers, and two redundant 32 cm transmitters each with an antenna for direct communications to Earth, in addition to science instruments.

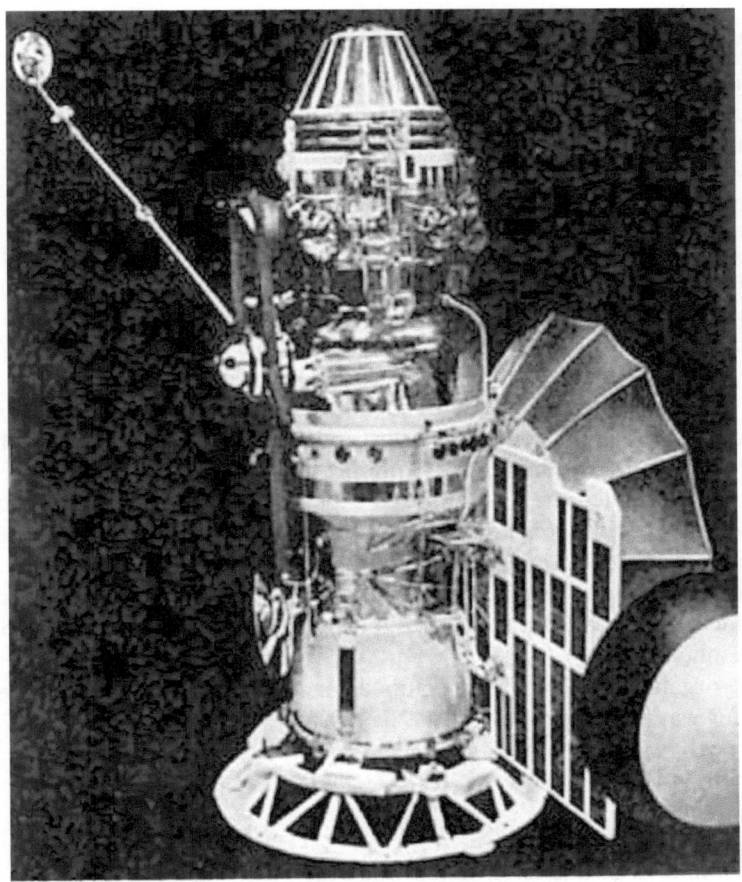

Figure 9.9 Zond 1.

Payload:

The payload for the test flight to Venus distance was probably similar to that of the lost test flight to Mars distance. The payloads for the missions to Venus launched on March 27 and April 2, 1964, were identical.

Zond 1 carrier spacecraft:

1. Radiation detector
2. Charged particle detector
3. Gas discharge and scintillation cosmic ray and gamma-ray detectors
4. Ion traps
5. Magnetometer
6. Micrometeoroid detector
7. Lyman-alpha atomic hydrogen detector

Zond 1 descent/landing capsule:

1. Temperature, pressure and density sensors
2. Atmospheric composition, acidity and electrical conductivity experiments
3. Gamma-ray surface composition detector and cosmic ray detector
4. Visible airglow photometer
5. Mercury level experiment

The atmospheric structure experiment consisted of two platinum wire resistance thermometers with ranges of -60°C to 460°C and 0 to 330°C, an aneroid barometer with a range of 0.13 to 6.9 bar, and a beta ray ionization chamber densitometer that was integrated with the thermometers and had a range of 0.0005 to 0.015 g/cc with a 5% error. The atmospheric composition, acidity and electrical experiments consisted of a set of gas analyzer cartridges with chemical and electrical tests for various gases including carbon dioxide, nitrogen, oxygen, and water vapor. The photometer was to search for airglow during the night landing. It was sensitive over the range 0.001 to 10,000 lux, and included the mercury level experiment to measure wave motion in a putative ocean. The anti-coincidence gas discharge and scintillation counter cosmic ray and gamma-ray detector was primarily to measure the surface composition of radioactive elements potassium, thorium and uranium from gamma-ray emissions on the surface, but it was also to be used during the interplanetary cruise to measure primary cosmic rays.

A micro-organism detector was planned for the 3MV Venus and Mars landing capsules, but was never included in the payloads.

Mission description:

The first test launch of this series failed when the third stage exploded. LOX leaking through a valve froze a fuel line which later broke. The loss of 3MV-1A No.4A so close to the imminent Venus launch window must have been disheartening, but the Soviets continued with preparations to launch the other two spacecraft.

The first attempt to launch the 3MV-1 No.5 spacecraft on March 1 was postponed owing to problems with the launcher during pre-launch tests. The second attempt on March 27 using the same vehicle failed when an electrical fault caused the fourth stage to lose attitude control and the engine did not restart for the escape burn. It was designated Cosmos 27 by the Soviets. This loss did have a very valuable result. For the first time a flight recorder had been added to the fourth stage telemetry system, and on its second pass the downlinked telemetry indicated a failure that was able to be traced to a generic problem in the I-100 control system circuitry. It required only 20 minutes of re-soldering to fix this for subsequent flights.

The third Venus spacecraft was dispatched successfully on April 2, 1964, but its initial trajectory was inaccurate and a midcourse maneuver was made the next day at a range of 564,000 km from Earth. It was the first midcourse maneuver carried out by a Soviet planetary spacecraft. Venera 1 and Mars 1 had both had this capability, but neither had been able to exercise it. However, Zond 1 was in serious trouble. A leak in the pressurized avionics section was detected right after launch due to a crack in the weld seam of the quartz dome for the Sun and star navigational sensors. The location of the leak was determined from analysis of how the escaping gas perturbed the spacecraft. After a week, the transmitters and other electronics failed when they were switched on as the pressure fell to about 5 millibars, which permitted coronal discharges to short out power lines. The ion engines also failed their test, operating erratically. Owing to sensible backup design, communications were able to continue using the entry system, and a second midcourse maneuver was made on May 14 at a distance of more than 13 million km from Earth. This resulted in a trajectory that would fly by the target at 100,000 km. In fact, the initial trajectory was probably so wide of the mark that even if the spacecraft had been fully functional, it would not have been able to adopt a collision course. Due to the pressure leak the Soviets did not reveal Venus as the target, merely announcing that the mission was a deep space engineering test, and named it Zond 1 rather than Venera 2. The leak was fatal, and on May 25 thermal control was lost and communications failed. The inert spacecraft passed Venus on July 19.

Results:

Zond 1 did return data on interplanetary plasma, including cosmic ray and Lyman-alpha measurements from the avionics module and proton measurements from the cosmic ray instrument in the lander capsule, but much of the data returned appears to have been lost.

TWO FRUSTRATING MISSIONS AT VENUS: 1965

Campaign objectives:

Nineteen months after their frustrating third campaign to Venus, the Soviets were ready with three more spacecraft for the late 1965 launch window. They had tried to reach this planet at every opportunity since February 1961, but after one test launch and seven launches they had nothing to show for it. Only two of the seven spacecraft survived their launch vehicles, and both of these failed in flight rather quickly. But the engineers reckoned they had fixed the problems that crippled Zond 1 and were encouraged by the success of Zond 3 at the Moon and its long interplanetary flight, so they prepared for the second 3MV Venus campaign with confident expectation.

Several 3MV spacecraft were left over from the November 1964 Mars campaign when only one had been launched during the window, flying as Zond 2. Another had been launched in July 1965 as Zond 3 for a test to Mars distance. Three 3MV Mars spacecraft, one configured with an entry probe (3MV-3 No.1) and the other two for flyby observations (3MV-4 No.4 and No.6), were modified for the Venus window in 1965. Their original target, Mars, accounts for their anomalous 'tail numbers'. Only

---

Spacecraft launched

First spacecraft: Venera 2 (3MV-4 No.4)
Mission Type: Venus Flyby
Country/Builder: USSR/OKB-1
Launch Vehicle: Molniya-M
Launch Date/Time: November 12, 1965 at 05:02:00 UT (Baikonur)
Mission End: February 10, 1966
Encounter Date/Time: February 27, 1966
Outcome: Failed in transit, communications lost.

Second spacecraft: Venera 3 (3MV-3 No.1)
Mission Type: Venus Atmosphere/Surface Probe
Country/Builder: USSR/OKB-1
Launch Vehicle: Molniya-M
Launch Date/Time: November 16, 1965 at 04:19:00 UT (Baikonur)
Mission End: February 16, 1966
Encounter Date/Time: March 1, 1966
Outcome: Failed in transit, communications lost.

Third spacecraft: 3MV-4 No.6 (Cosmos 96)
Mission Type: Venus Flyby
Country/Builder::   USSR/OKB-1
Launch Vehicle: Molniya-M
Launch Date/Time: November 23, 1965 at 03:22:00 UT (Baikonur)
Outcome: Failed to depart Earth orbit.

---

two were successfully dispatched. Venera 2 and 3 flew to the vicinity of their target and became the first truly successful interplanetary cruises since Korolev had begun launching planetary spacecraft in 1960. The long interplanetary cruise provided new confidence in the spacecraft, but the fact that they failed at or near their target made them agonizing disappointments. There was a fourth spacecraft, probably with an entry probe, but this was unable to be launched before the window closed.

Venera 2 and 3 were also the last planetary spacecraft to be built and launched by OKB-1 because in late 1965 Korolev had transferred responsibility for robotic lunar and planetary missions to NPO-Lavochkin. The next Venera spacecraft for the 1967 window would be built and launched under the leadership of Georgi Babakin.

Spacecraft:

The Venera 2 and 3 spacecraft were basically the same as Zond 2 and 3 but modified for the new target. The Venera 3 entry probe was essentially the same as that carried by Zond 1. By the time the mission was launched, there was strong evidence that the surface of Venus was hot, possibly 400°C. Although the surface pressure was not yet well determined, it was apparent that conditions were beyond the limits to which the 3MV probe was designed (77°C and 5 bar). As it was too late to make changes, Venera 3 was launched in full knowledge that its probe would provide only data on the atmosphere and would not survive the full descent to the surface.

Figure 9.10 Venera 2 (left) and Venera 3 (right).

Figure 9.11 Venera 3 probe.

Launch mass:	963 kg (Venera 2)
Launch mass:	958 kg (Venera 3)
Launch mass:	~950 kg (Cosmos 96)
Probe mass:	337 kg

Payload:

Venera 2 carrier spacecraft:

1. Lyman-alpha and oxygen spectrometer
2. Triaxial fluxgate magnetometer
3. Micrometeoroid detector
4. Charged particle detectors
5. Cosmic ray gas discharge and solid state detectors
6. Cosmic radio emission receivers for 20 to 2,200 kHz
7. Decimeter band radio solar plasma detector

The cosmic ray detectors now consisted of the gas discharge counters and silicon solid-state detectors. The decimeter band radiometer dish antenna was mounted on the ring between the avionics and instrument compartments.

Venera 2 flyby instrument module:

1. Facsimile imaging system
2. Ultraviolet spectrometer at 285 to 355 nm in the imaging system
3. Ultraviolet spectrometer for ozone at 190 to 275 nm
4. Infrared spectrometer at 7 to 20 and 14 to 38 microns

The camera system and ultraviolet spectrometers were identical to those carried by Zond 2 and 3. The camera was provided with a 200 mm lens. The Venus infrared spectrometer was similar to that of Mars 1 but designed to measure thermal radiation from the atmosphere and clouds. It covered two ranges in 150 increments each, the first using an InAn window and the second a LiF mirror. The instrument had a mass of 13 to 15 kg, was 50 cm in size, and was mounted outside the instrument module, coaxial with the imaging system, and included a visible photometer for reference. It could also make a spatial scan of the planet at the two fixed wavelengths of 9.5 and 18.5 microns.

Venera 3 carrier spacecraft:

1. Lyman-alpha and atomic oxygen photometers
2. Triaxial fluxgate magnetometer
3. Charged particle detectors
4. Cosmic ray gas discharge and solid state detectors
5. Decimeter band radio solar plasma detector

The cosmic ray instrument had an additional gas discharge counter on Venera 3, and both the micrometeoroid detector and the radio emission receivers were deleted.

Venera 3 descent/landing capsule:

1. Temperature, pressure and density sensors
2. Atmospheric composition, acidity and electrical conductivity experiments
3. Gamma-ray surface composition detector and cosmic ray detector
4. Visible airglow photometer
5. Mercury level motion experiment

The probe instruments were spares from the 1964 campaign. The photometer was included again since Venera 3 was to be a night-time landing. As with all of the 3MV missions, the probe also carried pennants of the Soviet Union.

Mission description:

The Venera 2 flyby spacecraft was successfully launched on November 12, 1965. It was intended to fly in front of the sunlit hemisphere of Venus and photograph it at a range of less than 40,000 km. The initial trajectory was so precise that no midcourse maneuver was required. The thermal system did not function well and the spacecraft began to overheat as it neared its target, causing problems with the communications system. An improper coating of the radiation domes was suggested as the cause. By February 10, which proved to be the final interrogation session, the temperature was considerably increased, the quality of communications was seriously degraded, and the command from Earth to initiate flyby observations was not acknowledged. After the flyby Venera 2 failed to respond to commands to download the flyby data, and on March 4 it was declared lost. It may very well have achieved its mission and been unable to transmit its results to Earth. The closest point of approach to the planet was at 02:52 UT on February 27, 1966, at a distance of 23,950 km.

Venera 3 was dispatched towards Venus on November 16, 1965. It performed satisfactorily during cruise and a midcourse correction on December 26 put it on an impact trajectory 800 km from the bull's-eye. However, the communications system failed on February 16, just seventeen days prior to arrival. The spacecraft may have released its entry probe automatically at 06:56 UT on March 1, 1966, but there was no telemetry from the capsule. Even so, the probe became the first human artefact to reach another planet, near the terminator on the night side somewhere between 20°S and 20°N and between 60°E and 80°E.

The post-mission investigation into the loss of Venera 2 and 3 revealed problems with the thermal control system in both spacecraft which had caused components in the communications system to overheat and fail.

The third spacecraft, 3MV-4 No.6, was launched on November 23. A broken fuel line caused one of the engine chambers in the third stage to explode shortly prior to stage shutdown, with the result that the fourth stage inherited an unstable attitude. It managed to achieve orbit, but the tumbling prevented it from restarting its engine for the escape maneuver. Written off as Cosmos 96, it re-entered on December 9.

A fourth spacecraft (probably 3MV-3 No.2) was to be launched at the very end of the window on November 26, 1965, but was scrubbed when a problem was found in the launch vehicle during pre-flight checks. The launch was abandoned because the vehicle could not be recycled before the window closed.

These were the last robotic interplanetary spacecraft launched by OKB-1. Out of a total of 39 launch attempts in a period of a little more than seven years, only Luna 2, Luna 3, and Zond 3 fulfilled their missions. Twenty lunar launch attempts gave eight successful launches, with only three spacecraft being fully successful. Eleven Venus launch attempts gave four successful launches, but unfortunately no spacecraft were successful. Out of six Mars launch attempts only two succeeded, but both spacecraft failed. Two 3MV test launches also failed.

Results:

The 1965 campaign produced no data from Venus. Some results were published on micrometeoroids, the interplanetary magnetic field, cosmic rays, low energy charged particles, solar wind plasma fluxes and their energy spectra.

10

Finally success at the Moon and Venus, but Mars eludes

TIMELINE: JAN 1966–NOV 1968

The race to put humans on the Moon heated up in 1966. Both the Soviets and Americans stepped up the pace of their robotic lunar missions in support of eventual manned missions, with each successfully sending landers and orbiters. In September 1967 the Soviets began automated tests of a version of the Soyuz spacecraft that was intended to fly cosmonauts on circumlunar missions. In September 1968 the Zond 5 mission flew this spacecraft around the Moon and returned to Earth with a biological payload and high quality photographs of Earth from deep space. Its success spurred the American decision to send Apollo 8 into lunar orbit in the hope that astronauts would beat cosmonauts to the vicinity of the Moon. The Soviets lost their chance in November, when Zond 6 flew a repeat of the circumlunar mission and crashed on its return. This left the way clear for Apollo 8 in December, whose success nullified the propaganda value to be gained from sending cosmonauts on a circumlunar mission.

On their first launch of 1966 the Soviets were finally successful with their Luna soft-lander series. Luna 9 became the first lander on the Moon on February 3. It was followed in December by a second lander, Luna 13. Immediately after the success of Luna 9 an orbital module was hastily assembled to replace the lander module, and the spacecraft converted to support the call by the manned program for information on the lunar gravity field and surface properties. With the Luna 10 mission launched on March 31, 1966, the Soviet Union became the first nation to put a spacecraft into orbit around the Moon. Additional orbiters were sent, and by the close of 1968 this series was concluded with Luna 14.

On May 30, 1966, America succeeded with its first attempt at a lunar soft landing. Surveyor 1 was a more sophisticated lander than Luna 9, as were the spacecraft of its Lunar Orbiter series, the first of which was successfully inserted into orbit around the Moon in August 1966. The purpose of the US landers and orbiters was precisely the same as their Soviet counterparts, to determine the properties of the Moon and to identify candidate landing sites for manned missions. All five Lunar Orbiters were successful, as were five of the seven Surveyor landers.

Launch date

1966

31 Jan	Luna 9 lander	Success, first lander on the Moon
1 Mar	Luna orbiter	Fourth stage failure
31 Mar	Luna 10 orbiter	Success, first orbiter of the Moon
30 May	Surveyor 1 lunar lander	Success, first US lander on the Moon
10 Aug	Lunar Orbiter 1	Success, first US orbiter of the Moon
24 Aug	Luna 11 orbiter	Successful orbiter, imager failed
20 Sep	Surveyor 2 lunar lander	Crashed on the Moon on Sep 22
22 Oct	Luna 12 orbiter	Success, returned images
6 Nov	Lunar Orbiter 2	Success
21 Dec	Luna 13 lander	Success

1967

5 Feb	Lunar Orbiter 3	Success
17 Apr	Surveyor 3 lunar lander	Success
4 May	Lunar Orbiter 4	Success
16 May	Luna orbiter test flight	Fourth stage early burnout
12 Jun	Venera 4 entry probe	Entry successful, didn't reach surface
14 Jun	Mariner 5 Venus flyby	Successful Venus flyby on Oct 19
17 Jun	Venera entry probe	Fourth stage failure
14 Jul	Surveyor 4 lunar lander	Lost contact minutes before landing
1 Aug	Lunar Orbiter 5	Success
8 Sep	Surveyor 5 lunar lander	Success
27 Sep	Zond Earth orbital test flight	Booster failed, unpiloted lunar Soyuz
7 Nov	Surveyor 6	Success
22 Nov	Zond Earth orbital test flight	Launch failed, unpiloted lunar Soyuz

1968

7 Jan	Surveyor 7 lunar lander	Success
7 Feb	Luna orbiter	Third stage early burnout
2 Mar	Zond 4 deep space test	Self-destructs on return, unpiloted lunar Soyuz
7 Apr	Luna 14 orbiter	Success
22 Apr	Zond circumlunar test	Second stage shutdown
14 Sep	Zond 5 circumlunar test	Success, returned on Sep 21
24 Aug	Zond 6 circumlunar test	Crashed on return on Nov 17

In the midst of all this lunar activity, the Soviets had their first fully successful planetary mission, launched as Venera 4 on June 12, 1967. A second launch on June 17 was unsuccessful. Venera 4 entered the atmosphere of Venus on October 18 and transmitted atmospheric data while descending by parachute until it fell silent at an altitude that was still far above the surface. The next day America made its second successful flyby mission of Venus, with Mariner 5.

Both the nations ignored the Mars launch opportunity in 1967. The US could not afford as many launches as the Soviets and focused on Venus in 1967. The Soviets, frustrated by their six failures at Mars, by their continuing difficulties with the 3MV spacecraft, and by the Mariner 4 flyby in 1965, decided to forego the 1967 window and create a more complex Mars spacecraft that would be dispatched by the Proton launcher on a mission to land on the planet. In contrast, the tantalizing performance of Venera 2 and 3 ensured that this 3MV line would continue for Venus, and after the success of Venera 4 it became evident that a landing on the surface of that planet was achievable.

THE YE-6M LUNAR LANDER SERIES: 1966

Campaign objectives:

Throughout 1965 there was a high level of frustration in the Soviet robotic lunar and planetary programs. In the period 1963–65, three of six Venus-type spacecraft were successfully dispatched, two of three Mars-type, and five of eleven soft landers for the Moon. Nevertheless, all of these spacecraft failed either in transit or at the target. Only one, Zond 3, a test launch to Mars distance, returned anything of scientific or propaganda value, and that was from its flyby of the Moon. On the plus side things were improving, because three missions in late 1965 reached their targets and failed only at the last minute, Venera 2 and Luna 7 and 8.

Spacecraft launched

First spacecraft:	Luna 9 (Ye-6M No.202/13)
Mission Type:	Lunar Lander
Country/Builder:	USSR/NPO-Lavochkin
Launch Vehicle:	Molniya-M
Launch Date/Time:	January 31, 1966 at 11:41:37 UT (Baikonur)
Landing Date/Time:	February 3, 1966 at 18:44:54 UT
Mission End:	February 6, 1966 at 22:55 UT
Outcome:	Success.
Second spacecraft:	Luna 13 (Ye-6M No.205/14)
Mission Type:	Lunar Lander
Country/Builder:	USSR/NPO-Lavochkin
Launch Vehicle:	Molniya-M
Launch Date/Time:	December 21, 1966 at 10:17:00 UT (Baikonur)
Landing Date/Time:	December 24, 1966 at 18:01:00 UT
Mission End:	December 28, 1966 at 06:13 UT
Outcome:	Success.

When responsibility for robotic lunar and planetary missions was transferred from OKB-1 to NPO-Lavochkin at the end of 1965, a dozen failed missions made Georgi Babakin decide to modify the Ye-6 lander as the Ye-6M. His changes produced an immediate success, with Luna 9 making the desired soft landing on February 3, 1966, and sending back the first pictures from the surface of another world. Once again the Soviets had beaten the US to a space exploration milestone. Western headlines proclaimed a new space lead for the USSR. Although several years behind schedule, largely owing to the difficulty in developing the upper stage for its launcher, the US succeeded at its first attempt at a lunar landing. Surveyor 1 touched down on June 2, 1966, and returned imagery of a far superior quality. In December 1966 the second and final lander in the Ye-6M series, Luna 13, was successful as well.

Spacecraft:

The Ye-6M spacecraft was identical to the Ye-6 with modifications to the landing shock absorbers and a new independent guidance system. The airbags that enclosed the lander were inflated after the retro-rocket had ignited, requiring relocation of the tank of the nitrogen with which to inflate the bags from one of the side modules onto the cruise stage itself, because the side modules were jettisoned prior to braking. No additional redundancy was introduced.

Luna 9 launch mass: 1,538 kg (*lander* 105 kg)
Luna 13 launch mass: 1,620 kg (*lander* 113 kg)

Figure 10.1 Luna 9 spacecraft.

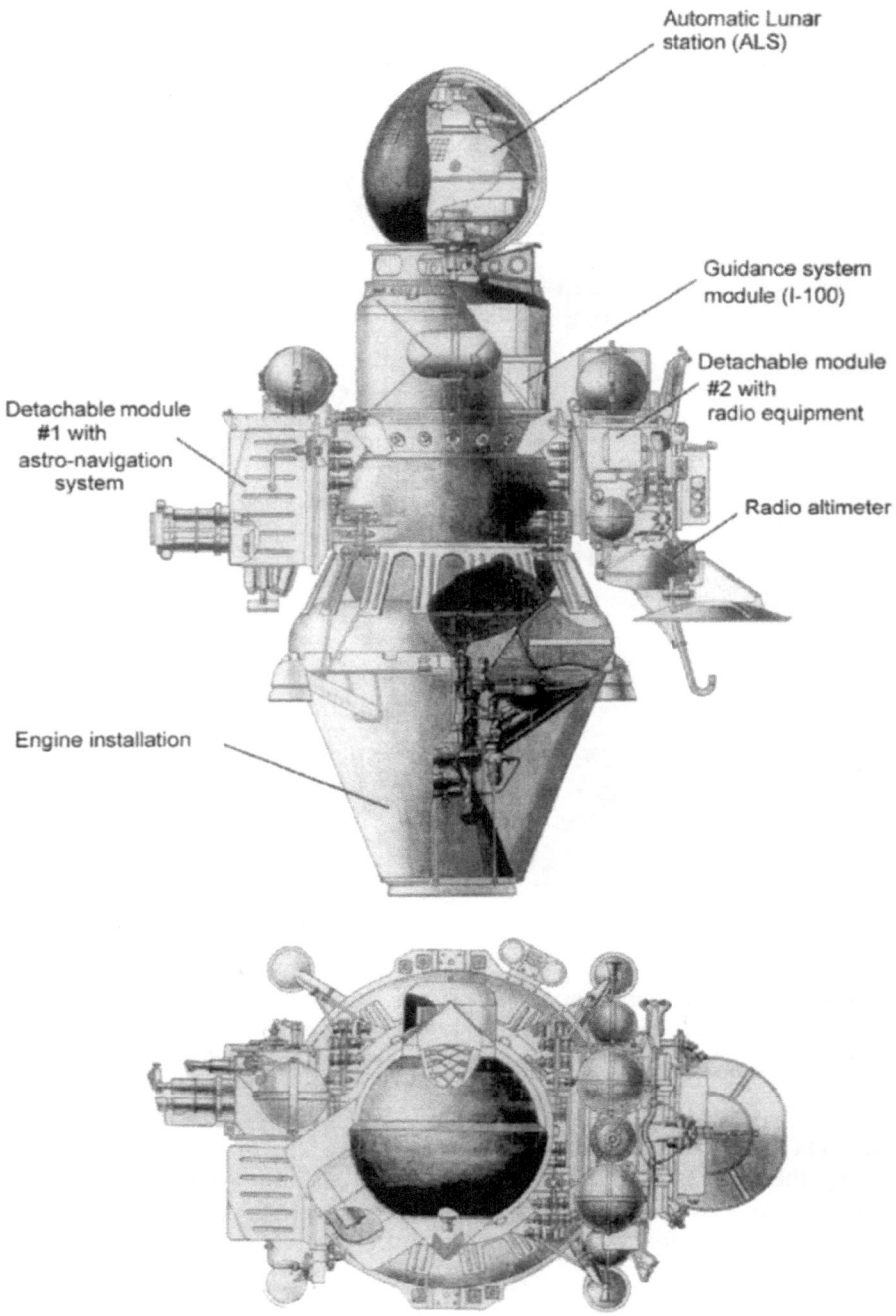

Figure 10.2 Luna 9 spacecraft diagram (courtesy Energiya Corp).

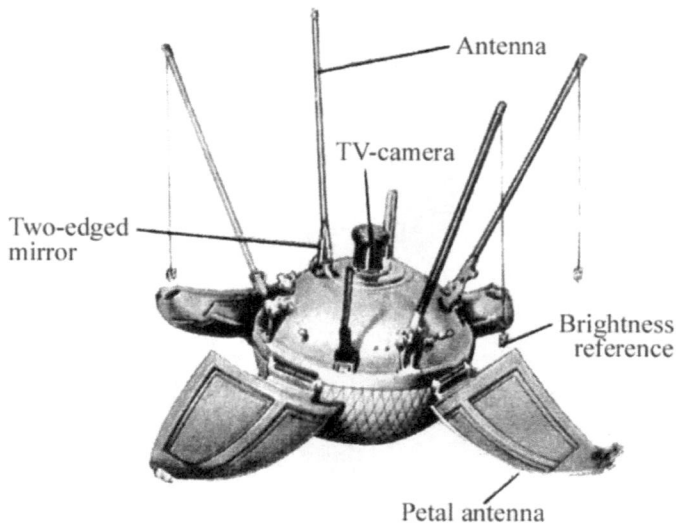

Figure 10.3 Luna 9 lander.

Figure 10.4 Luna 13 lander (from *Space Travel Encyclopedia*): 1. Transmitter antennae; 2. Receiver antennae; 3. Deployment arm; 4. Penetrometer; 5. Gamma-ray densitometer; 6. Panoramic stereo cameras; 7. Infra-red radiometers; 8. Stabilizing petals.

The Luna 9 lander had a mass of 105 kg, including 5 kg of scientific instruments. At 112 kg, the Luna 13 lander carried an increased scientific payload including two panoramic cameras for stereoscopic imaging and a pair of spring-loaded deployable 1.5 meter long booms for testing soil mechanics.

Figure 10.5 Luna 9 lander with single camera, and Luna 13 lander with dual-camera (inset).

Payload:

Luna 9:

1. Panoramic camera
2. Radiation detector

The scanning photometer camera system of Luna 4 to 8 was improved. It weighed only 1.5 kg, drew only 2.5 W, and had higher resolution. It used a tilting mirror in a revolving turret to produce a 29 × 360 degree panorama of 6,000 lines. The sensitivity could be adjusted by command, and it could operate from 80 to 150,000 lux. A panorama required approximately 100 minutes to transmit as 250 Hz analog video on the 183.538 MHz channel. Optical calibration and tilt-indication targets were suspended from the pop-up antennas, and three short poles carried dihedral mirrors to provide stereo images for small areas of the surface, to measure distances, and to locate the horizon and tilt more precisely. The radiation detector measured solar corpuscular radiation both in flight and on the lunar surface.

Luna 13:

1. Dual panoramic cameras for stereo
2. Radiation detector
3. Infrared radiometer for soil temperature
4. Penetrometer for soil strength and bearing capacity
5. Gamma-ray radiation/backscatter densitometer
6. Three axis accelerometer for surface mechanics on landing

The penetrometer, which had a 5 cm long cone, was mounted at the end of one of the deployable booms and a 65 N explosive charge drove it into the ground in order to measure the mechanical soil properties. The gamma-ray backscatter densitometer was mounted at the end of the other boom to measure soil density.

Mission description:

Launched on January 31, 1966, the Luna 9 spacecraft flew flawlessly across cislunar space, made its braking maneuver and ejected its capsule, which bounced and rolled to a halt at 18:45:04 UT on February 3 at 7.08°N 295.63°E in the Ocean of Storms. After the four petals opened outward and stabilized the capsule, the spring-loaded antennas were commanded to deploy, with one evidently failing. Five minutes after touchdown the television camera was activated to show the first ground-level view of the lunar surface. At that time the Sun was just 3.5 degrees above the horizon and much of the ground was in shadow. In one of the ironies of the Cold War, the British using the Jodrell Bank radio telescope were the first to publish pictures from Luna 9, after intercepting and readily recognizing the transmission as a fax machine signal. Although the Soviets had published their frequencies and had enlisted the assistance of Jodrell Bank to track previous missions, they were understandably upset to have their accomplishment 'scooped' in the world's press, particularly as the aspect ratio was incorrectly set. The US intelligence station at Asmara, Ethiopia, also intercepted the images but this was not made known at the time.

Luna 9 came to rest near the rim of a 25 meter diameter crater and was tilted at 15 degrees. Over the next few hours it settled to a tilt of 22.5 degrees, enabling stereo images to be made of nearby features. Over seven communications sessions lasting a total of 8 hours and 5 minutes four panoramas were transmitted, the final one with the Sun approaching an elevation of 40 degrees. The last contact was at 22:55 UT on February 6, as the battery depleted.

The second Ye-6M spacecraft, Luna 13, landed at 18:01:00 UT on December 24, 1966, at 18.87°N 297.95°E between the craters Seleucus and Krafft in the Ocean of Storms. In the act of bouncing and rolling, the accelerometer recorded data on soil density to depths of about 20 cm. It deployed two booms to measure soil density and surface radioactivity. The television system provided imagery at various times over the next 2 days, but the failure of one of the two cameras precluded stereo imagery. The depletion of the battery terminated operations at 06:13 UT on December 28.

Results:

Luna 9:

Nine images were returned by Luna 9, including five that were assembled to provide a panoramic view of the surface in the vicinity of the lander. The radiation detector measured a daily dosage of 30 millirads, which would not be hazardous to humans. The successful landing was clear evidence that the lunar surface was sufficiently dense to support a future manned spacecraft.

Figure 10.6 Portion of a Luna 9 panorama.

Figure 10.7 Portion of a Luna 13 panorama.

Luna 13:

Only one camera worked on Luna 13, returning five 220-degree panoramas in which the Sun was at increasing elevations. The soil density was found to be approximately 0.8 g/cc, much less than lunar bulk density and terrestrial analogs, but sufficient to support heavy landers. The radiation detector confirmed the 30 millirads/day reading by Luna 9. The infrared radiometer recorded surface temperature as a function of solar elevation, measuring a temperature of 117°C at local noon. It was decided that the first cosmonauts to land on the Moon would do so in the Ocean of Storms.

THE YE-6 LUNAR ORBITER SERIES: 1966–1968

Campaign objectives:

OKB-1 started development work on the Ye-7 lunar orbiter at the same time as the Ye-6 soft lander, but it progressed more slowly. After NPO-Lavochkin took over the robotic program the Soviets became anxious to upstage the American orbiter, which was scheduled for its first launch in mid-1966. They also needed to acquire close up imagery of potential landing sites and information on the environment in lunar orbit for the manned lunar program. An incentive presented itself in early 1966 when the long duration manned Voskhod 3 flight that was to have coincided with the opening of the 23rd Congress of the Communist Party in April (the first for new Communist Party leader Leonid Brezhnev) was canceled and there was a requirement for a new space spectacular. The Ye-7 was not ready, but Babakin offered to produce a lunar satellite by replacing the lander of the Ye-6 with a pressurized module carrying a payload of readily available instruments. The first orbiter, the Ye-6S, was cobbled together in less than a month. It is possible that the orbiter module was adapted from an Earth satellite of the Cosmos series.

After a launch failure on March 1, 1966, a backup spacecraft was prepared and successfully dispatched to the Moon on March 31, fortunately for Babakin just in time to satisfy the political objective. By becoming the first lunar orbiter, Luna 10 marked another milestone for the Soviet space program. In a moment of theater, it played a recording of the '*Internationale*' to the Party Congress.

The US sent its first orbiter to the Moon a little over 4 months later. This project was successful on the first try, and Lunar Orbiter 1 sent back the first pictures from lunar orbit. With landers and orbiters returning high quality data, the US was finally catching up.

Spacecraft launched

First spacecraft:	Ye-6S No.204 (Cosmos 111)
Mission Type:	Lunar Orbiter
Country/Builder:	USSR/NPO-Lavochkin
Launch Vehicle:	Molniya-M
Launch Date/Time:	March 1, 1966 at 11:03:49 UT (Baikonur)
Outcome:	Failed to leave Earth orbit.

Second spacecraft:	Luna 10 (Ye-6S No.206)
Mission Type:	Lunar Orbiter
Country/Builder:	USSR/NPO-Lavochkin
Launch Vehicle:	Molniya-M
Launch Date/Time:	March 31, 1966 at 10:47:00 UT (Baikonur)
Encounter Date/Time:	April 3, 1966 at 18:44 UT
Mission End:	May 30, 1966
Outcome:	Success.

<u>Third spacecraft:</u>	Luna 11 (Ye-6LF No.101)
Mission Type:	Lunar Orbiter
Country/Builder:	USSR/NPO-Lavochkin
Launch Vehicle:	Molniya-M
Launch Date/Time:	August 24, 1966 at 08:03:00 UT (Baikonur)
Encounter Date/Time:	August 28, 1966 at 21:49 UT
Mission End:	October 1, 1966
Outcome:	Successful orbiter, but no images returned.
<u>Fourth spacecraft:</u>	Luna 12 (Ye-6LF No.102)
Mission Type:	Lunar Orbiter
Country/Builder:	USSR/NPO-Lavochkin
Launch Vehicle:	Molniya-M
Launch Date/Time:	October 22, 1966 at 08:42:00 UT (Baikonur)
Encounter Date/Time:	October 25, 1966
Mission End:	January 19, 1967
Outcome:	Success.
<u>Fifth spacecraft:</u>	Ye-6LS No.111 (Cosmos 159)
Mission Type:	Lunar Orbiter Test Mission
Country/Builder:	USSR/NPO-Lavochkin
Launch Vehicle:	Molniya-M
Launch Date/Time:	May 16, 1967 at 21:43:57 UT (Baikonur)
Outcome:	Earth orbit test mission, lower orbit than desired.
<u>Sixth spacecraft:</u>	Ye-6LS No.112
Mission Type:	Lunar Orbiter
Sponsoring Agency:	USSR/NPO-Lavochkin
Launch Vehicle:	Molniya-M
Launch Date/Time:	February 7, 1968 at 10:43:54 UT (Baikonur)
Outcome:	Stage 3 failure. Did not reach Earth orbit.
<u>Seventh spacecraft:</u>	Luna 14 (Ye-6LS No.113)
Mission Type:	Lunar Orbiter
Country/Builder:	USSR/NPO-Lavochkin
Launch Vehicle:	Molniya-M
Launch Date/Time:	April 7, 1968 at 10:09:32 UT (Baikonur)
Encounter Date/Time:	April 10, 1968 at 19:25 UT
Mission End:	June 24, 1968
Outcome:	Success.

Having achieved lunar orbit ahead of the US and provided the required space spectacular for Moscow, Babakin resumed work on the Ye-7 orbiter whose task was lunar photography. When it was decided to use the Ye-6 cruise stage to make the midcourse and orbit insertion maneuvers, the orbiter spacecraft became the Ye-6LF.

Luna 10 had significantly deviated from its predicted path in lunar orbit, and radio tracking had shown the Moon to have an irregular gravity field. As a manned lunar lander would require a precise orbit if it were to land at a preselected point, in addition to lunar photography the Ye-6LF was given the task of accurately mapping the lunar gravity field.

Two of these spacecraft flew as Luna 11 and 12, and although both achieved lunar orbit an attitude stabilization problem prevented Luna 11 from providing any useful imagery. Even with new tracking data, the navigators could not predict their orbits with the desired accuracy. A second modification to the orbiter module produced the Ye-6LS that was to acquire more precise lunar gravity field data and also test deep space communications for the manned lunar program. After a test mission in Earth orbit and a lunar attempt that was lost to a launch vehicle failure, the third Ye-6LS was successful as Luna 14.

Spacecraft:

These orbiters all used the Ye-6 cruise stage with the lander replaced by an orbiter module. An interesting feature of the orbiters is that they carried no solar panels for recharging batteries, with the result that their operating time was defined by battery life. Lunar orbit insertion was a much smaller burn than the braking maneuver for a landing mission. In the case of the Ye-6S Luna 10, the orbiter module was released on its own in a spin stabilized condition. For the Ye-6LF and Ye-6LS, however, the orbiter modules of Luna 11, 12 and 14 were retained so that they could be stabilized to perform photography. The Ye-6S orbiter module was 1.5 meters long, 0.75 meters in diameter and massed 248.5 kg. It carried two radios transmitting at 183 and 922 MHz. Including the cruise stage and a large conical equipment module the Ye-6LF orbiters were 2.7 meters long and 1.5 meters in diameter. The Ye-6LS was similar to the Ye-6LF but with an improved navigation system to more precisely measure the orbit and apparatus to test a communication system intended for the manned lunar program.

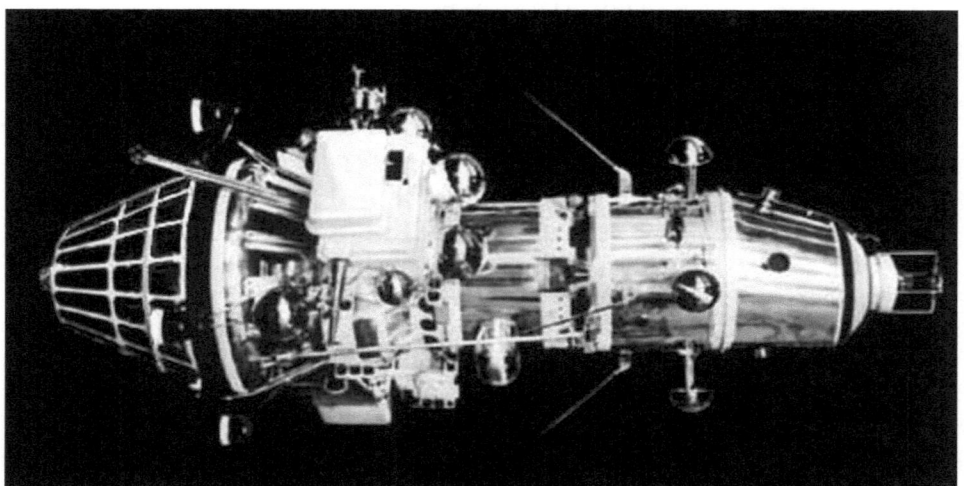

Figure 10.8 Luna 10 spacecraft.

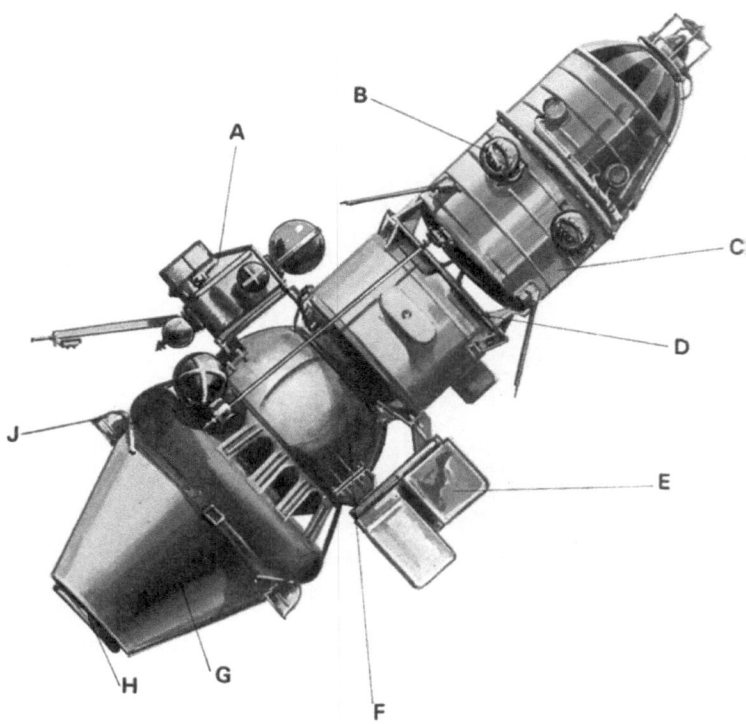

Figure 10.9 Luna 10 Ye-6S spacecraft (from *Robot Explorers*): A. Attitude control and radar altimeter; B. Orbiter omni antenna; C. Orbiter module; D. Avionics and communications module; E. Attitude reference sensors; F. Propellant tank; G. Propulsion system; H. Engine nozzle; J. Attitude control jets.

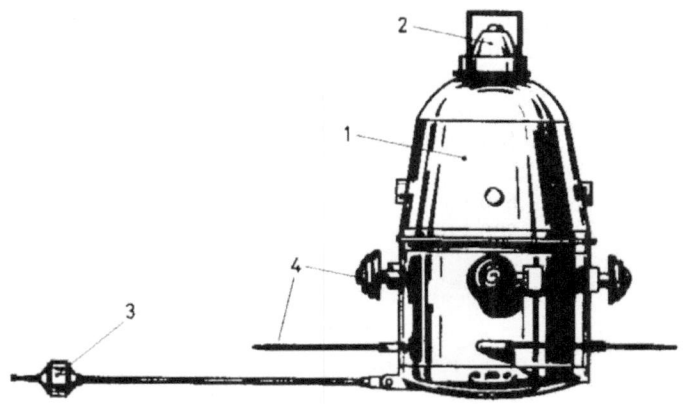

Figure 10.10 Luna 10 orbiter module (from *Space Travel Encyclopedia*): 1. Pressurized instrument module; 2. Radiometer; 3. Magnetometer; 4. Antennas.

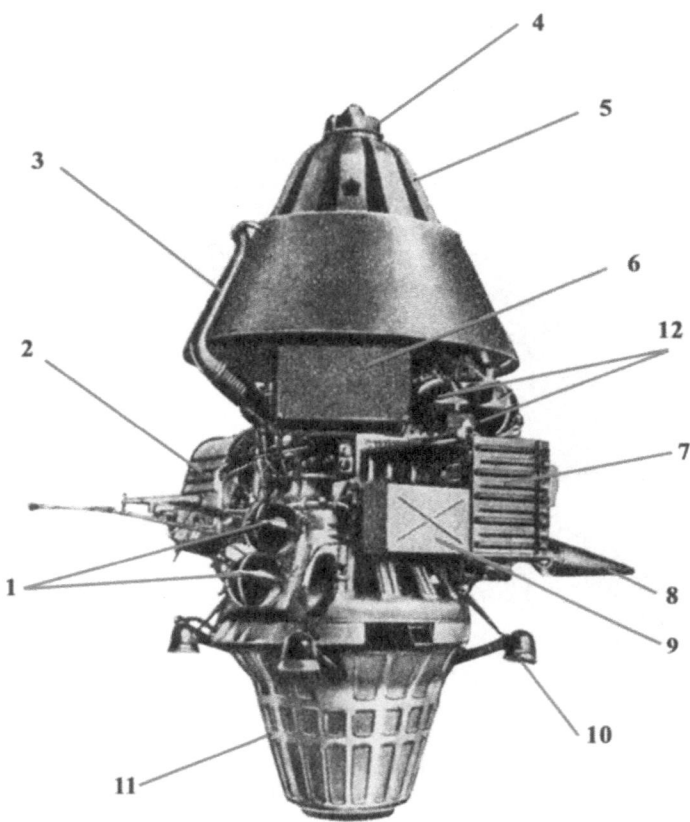

Figure 10.11 Luna 11 and 12 Ye-6LF spacecraft: 1. Helium tanks for attitude control pressurization; 2. Imaging system; 3. Thermal control radiator; 4. Radiometer; 5. Instrument module; 6. Battery; 7. Attitude control sensors; 8. Omnidirectional antenna; 9. Attitude control avionics; 10. Attitude control engines; 11. Main engine; 12. Propellant tanks.

Luna 10 launch mass:	*1,584 kg (orbiter module 248.5 kg)*
Luna 11 launch mass:	1,640 kg
Luna 12 launch mass:	1,620 kg
Luna 14 launch mass:	1,700 kg

Payload:

Ye-6S Luna 10:

1. Boom-mounted triaxial fluxgate magnetometer
2. Low energy x-ray spectrometer

3. Gamma-ray spectrometer
4. Gas discharge counters
5. Ion trap solar plasma detectors
6. SL-1 radiometer
7. Micrometeoroid detector
8. Infrared radiometer
9. Gravitational field mapping experiment (using spacecraft tracking)

The Ye-6S flown as Luna 10 had seven instruments developed for the Ye-7, but not the camera. The magnetometer was on a 1.5 meter long boom. Low energy x-ray and gamma-ray spectrometers were to measure the composition of the lunar surface; piezoelectric sensors were to measure micrometeoroid fluxes; an infrared radiometer was to measure thermal radiation from and the temperature of the lunar surface; gas discharge counters were to measure solar and cosmic rays in the lunar environment and soft electrons in a lunar ionosphere; ion traps were to measure electrons and ions in the solar wind and search for a lunar ionosphere; and the SL-1 radiometer was to measure the lunar radiation environment. Another key investigation was to measure the lunar gravity field by radio tracking of the spacecraft. There was no imaging on this first hastily prepared orbiter.

Ye-6LF Luna 11 and 12:

1. Facsimile imaging system
2. Low energy x-ray spectrometer
3. Gamma-ray spectrometer
4. SL-l radiometer
5. Micrometeoroid detectors
6. Ultraviolet spectrometer
7. Long-wave radio astronomy experiment
8. Gravitational field mapping experiment (using spacecraft tracking)
9. Lunar rover wheel drive technology experiments

The imaging system was similar to the facsimile film camera used by Zond 3 the previous year. At the altitude at which the pictures were to be shot, an image would encompass an area of 25 square kilometers and the 1,100 scan lines would provide a maximum surface resolution of 15 to 20 meters. Two cameras were carried on each mission. The ultraviolet reflectance spectrometer was to measure the structure of the surface. No data was ever published on the composition of the lunar surface or of the magnetic field, radiation and micrometeoroids in the lunar environment. Nor was the analysis of the irregular gravity field published. In addition to scientific instruments, the spacecraft carried technology tests of lubricants for operating gears and bearings in vacuum to qualify them for use on lunar rovers.

Ye-6LS Luna 14:

1. Lunar communications system test
2. Cosmic ray detector

3. Solar wind plasma sensors
4. Radiation dosimeter
5. Gravitational field mapping experiment (using spacecraft tracking)
6. Lunar rover wheel drive technology experiment

The main goal of Luna 14 was to test the spaceborne and ground segments of the new communication system for the manned lunar program. Other objectives were to continue to investigate the lunar radiation and plasma environment and to use a new navigation system to more precisely map the lunar gravity field and librations. The spacecraft also carried more engineering tests of lunar rover motors including gears, ball bearings and lubricants for vacuum operation.

Mission description:

On March 1, 1966, the first spacecraft, Ye-6S No.204, was stranded in parking orbit when the fourth stage lost roll control during the unpowered coast and was unable to fire its engine for the escape burn. It was designated Cosmos 111 and re-entered the atmosphere 2 days later.

Four weeks later on March 31, 1966, Ye-6S No.206 was successfully dispatched as Luna 10. It made a midcourse correction the following day and then at 18:44 UT on April 3 it became the first spacecraft to enter orbit around another celestial body, achieving a 350 × 1,017 km orbit inclined at 72 degrees to the lunar equator with a period of 2 hours 58 minutes. The cruise stage then released the orbiter module. On April 4 the '*Internationale*' was played to the 23rd Congress of the Communist Party of the Soviet Union. In fact what the Congress heard was a rehearsal playback from the previous evening, because at a second rehearsal on the morning of the meeting it was observed that one of the notes had gone missing. Luna 10 operated for 56 days, 460 lunar orbits and 219 radio transmissions before the battery drained and contact was lost on May 30, 1966.

Spacecraft Ye-6LF No.101 was launched on August 24, 1966, as Luna 11, and at 21:49 UT on August 28 it entered a 160 × 1,193 km lunar orbit that was inclined at 27 degrees with a period of 3 hours. Coming 2 weeks after the first US orbiter, there was every expectation in the West that the Soviet mission would send back images. The transmissions had been recoded to prevent interception by the likes of Jodrell Bank, but no pictures were forthcoming. An attitude control thruster had failed and prevented aiming either the camera or the ultraviolet instrument at the lunar surface. It was suspected that something had become lodged in the nozzle of the thruster. By sheer bad luck, the x-ray and gamma-ray spectrometers also failed. The spacecraft was placed into a spin stabilized mode and the other experiments apparently worked satisfactorily. After 38 days, 277 orbits and 137 radio transmissions the batteries ran out on October 1, 1966, and the Soviets, without mentioning imaging, reported that the mission was complete.

Ye-6LF No.102, Luna 12, entered a 3 hour 25 minute 103 × 1,742 km lunar orbit inclined by 36.6 degrees on October 25, 1966. It did not suffer the problems of its predecessor except for the ultraviolet spectrometer, which again failed. The primary objectives were the lunar photography that Luna 11 had been unable to provide and to continue to chart the gravitational field. On October 29 the spacecraft transmitted its first photographs. In this regard the Soviets were 2 months behind the Americans. A total of 40 images were returned by each of the two cameras. Once the spacecraft had finished imaging, it was placed into a spin stabilized mode and the other tasks were successfully performed, including testing electric motors for a rover. Even with the new gravity maps developed by tracking Luna 11, the orbit of Luna 12 deviated surprisingly far from that predicted. Its perilune dropped by 3 to 4 km/day relative to the prediction, and the failure of one of the attitude control thrusters made it difficult to raise the perilune to compensate. Finally, on January 19, 1967, after 85 days, 602 orbits and 302 communication sessions, transmissions ceased.

The next mission to be launched was a test flight of the second modification to the orbiter. The Ye-6LS No.111 spacecraft was to be launched into a highly elliptical Earth orbit with an apogee near 250,000 km in order to perfect a means of accurately measuring and adjusting an orbital trajectory to compensate for gravity anomalies, but the fourth stage shut down prematurely leaving the spacecraft in a lower 260 × 60,710 km orbit. Despite the low apogee, it was probably put through its intended operations. Designated Cosmos 159, it re-entered on November 11, 1967. The Ye-6LS No.112 spacecraft failed to achieve parking orbit when the third stage ran out of propellant early at the 524 second mark as a result of excessive fuel consumption through the turbine gas generator. The last spacecraft of this type, Ye-6LS No.113, was successfully dispatched as Luna 14 and at 19:25 UT on April 10, 1968, it entered a 160 × 870 km lunar orbit at 42 degrees. It marked the end of the second generation of Luna missions.

Results:

Luna 10 conducted an extensive study of the Moon from lunar orbit. Its path varied much more than the Soviets expected. This was due to a very uneven gravity field that featured localized 'mass concentrations' (mascons) below the surface. Luna 10 established the importance of charting the lunar gravity field, and also of providing spacecraft with robust propulsion for precise control of their orbital trajectories. The Soviets discovered this well before the Americans, who were generally given credit because their space program results were more widely published. Luna 10 found the Moon to have no detectable atmosphere, the lunar surface to have large expanses of basalt but few, if any, granitic provinces, and measured the amount of potassium, uranium, and thorium in the crust. It mapped a magnetic field whose strength was 0.001% of Earth, and found it not likely to be intrinsic. It showed that the Moon had no trapped radiation belts like those of Earth. The micrometeoroid flux and cosmic radiation in lunar orbit was also measured.

The x-ray and gamma-ray spectrometers on Luna 11 failed and imaging was not possible, but it was able to make observations of solar radiation and also conducted a successful test of lunar rover reduction gear operation in vacuum. The first images

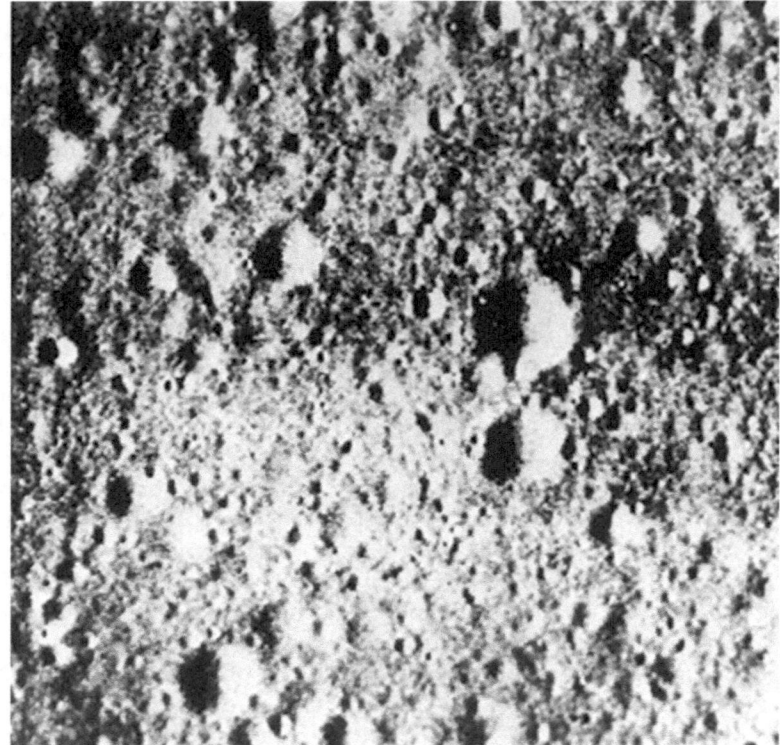

Figure 10.12 Luna 12 image taken from 250 km of an area of 25 square kilometers south of the crater Aristarchus.

were obtained by Luna 12. On October 29, 1966, it transmitted images of the Sea of Rains and the crater Aristarchus with resolution of 15 to 20 meters. Owing to their low quality, only the first few images were released. Ironically, the Soviets used the more extensive and far better imagery from the Lunar Orbiter series to select landing sites. Furthermore, like the Americans they ultimately chose three sites, one each in the Sea of Tranquility, the Central Bay and the Ocean of Storms. The radio tracking of Luna 11 and 12 revealed the need for even better orbital tracking experiments to provide a more precise lunar gravity map. Luna 12 also detected x-ray fluorescence of the surface induced by the solar wind, and this provided a means of measuring the composition of the surface. Radiation fields and micrometeoroid flux were measured in lunar space and the engineering test of reduction gears was successful.

The mission of Luna 14 passed almost without comment, perhaps because to have described what it was doing would have revealed too much about the manned lunar program. It was assumed in the West at the time to be a failed photographic mission, but it is now known to have mapped the figure of the Moon and its gravity field to a high degree of precision, to have provided data on the propagation and stability of radio communications to the spacecraft at different orbital positions, measured solar wind plasma and cosmic rays in lunar orbit, measured the librational motion of the Moon, and determined the Earth-Moon mass ratio. The communications system for the manned lunar program was also successfully tested. Further engineering tests of rover motors were conducted to select the materials and lubricants that operated best in vacuum for the systems intended for roving vehicles.

THE FIRST SUCCESS AT VENUS: 1967

Campaign objectives:

By the end of 1965 the Soviets had failed in a total of sixteen launches to Venus and Mars. Ten of these failed attempts were aimed at Venus, including the most recent Venera 2 and 3 missions that had come so close to achieving their goals. Adding to the frustration was the fact that by this time the US had succeeded with close flybys at Venus in 1962 and at Mars in 1964. Nevertheless, the Soviets were encouraged by their near successes, and were determined to push on. Realizing that the US was to attempt another Venus flyby mission in 1967, the Soviets wanted to outdo them with two missions to pierce the cloudy veil of the planet and obtain new information on its mysterious atmosphere and surface.

Spacecraft launched

First spacecraft:	Venera 4 (1V No.310)
Mission Type:	Venus Atmosphere/Surface Probe
Country/Builder:	USSR/NPO-Lavochkin
Launch Vehicle:	Molniya-M
Launch Date/Time:	June 12, 1967 at 02:39:45 UT (Baikonur)
Encounter Date/Time:	October 18, 1967
Outcome:	Successful.
Second spacecraft:	Cosmos 167 (1V No.311)
Mission Type:	Venus Atmosphere/Surface Probe
Country/Builder:	USSR/NPO-Lavochkin
Launch Vehicle:	Molniya-M
Launch Date/Time:	June 17, 1967 at 02:36:38 UT (Baikonur)
Outcome:	Failed to depart Earth orbit.

After Venera 2 and 3 the robotic planetary program was transferred from OKB-1 to NPO-Lavochkin. Beginning in April 1965 Babakin decided not to send any more flyby missions to Venus after the 1965 campaign and Lavochkin began to revise the 3MV spacecraft for the 1967 window for this planet, concentrating heavily on entry and landing. Working from 3MV blueprints supplied by OKB-1 and insight drawn from the Venera 2 and 3 experience, Babakin's engineers devised improvements to the thermal control and other systems. Lavochkin did more ground testing and built two new test facilities, one a thermal vacuum chamber completed in January 1967 to test the spacecraft under simulated flight conditions and the other a centrifuge rated at 500 G to test the entry and descent system. The first test of an entry probe in this chamber at the 350–450 G load expected for high angle Venus entries near 11 km/s destroyed its internal components. As the earlier descent capsules would certainly not have worked, the design had to be modified. This revitalized effort was rewarded immediately with the USSR's first truly successful planetary mission in 8 years of trying, with Venera 4 yielding in-situ data on the atmosphere of Venus. It began a new and much more fruitful era in the Soviet investigation of this planet.

Spacecraft:

Carrier spacecraft:

These spacecraft were the first 3MV for Venera missions built by NPO-Lavochkin which, in particular, greatly improved the thermal control system that had caused so much trouble with Venera 2 and 3. The hemispherical fluid radiators on the ends of the solar panels were deleted and a new system of heat transfer pipes located behind the parabolic antenna, which itself served as a radiator since it faced in the opposite direction to the solar panels. Liquid coolant was abandoned in favor of gas coolant. The communication system was also improved and the omnidirectional antenna was replaced by low gain spiral cone antennas mounted on booms connected to the solar panels and angled in flight to keep Earth in the radiation pattern. As previously, the spacecraft had to be turned to aim its high gain antenna at Earth, but this was only during scheduled communications sessions and operations at Venus.

Like its predecessors, Venera 4 was 3.5 meters tall, the solar panels had a span of 4 meters and the parabolic high gain antenna was 2.3 meters in diameter. The panels measured 2.5 square meters but, as previously, were sparsely populated with cells. The noticeable difference between Venera 4 and its predecessors in the 3MV series were the change in the solar panels to a more rectangular shape and the absence of the hemispherical radiators.

Entry vehicle:

For the 1967 mission the entry capsule was strengthened to resist stresses as high as 350 G and given an internal damper to reduce shock effects during entry and landing. At 1 meter in diameter it was 10 cm larger than the previous probes and nearly spherical with an ablative surface and a covered opening in the rear hemisphere for deployment of the parachute and antennas. It was the first of a series of entry probes which would be progressively better suited to survive the descent down to the surface. The internal mass distribution was bottom heavy to ensure the proper pointing on entry and aerodynamic stability during the descent. It was pre-cooled to -10°C by a system in the main module prior to separation, and operated a re-circulating fan thereafter. The capsule was intended to transmit atmospheric data

Figure 10.13 Venera 4 spacecraft, front and back views. These publicity photos do not show the thick ablative material on the entry system or the thermal blankets.

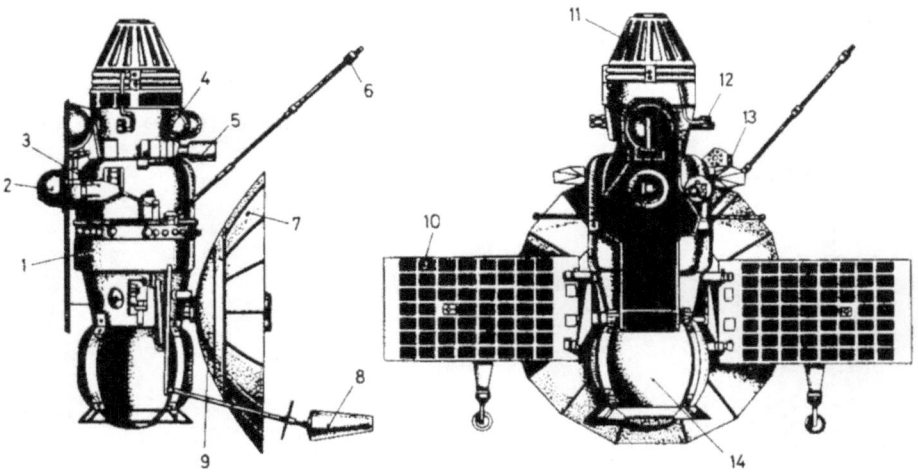

Figure 10.14 Venera 4 spacecraft diagram (from *Space Travel Encyclopedia*): 1. Carrier vehicle; 2. Star sensor; 3. Sun sensor; 4. Attitude control gas tanks; 5. Earth sensor; 6. Magnetometer; 7. Parabolic antenna; 8. Omnidirectional spiral antennas; 9. Thermal radiator; 10. Solar panels; 11. Propulsion system; 12. Attitude control microengines; 13. Cosmic ray detector; 14. Entry vehicle.

and radar data on descent, survive the impact and make measurements on the surface. The 28 amp-hour battery, which was rechargeable by the spacecraft during the cruise, could sustain 100 minutes of independent operation. The capsule design pressure was 10 bar with a margin up to about 18 bar, and the maximum survivable temperature for the parachute was 400°C.

The Venera probes were targeted to the center of the planetary disk as seen from Earth for optimum communications directly back to home. A helical antenna on top of the descending capsule was used to direct a radiation pattern to the zenith, and the telemetry was sent at 1 bit/s on 922.8 MHz using a pair of redundant transmitters. Measurements were sent back every 48 seconds. If the capsule were to splash down in an ocean, which few people believed was likely, it would float and a 'sugar seal' would release a semaphore signal to signify this fact.

Figure 10.15 (left) shows the entry vehicle without its upper insulation layers. The two ports are for testing the insulation system on this engineering model. Inside the thick, porous and lightweight ablative material is the descent capsule itself shown in Figure 10.15 (right). Hanging out over the side are the radio altimeter antennas that spring out when the parachute deploys. In accordance with international regulations, the capsule was sterilized prior to launch.

Five levels of redundancy were provided to ensure separation from the spacecraft. First by direct command from Earth, second by the on board sequencer, third by the triggering of a G switch on atmospheric entry, fourth by a sensor activated if Earth communications were interrupted by reorientation on entry and, as a last resort, the bands attaching the capsule to the spacecraft would burn through during initial entry.

Launch mass: 1,106 kg
Entry vehicle mass: 383 kg

Figure 10.15 Venera 4 entry system and enclosed descent capsule.

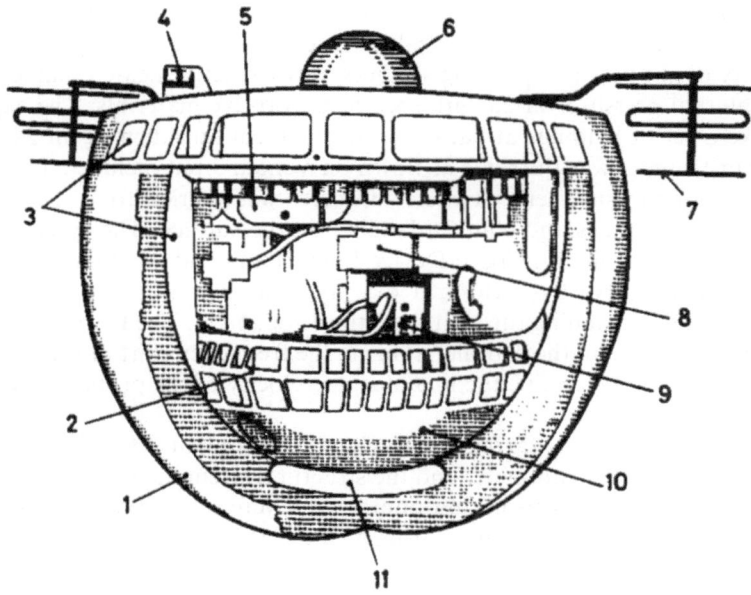

Figure 10.16 Venera 4 descent capsule diagram (from *Space Travel Encyclopedia*): 1. Outer heat shield; 2. Structural frame; 3. Probe walls; 4. Altimeter deployment system; 5. Heat exchanger; 6. Communication antenna; 7. Altimeter antenna; 8. Avionics unit; 9. Battery; 10. Insulation; 11. Shock absorber.

Payload:

Carrier spacecraft:

1. Triaxial fluxgate magnetometer
2. Solar wind charged particle detector
3. Lyman-alpha and atomic oxygen photometers
4. Cosmic ray gas discharge and solid state detectors

It had the same instruments as the Venera 2 and 3 cruise modules, except that the cosmic ray instrument included a second gas discharge counter of a different type.

Descent/landing capsule:

1. Temperature, pressure and density sensors
2. Atmospheric chemical gas analyzers
3. Radio altimeter
4. Doppler experiment

The temperature, pressure and density sensors were the same as on Venera 3. The gas analyzers used eleven cells to measure carbon dioxide, molecular nitrogen, molecular oxygen and water vapor. The composition was identified by how the atmosphere reacted with the material in each cell, such as by the electrical conductivity of chemically absorbing surfaces; or by reactive heated filaments; or by how the internal pressure varied with specific absorptive materials. The experiment was to take a set of readings at parachute deployment and then again 347 seconds later. The instrument was the same as flown on Venera 3 but included a hydrometer for water vapor measurement. A radio altimeter was carried for the first time to obtain absolute altitudes and confirm landing on the surface. The system was built by the Research Institute for Space Device Engineering and adapted from one used in aircraft. To conserve bandwidth, it did not issue continuous data, but only a semaphore to indicate falling through the altitude of 26 km. The Doppler experiment required no hardware on the capsule, utilizing the frequency shift of the carrier wave of the transmitter to determine the line of sight velocity of the probe as it descended through the atmosphere.

Some of the instruments carried by previous probes had to be deleted in order to release mass for the radio altimeter and the structural strengthening. The gamma-ray instrument, wave motion sensor, and photometer were sacrificed. But, as always, it carried a medallion with the coat of arms of the USSR and a bas relief of Lenin.

Mission description:

The first spacecraft was launched successfully towards Venus on June 12, 1967, and became Venera 4. The second was stranded in parking orbit on June 17, when the fourth stage did not ignite because the turbopump had not been pre-cooled. It was named Cosmos 167 by the Soviets and re-entered 8 days later. Venera 4 performed well during cruise, reorienting itself every few days to point its high gain antenna at Earth for a communication session. A midcourse correction was made on July 29 at a range of 12 million km from Earth. It arrived at Venus on October 18 and released the entry capsule at 04:34 UT, at which time it was 44,800 km over the night side. The carrier spacecraft sent measurements on the upper atmosphere and ionosphere until it broke up in the atmosphere. The capsule entered the atmosphere at 10.7 km/s and slowed through a peak deceleration of 350 G. At a pressure level of 0.6 bar and a speed of 300 m/s it shed the rear cover and deployed the 2.5 square meter drogue parachute. Several seconds later it deployed the 55 square meter main parachute and radio altimeter antennas. The instruments were turned on at 55 km altitude, at which time the rate of descent was 10 m/s. The mechanical commutator interrogated each instrument in turn and fed the data to the transmitter. It transmitted for 93 minutes on its parachute descent before falling silent. It reached the surface at 19°N 38°E, in darkness near the morning terminator. It was 4:40 Venus solar time and the solar zenith angle was 110 degrees. Including three intended test flights, this was the first successful Soviet planetary mission after twenty attempts, and the first successful entry probe by either spacefaring nation.

Jodrell Bank reported receiving signals from the surface, not realizing that these had been sent during the descent. Thinking the capsule had reached the surface in an

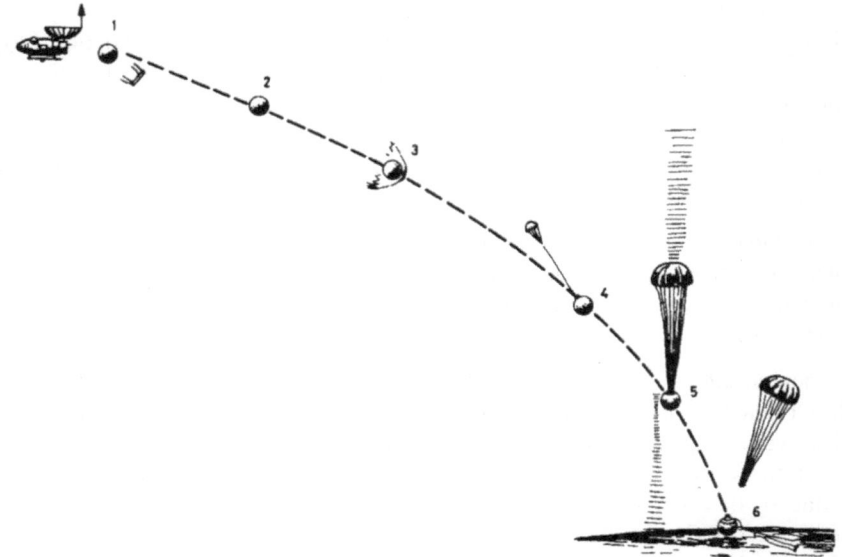

Figure 10.17 Venera 4 descent sequence (from *Space Travel Encyclopedia*): 1. Separation; 2. Unstabilized free flight; 3. Entry and stabilization; 4. Braking parachute deployed; 5. On main parachute, transmitter and altimeter on, acquiring and transmitting data on descent; 6. Surface impact, main chute release.

operational state, the Soviets reported that it had landed. But it slowly became clear that this could not be the case. The data from Mariner 5, which flew by Venus a day after Venera 4 arrived, indicated that the surface temperature was much higher than the final measurement reported by the entry probe. A series of meetings by Soviet and American scientists conducted over the next 2 years decided that the probe had succumbed to the increasingly hostile environment and had been disabled while still far above the ground. Nevertheless, as the first mission to transmit data from within a planetary atmosphere it achieved a major scientific milestone. The data return was significant and demonstrated just how hostile was the environment of Venus. It was evident that future probes would have to be further strengthened.

Results:

During its descent the Venera 4 entry probe returned more than 23 sets of readings by the atmospheric structure experiment. They began at an altitude of 55 km, and the atmospheric temperature was measured over the entire 93 minute descent. The initial temperature was 33°C and it increased to 262°C. The initial pressure reading was 0.75 bar, and the instrument reached its limit of 7.3 bar long before the probe ceased to transmit. Using atmospheric models constructed from the data at the time, it was concluded that the signal was lost at an altitude of 24 km. Atmospheric density was obtained by plugging the temperature and pressure data into the hydrostatic equation and the result tested against the parachute descent characteristics. Doppler data (i.e. changes in received master oscillator frequency) provided altitude profiles of wind speed and direction, both horizontal and vertical, but the measurement errors were large.

The atmospheric composition experiment showed the atmosphere to be composed mainly of carbon dioxide:

carbon dioxide	90 ± 10%
molecular nitrogen	less than 2.5%
molecular oxygen	0.4 to 1.6%
water vapor	0.05 to 0.7%

The percentage of carbon dioxide was initially disputed because the expectation was that at least 50% of the atmosphere would be molecular nitrogen, and American scientists were skeptical. But later missions would prove Venera 4 correct. The arid nature of the atmosphere was also unexpected. The model of Venus as a watery world had to be completely scrapped.

The aircraft-derived radio altimeter was designed to send a signal semaphore at an altitude of 26 km, but it had not been adequately adapted for the Venus mission and actually sent its signal at twice that altitude, 52 km. This was a principal cause of the confusion over whether Venera 4 had reached the surface or not. Atmospheric data and Doppler measurements showed that the probe had descended through about 28 km during transmission and the altimeter semaphore indicated that the top level was 26 km. The last measured temperature of 262°C and the derived pressure of ~18 bar were about what was expected at the surface at the time. However, measurements of the planet's microwave brightness made by terrestrial radio telescopes had indicated values of about 325°C. The chemical analysis by Venera 4 showing the dominance of carbon dioxide required a reanalysis of the radio-telescope microwave brightness based on atmospheric models with less carbon dioxide. A new analysis in 1967 explained some of the unusual features of the microwave spectrum of Venus as due to carbon dioxide, and resulted in surface conditions of about 427°C and 75 bar that were inconsistent with the Venera 4 probe having reached the surface. Atmospheric models based on Mariner 5 data also showed far higher temperatures and pressures at the surface. One suggestion was that Venera 4 had landed on a large mountain, but Carl Sagan pointed out that radar studies of the planet had found no such large edifice. Extrapolation of the Venera 4 atmospheric profile indicated conditions at the surface at the impact site to be 500°C and 75 bar. Eventually the data was reconciled by Avduevsky, Marov, and Rozhdestvensky (1969) using an adiabatic model of the Venusian atmosphere which confirmed loss of signal at 18 ± 2.5 bar at an altitude of about 24 km and extrapolated conditions at the surface as 442°C and 90 bar.

The signal ceased near the pressure limit of the capsule, but it is possible that the probe exhausted its battery near the 18 bar level after 93 minutes of operation. In any case the capsule would have been crushed and thereafter the parachute would have burned, leaving the capsule to free fall to the surface.

Prior to breaking up, the main spacecraft provided the first in-situ measurements of the close-in magnetic field, thermosphere, ionosphere, and solar wind interaction. In 1962 Mariner 2 had flown past Venus at 34,773 km, which was too great a range to detect a magnetic field or magnetospheric signature. Venera 4 found no intrinsic planetary magnetic field. The low fields detected were due to interaction of the solar wind with the ionosphere. No radiation belts were found, and an extended corona of atomic hydrogen was discovered reaching 10,000 km into space from the planet.

THE ZOND CIRCUMLUNAR SERIES: 1967–1970

Campaign objectives:

After the August 1964 government declaration that divided the work on the Soviet manned lunar program between Chelomey's Proton-launched circumlunar program and Korolev's N-1 launched lander program, Korolev continued throughout the year to argue for both to be consolidated under his OKB-1. In addition to organizational and economic arguments, he offered the prospect of reaching the Moon much sooner with a spacecraft already in an advanced stage of design, whereas Chelomey would have to start from scratch. For the circumlunar mission Korolev proposed a stripped down form of his Earth orbital Soyuz complex which, as it happened, was originally conceived with lunar missions in mind. The anxiety to get a Soviet cosmonaut to the Moon first, and Korolev's persuasiveness, won him a partial victory in October 1965 when the government approved his 7K-L1 lunar Soyuz for circumlunar flights that would be launched by Chelomey's Proton using OKB-1's Block D as an upper stage. The N-1 was neither ready nor appropriate for circumlunar missions, it was for the manned lunar landing program. Although rivals, Korolev and Chelomey appeared to work together well in implementing this circumlunar plan. Meanwhile Korolev would continue to develop the hardware for the manned lunar landing program: the N-1 launcher, the LOK lunar orbital version of the Soyuz, and the LK lunar lander.

Spacecraft launched

First spacecraft: 7K-L1 No.4L
Mission Type: Circumlunar and Return Test Flight
Country/Builder: USSR/TsKBEM
Launch Vehicle: Proton-K
Launch Date/Time: September 27, 1967 at 22:11:54 UT (Baikonur)
Outcome: Booster destroyed.

Second spacecraft: 7K-L1 No.5L
Mission Type: Circumlunar and Return Test Flight
Country/Builder: USSR/TsKBEM
Launch Vehicle: Proton-K

Launch Date/Time:	November 22, 1967 at 19:07:59 UT (Baikonur)
Outcome:	Second stage destroyed.

Third spacecraft: Zond 4 (7K-L1 No.6L)
Mission Type:	Test Flight to Lunar Distance and Return
Country/Builder:	USSR/TsKBEM
Launch Vehicle:	Proton-K
Launch Date/Time:	March 2, 1968 at 18:29:23 UT (Baikonur)
Return Date/Time:	March 9, 1968
Outcome:	Spacecraft self-destructed while on parachutes.

Fourth spacecraft: 7K-L1 No.7L
Mission Type:	Circumlunar and Return Test Flight
Country/Builder:	USSR/TsKBEM
Launch Vehicle:	Proton-K
Launch Date/Time:	April 22, 1968 at 23:01:27 UT (Baikonur)
Outcome:	Failed due to second stage shutdown.

Fifth spacecraft: Zond 5 (7K-L1 No.9L)
Mission Type:	Circumlunar and Return Test Flight
Country/Builder:	USSR/TsKBEM
Launch Vehicle:	Proton-K
Launch Date/Time:	September 14, 1968 at 21:42:11 UT (Baikonur)
Encounter Date/Time:	September 18, 1968
Return Date/Time:	September 21, 1968 at 16:08 UT
Outcome:	Successful recovery in Indian Ocean.

Sixth spacecraft: Zond 6 (7K-L1 No.12L)
Mission Type:	Circumlunar and Return Test Flight
Country/Builder:	USSR/TsKBEM
Launch Vehicle:	Proton-K
Launch Date/Time:	November 10, 1968 at 19:11:31 UT (Baikonur)
Encounter Date/Time:	November 14, 1968
Return Date/Time:	November 17, 1968
Outcome:	Spacecraft crashed on return landing in Kazakhstan.

Seventh spacecraft: 7K-L1 No.13L
Mission Type:	Test Flight to Lunar Distance and Return
Country/Builder:	USSR/TsKBEM
Launch Vehicle:	Proton-K
Launch Date/Time:	January 20, 1969 at 04:14:36 UT (Baikonur)
Outcome:	Upper stages failed.

Eighth spacecraft: Zond 7 (7K-L1 No.11)
Mission Type:	Circumlunar and Return Test Flight
Country/Builder:	USSR/TsKBEM
Launch Vehicle:	Proton-K
Launch Date/Time:	August 7, 1969 at 23:48:06 UT (Baikonur)
Encounter Date/Time:	August 11, 1969
Return Date/Time:	August 14, 1969
Outcome:	Successful recovery in Siberia.

Ninth spacecraft:	Zond 8 (7K-L1 No.14)
Mission Type:	Circumlunar and Return Test Flight
Country/Builder:	USSR/TsKBEM
Launch Vehicle:	Proton-K
Launch Date/Time:	October 20, 1970 at 19:55:39 UT (Baikonur)
Encounter Date/Time:	October 24, 1970
Return Date/Time:	October 27, 1970 at 13:55 UT
Outcome:	Successful recovery in Indian Ocean.

It was decided to make the test flights of the lunar versions of the Soyuz in the Zond series, beginning with the 7K-L1 circumlunar model launched by the Proton. Approved in December 1966, the plan for the L1 program called for four automated tests prior to the first manned circumlunar flight, which was scheduled for launch on June 26, 1967. The follow-on series was intended to test the 7K-LOK lunar orbital model launched by the N-1 lunar rocket, the Soviet counterpart to the Saturn V. This extensive flight testing in the manned program was unique to the Soviets. Ironically, Korolev generally eschewed full-up piloted flight tests in the manner of Apollo, but he abhorred the extensive ground testing the US conducted in all its programs. As a result, Soviet spacecraft generally had a much higher degree of automation than US spacecraft, often to the chagrin of the cosmonauts, and the lack of ground testing led to poor performance in test flights and consequent failures and delays.

The success of Zond 5 in circumnavigating the Moon, returning to Earth and then being recovered in September 1968 set off celebrations in the USSR. It was the first mission of either nation to achieve this feat. Shortly afterwards the Soviets revealed that it was an automated flight of a Soyuz manned capsule. This set off alarms in the US, because the way seemed clear for a cosmonaut to fly the same mission. There were windows in November and mid-December which would enable the Soviets to steal a march on Apollo 8, whose launch window was in late December. (A window for the northerly Baikonur site opened slightly earlier than for the lower latitude site in Florida.) The Soviets used the November window for another test, but Zond 6 failed catastrophically during landing. Unaware of this setback the US expected a Soviet manned circumlunar mission in December, but there was no such launch. The way was now clear for Apollo 8. The Soviets had insisted on four successful automated flights of Zond prior to a manned mission. There had been an internal debate on whether it would be justified to make an attempt as reckless as the Soviets felt the Americans were making with a mission to orbit the Moon using the first manned Saturn V. The failure of Zond 6 rendered this debate moot. After Apollo 8 the Soviets doggedly continued to perfect Zond for the circumlunar mission, but switched their focus and rushed tests of the N-1 launcher.

Although most Zond test flights were designed to provide information on the techniques and technologies needed to fly cosmonauts to the Moon and back safely, they also provided information of scientific interest. Instrumentation flown on these missions gathered data on micrometeoroid flux, solar and cosmic rays, magnetic fields, radio emissions, and the solar wind. Biological payloads were also flown and many excellent photographs of the Moon and Earth were taken.

Spacecraft:

The Soyuz 7K-L1 was a version of the 7K-LOK lunar orbital spacecraft modified to perform a circumlunar mission. Lacking an orbital module, it would have carried two rather than three cosmonauts. It had redesigned instrument panels for lunar missions and a thicker heat shield to handle the faster re-entry from a lunar return. There were a number of other design differences for the circumlunar missions.

Zond 4 is notable for being the first Soviet spacecraft to possess a computer. The 34 kg Argon 11 used integrated circuits, drew 75 W, was capable of 15 operations, and was provided with 4K of instruction ROM and 128 words of RAM.

Zond launch mass: ~5,375 kg

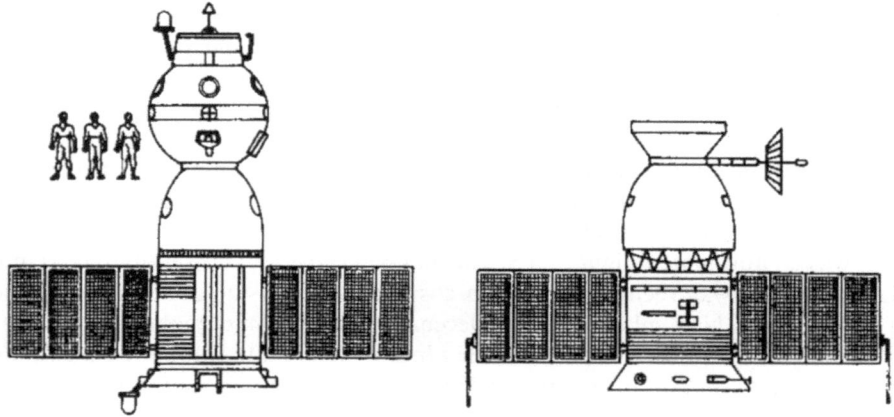

Figure 10.18 Comparison of Soyuz at left and Zond at right (from *Space Travel Encyclopedia*).

Payload:

Zonds 5 to 7:

1. Imaging system (color on Zond 7)
2. Cosmic ray detectors
3. Micrometeoroid detectors (Zond 6 only)
4. Biology payload

Figure 10.19 Zond 4 to Zond 8 spacecraft (courtesy Energiya Corp).

Zond 8:

1. Imaging system
2. Solar wind collector

The imaging systems on Zond 5, 6 and 8 carried a 400 mm camera for 13 × 18 cm frames of isopanchromatic film. Zond 7 carried a 300 mm camera using 5.6 × 5.6 cm film, both panchromatic and color. The solar wind collector utilized aluminum foil targets similar to the solar wind collectors that Apollo astronauts set up on the lunar surface, except that in the case of the Zond missions they were on the exterior of the descent capsule.

Mission description:

The constraints for circumlunar flights with launch and recovery in the high latitudes of the Soviet Union were fairly severe, yielding only 5 or 6 launch opportunities per year and not always spread out evenly over the year. Hence some Zond tests were made out to lunar distance without the Moon being available. Before the automated circumlunar series, there were two Earth orbital test flights. The first, launched on March 10, 1967, was only the fifth flight of the Proton and the first with Korolev's Block D fourth stage. It successfully placed the Soyuz into an orbit with an apogee far from Earth and telemetry tests were carried out. No recovery was attempted. The mission, designated Cosmos 146, was an auspicious start. The schedule for the first manned circumlunar flight in June 1967 seemed feasible. But then everything went wrong. The second flight test on April 6, 1967, went awry when the Block D failed.

Worse, on April 23 the first manned Earth orbital flight of the Soyuz failed when it crashed upon landing, killing cosmonaut Vladimir Komarov. His brief orbital flight had been plagued with serious problems and in attempting an emergency landing the parachute became entangled. Both the Zond and Soyuz programs stood down while the common issues were addressed. By September the Zond program was ready to resume.

The first two circumlunar mission attempts fell victim to the Proton launcher. In the first on September 27, 1967, one of the six engines on the booster failed to ignite owing to a blocked propellant line and the rocket was destroyed 97 seconds into the flight. Pad engineers had not removed a cover before flight. On the second launch on November 22, 1967, one of the four engines on the second stage failed to ignite at 125 seconds into the flight and the rocket was destroyed at the 130 second mark.

The next launch had been scheduled for a window in April 1968, but the Soviets were anxious to finish testing and beat the US to the Moon so it was decided to go early, without the Moon, and test the spacecraft to lunar distance with the high speed re-entry. On March 2, 1968, Zond 4 was launched into a highly elliptical orbit with an apogee at 354,000 km. It suffered many in flight system failures. An erratic star tracker in the attitude control system complicated the flight, but engineers managed to navigate the spacecraft back to Earth. The sensor failed again in the automated re-entry sequence, resulting a ballistic rather than the intended guided descent. While descending on its parachute over the Gulf of Guinea the capsule realized that it was off course and self-destructed using a mechanism installed in part to ensure that the Americans could not recover the spacecraft in the event of a badly targeted re-entry.

The next Zond launch attempt was to have been a circumlunar flight, but it failed ignominiously on April 22, 1968, when the spacecraft emergency escape system was erroneously triggered at the 194 second mark, shutting down the second stage of the launcher and carrying the Zond spacecraft clear. It was recovered 520 km from the pad. Another planned Zond launch went awry on July 14, 1968, when the oxygen tank in the fourth stage ruptured on the pad during pre-launch testing, destroying the Block D stage and killing three engineers. Fortunately, the lower stages, all loaded with propellant, did not explode. Although undamaged, spacecraft 7K-L1 No.8 was discarded.

After this frustrating series of failures, Zond 5 became the first fully successful circumlunar flight. Launched on September 14, 1968, it achieved its mission despite suffering several technical problems. The Earth sensor had been mounted incorrectly owing to an error in the documentation. The star-tracker attitude control system was rendered ineffective when its optical surfaces became contaminated by sublimating thermal protection material. Worse, the backup system was mistakenly turned off. But engineers were able to control the spacecraft's attitude in an awkward and slow process using the Sun sensors. Two midcourse maneuvers were conducted, and the spacecraft flew around the Moon on September 18 with a closest point of approach at an altitude of 1,950 km. Jodrell Bank announced the achievement and reported the spacecraft heading back to Earth. Only at this point did the Soviets actually admit to having launched the mission! On the voyage home a second attitude control sensor failed. On September 21 the spacecraft re-entered at 11 km/s at an overly steep angle on a ballistic rather than the planned guided 'skip' re-entry. Fortunately, after the loss of Zond 4, the self-destruction system had been deleted. After a 6 minute ride through the atmosphere with peak stresses of 16 G and 13,000°C, the descent capsule parachuted into the Indian Ocean at 16:08 UT near 32.63°S 65.55°E, 105 km from the nearest Soviet tracking ship. This was the first water recovery for the Soviets. The capsule was offloaded at Mumbai, India, for return to Moscow by aircraft. This boosted Soviet confidence in their human lunar program, but the flight had suffered serious non-fatal failures.

Zond 6 was successfully launched on November 10, 1968. Attitude control again became a problem when the high gain antenna, with the main star tracker attached, failed to deploy. The backup star tracker system and a lower gain antenna were used instead. As the spacecraft flew around the Moon on November 14 at an altitude of 2,420 km it photographed the near and far sides. On its way home engineers had to reorient the spacecraft to try to control the temperature in a hydrogen peroxide tank by exposing it to direct sunlight and unfortunately this also heated and deformed the hatch seal, causing the cabin to depressurize. The biology payload was killed and the altimeter damaged. In preparation for re-entry on November 17 the service module separated as intended but the high gain antenna remained attached to the front of the capsule. The vehicle entered the atmosphere at 11 km/s and after bleeding off some energy it skipped back into space at 7.6 km/s as intended, at which time the antenna detached. All went well with the 'second dip' until at an altitude of 5,300 meters the altimeter failed to function properly and commanded the parachute to be jettisoned. Some film was recovered from the wreckage, including the first color pictures of the Moon, before explosives engineers detonated the 10 kg of TNT that had been carried on board the capsule for the radio-command destruct system.

Of two circumlunar missions, Zond 5 had failed to perform the 'skip' maneuver during re-entry and had made a ballistic descent and splashed into the Indian Ocean, and Zond 6 had flown the intended trajectory only to fail on its parachute within the Baikonur boundary only 16 km from the launch pad. Undeterred, the Soviets made a valiant effort to prepare a mission for the December opportunity. Cosmonauts in training for a circumlunar flight argued to be allowed to fly. A vehicle was rolled out to the pad on December 1, but the spacecraft suffered so many technical problems that the window closed before approval to launch could be obtained. Even after the successful flight of Apollo 8 later in the month, the Soviets continued with the Zond tests. The hardware was available and the obvious thing was to push for success and realize the benefits of the investment. There was no precedent for actually stopping a Soviet space project with this much visibility. There remained the hope of making a manned circumlunar flight, although this prospect faded as the focus switched to the N-1 launcher.

But in the next 8 months, while the Americans reveled in the success of Apollo 8, 10 and 11, three more set backs were to plague the Soviet manned lunar program. Anxious to solve the problems with the Zonds, another test flight to lunar distance was planned before the circumlunar windows opened in summer. The launch of 7K-L1 No.13L failed in January 1969 due to problems with the second and third stages.

Figure 10.20 Proton launcher ready on the pad with Zond 5.

One of four second-stage engines shut down early, and then during its phase of the ascent the third stage suffered a breakdown in its fuel feed system and cut off. Even more significant for the manned program, in February and July of 1969 the first two test launches of the N-1 lunar rocket also failed.

Nine months after the Zond 6 disaster, and shortly after Apollo 11 succeeded and the Soviet robotic lander Luna 15 crashed, Zond 7 was launched towards the Moon on August 7, 1969. The next day, it conducted a midcourse maneuver and took color pictures of Earth. On August 11 the spacecraft flew around the Moon at an altitude of 1,985 km and performed two sessions of color photography of both the Moon and Earth. It returned to Earth on August 14, made the intended 'skipping' re-entry and landed successfully in a preselected area south of Kustanai in Kazakhstan.

The plan had called for four test flights prior to a manned circumlunar mission, but of the four only Zond 5 and 7 were fully successful. This poor success rate led to

Figure 10.21 Picture of Earth from Zond 5.

a proposal to fly one more test flight and then a manned mission to celebrate Lenin's birthday in April 1970. The supporters argued that the flight would contribute to the program by demonstrating to the world that the USSR was capable of manned lunar missions. But the government rejected it because it would look second rate. To have sent cosmonauts on a circumlunar mission ahead of Apollo 8 would have been one thing, but to do so after Apollo 11 was something else. At the political level, benefit was now being determined relative to American achievements. Approval was given only for the final automated test flight.

Zond 8 was launched on October 20, 1970. The next day it took pictures of Earth

Figure 10.22 Photograph of the lunar surface and Earth from Zond 7.

from 64,480 km, and the day after that it made a midcourse maneuver at a range of 250,000 km. It transmitted images of Earth for 3 days during its outbound flight and conducted two imaging sessions as it passed behind the Moon on October 24 at an altitude of 1,110 km. After two midcourse correction maneuvers on the return leg, it made a ballistic re-entry over the northern hemisphere on a southbound trajectory to sustain communication in most of the re-entry sequence. All previous Zonds had re-entered over the southern hemisphere, heading north. The capsule splashed into the Indian Ocean at 13:55 UT on October 27, approximately 24 km from the target point 730 km southeast of the Chagos Islands. It was recovered 15 minutes later by the Soviet oceanographic vessel *Taman* for return to Moscow by way of Bombay, India.

Results:

These were primarily engineering test flights, but they also carried payloads which provided scientific results.

Zond 5: High quality photographs of Earth were taken on the way home at a range of 90,000 km. They were useful because the film would be able to be processed on Earth rather than scanned on board and transmitted by radio. A biological payload of turtles, wine flies, fly eggs, meal worms, plants, seeds, bacteria, and other

Figure 10.23 Photograph of Earth from Zond 7.

living matter was recovered. The turtles had lost significant body mass and exhibited other metabolic anomalies. The fly eggs had not produced the expected number of adults, and the next generations of these showed a large increase in mutations.

Zond 6: The crash broke the film canister, but some film was recovered including images of the lunar limb and far side features taken at ranges of approximately 3,300 to 11,000 km. Some stereo pairs were also obtained. Only a few of its images have been published.

Zond 7: Color photography of both Earth and the Moon. A biological payload of turtles, wine flies, meal worms, plants, seeds, bacteria, and other living matter was successfully recovered.

Zond 8: Obtained photographs of Earth and the Moon from distances of 9,500 and 1,500 km.

To sum up, although the photography was excellent the science results from these

Figure 10.24 Picture from Zond 8.

test flights were minimal. Data on solar wind and cosmic rays was obtained but not published. The seed samples on recovered flights all showed chromosomal damage but of the animals only the turtles on Zond 5 showed any ill-effects.

11

Robotic achievements in the shadow of Apollo

The flight of Apollo 8 in December 1968 marked the beginning of the end for the Soviet Union's campaign to put cosmonauts on the Moon. The Zond circumlunar flight test series had been plagued by problems. Even the successful flight of Zond 5 suffered so many subsystem anomalies that engineers were very reluctant to trust a spacecraft to a manned mission. The crash of Zond 6 made beating Apollo 8 to the Moon almost impossible, and the circumlunar program endured a further setback on January 20, 1969, when the next Zond test flight fell victim to yet another launcher failure. Any chance that cosmonauts could reach the Moon in competition with the Americans was dealt a severe blow on February 21, 1969, when the counterpart of the Saturn V, the N-1, failed spectacularly on its maiden flight. It had been intended to deliver a modified Zond into lunar orbit. The second attempt to qualify the N-1 on July 3, 1969, less than a fortnight ahead of the launch of Apollo 11, resulted in the biggest explosion in the history of rocketry and destroyed the pad facilities. The last of the scheduled Zond flight tests, Zond 7, was a success in August, 1969, but by then the race was over. Instead of following up with a manned circumlunar mission the Soviets added another automated flight, which flew successfully in October 1970 as Zond 8. After two further attempts to qualify the N-1 in June 1971 and November 1972 also failed, the manned lunar program was canceled.

However, the Soviets countered the Apollo program with a series of robotic lunar missions using a new, large spacecraft that was originally designed to land a rover for a cosmonaut to employ on the lunar surface. When in late 1968 and early 1969 it became clear that the Americans were likely to beat them to the Moon, the Soviets opted to use this robotic landing system to try and upstage Apollo by being the first to return a lunar sample to Earth.

While the sample return system was being developed for use with the lander, the first launch of the new lander with a rover was attempted on February 19, 1969, but it was lost when the payload shroud failed shortly after the Proton launcher lifted off. The first sample return spacecraft was launched on June 14, but lost when the fourth stage failed. Rushing to beat Apollo 11 to the Moon, another sample return mission was launched on July 13, 1969, ten days after the devastating N-1 explosion and 3 days before Apollo 11 was launched. This spacecraft, Luna 15, successfully reached lunar orbit 2 days ahead of Apollo 11 and Westerners, uninformed of its intentions, viewed it with suspicion. Shortly after the lunar module of Apollo 11 set down on the Moon on July 20 and its astronauts made their historic moonwalk, the Soviet spacecraft crashed attempting to land in the Sea of Crises, some distance east of the Apollo 11 site in the Sea of Tranquility. The next three attempts through the middle of 1970 to return samples from the Moon with this type of spacecraft were all lost to launch vehicle failures.

Launch date

1968

21 Dec	Apollo 8 lunar orbiter	Success, first men to orbit the Moon

1969

5 Jan	Venera 5 entry probe	Entry successful, didn't reach surface
10 Jan	Venera 6 entry probe	Entry successful, didn't reach surface
20 Jan	Zond Earth orbital test flight	Second stage failed
19 Feb	Luna rover	Launcher shroud failed
21 Feb	N-1 Moon Rocket test	First stage failed in flight
25 Feb	Mariner 6 Mars flyby	Success on Jul 31
27 Mar	Mariner 7 Mars flyby	Success on Aug 4
27 Mar	Mars orbiter	Third stage exploded
2 Apr	Mars orbiter	Booster exploded
18 May	Apollo 10 lunar orbit test	Success
14 Jun	Luna sample return	Fourth stage failed
3 Jul	N-1 Moon Rocket test	First stage exploded at liftoff
13 Jul	Luna 15 sample return	Crashed on Moon on Jul 21
16 Jul	Apollo 11 lunar landing	Success, first men on Moon on Jul 20
7 Aug	Zond 7 circumlunar test	Success, returned to Earth on Aug 14
23 Sep	Luna sample return	Fourth stage failed
22 Oct	Luna sample return	Fourth stage failed
14 Nov	Apollo 12 lunar landing	Success

1970

6 Feb	Luna sample return	Second stage premature shutdown
11 Apr	Apollo 13 lunar landing	Explosion damage enroute, safe return

In early January 1969 the Soviets followed up their 1967 success at Venus with two launches of spacecraft similar to Venera 4 but modified to descend through the atmosphere more rapidly, and thereby provide data from deeper levels than before. Although Venera 5 and Venera 6 worked well, they imploded far above the surface.

The Soviets were ready with a new spacecraft designed for Mars in March 1969. Like the new Luna for delivering rovers and sample return spacecraft, these were heavy spacecraft that needed the Proton launcher. They had been designed to be able to enter orbit around Mars and dispatch a lander, but for the 1969 window they were to release a probe to get the data on the atmosphere that was required to design that lander. When the probe was deleted owing to development and test problems it was decided to equip the two spacecraft for this window as orbiters. Neither survived its launcher. These spacecraft and their launch attempts were virtually unknown in the West until after the Cold War. Blissfully unaware of this potentially overwhelming competition, the US dispatched two more flyby missions to Mars in 1969, Mariners 6 and 7, both of which were successful.

FOLLOWING UP AT VENUS: 1969

Campaign objectives:

In 1967 Venera 4 worked as well as could be expected considering the unknown environment to which it was sent. The US had only managed a flyby in this launch window for Venus, while the first successful Soviet planetary mission had achieved the impressive technical challenge of descending through the planetary atmosphere and sending back critical data on its characteristics. Knowing the US had no plans to return to Venus because it desired to focus on Mars, the USSR set out to develop an unassailable lead in the investigation of Venus.

NPO-Lavochkin built two new 3MV spacecraft for the 1969 launch opportunity. Venera 4 was high above the surface when it fell silent after 93 minutes of descent, either when it was crushed or when the battery ran out. It was decided that the new capsules must fall more rapidly in order to reach a deeper level in the time allowed by the battery. The capsules were similar to that of Venera 4, but modified to endure a higher entry velocity and with a smaller parachute for a faster descent.

Spacecraft launched

First spacecraft:	Venera 5 (2V No.330)
Mission Type:	Venus Atmosphere/Surface Probe
Country/Builder:	USSR/NPO-Lavochkin
Launch Vehicle:	Molniya-M
Launch Date/Time:	January 5, 1969 at 06:28:08 UT (Baikonur)
Encounter Date/Time:	May 16, 1969
Outcome:	Successful.
Second spacecraft:	Venera 6 (2V No.331)
Mission Type:	Venus Atmosphere/Surface Probe
Country/Builder:	USSR/NPO-Lavochkin
Launch Vehicle:	Molniya-M
Launch Date/Time:	January 10, 1969 at 05:51:52 UT (Baikonur)
Encounter Date/Time:	May 17, 1969
Outcome:	Successful.

Taken together, data from the Venera 4 and Mariner 5 missions in 1967 and from terrestrial measurements of the planet's radio brightness indicated that the surface pressure on Venus greatly exceeded the design tolerance of the descent capsule. But the debate about whether Venera 4 had reached the surface raged on for 2 years. In 1968 Soviet and American scientists met first in Tucson in March, then again at the COSPAR meeting in Tokyo in May, and for a third time at a symposium in Kiev in October. The result was general agreement that the surface conditions were 427°C and 90 bar. But this long debate did not reach consensus soon enough to influence the short construction schedule available for the 1969 probes. Knowing only that the pressure exceeded 18 bar, NPO-Lavochkin increased the tolerance to 25 bar for the new missions. By the time Venera 5 and 6 were launched, however, it was accepted that the surface pressure was much greater. With no time for further improvement, the missions were treated simply as an opportunity to obtain data using more precise instruments while new higher pressure designs were created for the next window.

Both missions were successful in descending through the atmosphere and, just as expected, the capsules imploded at altitude. The Soviet media had played down any expectations of reaching the surface. Venera 5 and 6 firmly set the stage for the next mission, which would attempt to reach the surface. This was the first 100% success rate for multiple launches and the first 100% success rate for multiple spacecraft.

Spacecraft:

The carrier vehicle was essentially the same as Venera 4, but the descent probe was strengthened to handle the higher velocity of approach in 1969, which would impose a higher deceleration load of 450 G, and also to withstand a pressure of 25 bar. This used up the buoyancy mass allocation. A lower velocity of 210 m/s was preset for the deployment of the pilot parachute, with smaller parachutes employed to descend more quickly and obtain measurements nearer to the surface before either the battery expired or the internal temperatures became lethal. The pilot parachute was reduced in size from 2.2 to 1.9 square meters, and the main chute from 55 to just 12 square meters.

Figure 11.1 (right) shows the Venera 6 spacecraft folded in launch configuration. The entry system is at the bottom covered with its dark ablation material. Most 3MV pictures are taken with the probe painted white and bearing the letters 'CCCP'. The antenna and solar panels are folded to fit into the launcher shroud and the spiral gas cooling pipes are visible on the back of the high gain antenna dish which, by facing in the opposite direction to the solar panels, acts as a radiator. The Venera 5 and 6 spacecraft were identical.

Launch mass: 1.138 kg (*probe* 405 kg)

Figure 11.1 Venera 5 on display and Venera 6 folded for launch.

Payload:

Carrier spacecraft:

1. Solar wind charged particle detector
2. Lyman-alpha and atomic oxygen photometers
3. Cosmic ray gas discharge and solid state detectors

These are the same as on Venera 4.

Descent/landing capsule:

1. Temperature, pressure and density sensors
2. Atmospheric chemical gas analyzers
3. Visible airglow photometer
4. Radio altimeter
5. Doppler experiment

Atmospheric density was measured during entry by a combination of Doppler tracking and the probe's accelerometers. In the parachute descent the atmospheric structure experiment measured temperature, pressure and density. This instrument was improved over Venera 4 with three platinum wire resistance thermometers for more precision, two redundant sets of three aneroid barometers covering the ranges 0.13 to 6.6 bar, 0.66 to 26 bar and 1 to 39 bar, and a tuning fork densitometer for the

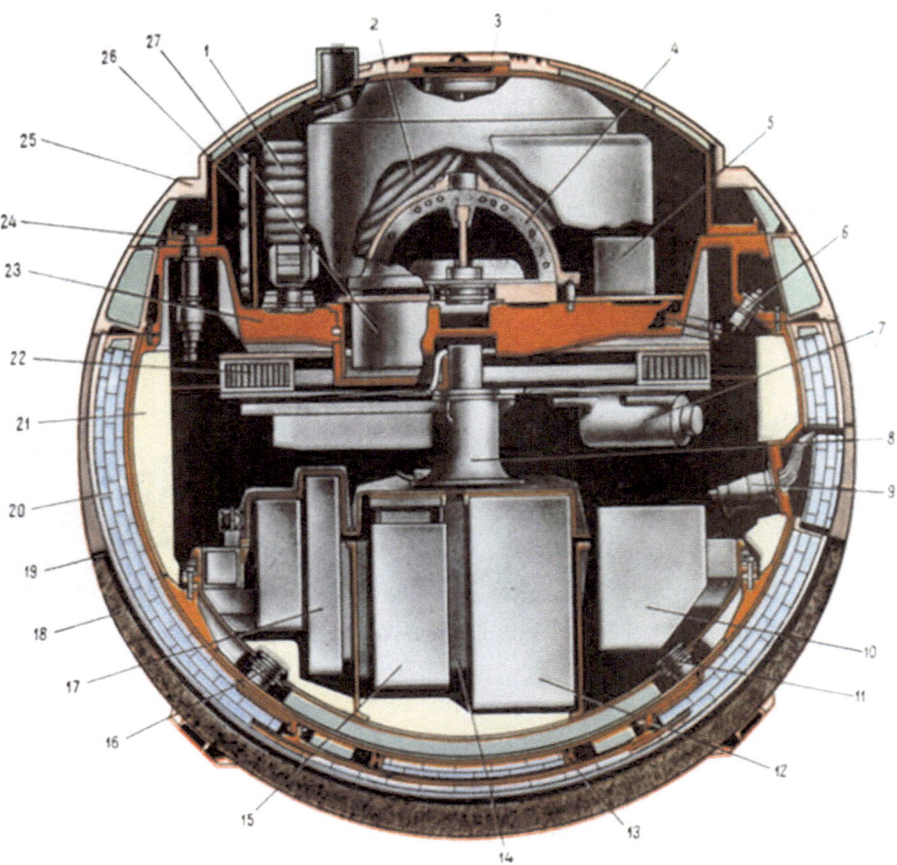

Figure 11.2 Venera 5 and Venera 6 entry capsule diagram: 1. Drogue parachute; 2. Main parachute; 3. Pyro piston lid; 4. Transmitter antenna; 5. Density sensor; 6. Gas fill valve; 7. Dehumidifier; 8. Thermal control fan; 9. Pressurization valve; 10. Commutation block; 11. Accelerometer; 12. Transmitter; 13. Oscillation damper; 14. Battery; 15. Redundant transmitter; 16. Accelerometer; 17. Timer; 18-20. External insulation; 21. Internal insulation; 22. Thermal control system; 23. Lander cover; 24. Pyro piston; 25. Parachute compartment lid; 26. Radio altimeter antenna; 27. Gas analyzer.

wider range of 0.0005 to 0.040 g/cc. The atmospheric gas analyzers were improved and reconfigured to benefit from the experience of Venera 4. A getter was added to measure the total inert gases including molecular nitrogen, and refinements were introduced to make better measurements of molecular oxygen and water. Accuracies were improved by transmitting a pressure reading at the same time as a composition analysis. Venera 5 would make its composition measurements at higher altitudes and Venera 6 at lower altitudes. A number of improvements were made to the radar altimeter to avoid the problems on Venera 4, and three semaphores were provided for altitudes of 45, 35, and 25 km at an accuracy of about 1.3 km. Even though these were night-time landings, a visible photometer was included to measure light levels during the descent because the dark hemisphere of the planet was known to exhibit an airglow phenomenon of unknown origin.

Mission description:

Venera 5 was launched on January 5, 1969, performed a course correction maneuver on March 14 when 15.5 million km from Earth, and reached Venus on May 16. The entry capsule was released at a distance of 37,000 km and at 06:01 UT it entered the atmosphere at 11.17 km/s at an approach angle of 65 degrees. During the parachute descent at 3°S 18°E on the night side, it provided a set of instrument readouts every 45 seconds for 53 minutes. The transmission ceased at an altitude of about 18 km when the pressure exceeded 27 bar. At that time the external temperature was 320°C and internal temperatures had reached 28°C. It was 4:12 Venus solar time and the solar zenith angle was 117 degrees.

Venera 6 was launched on January 10, 1969, conducted a midcourse correction on March 16 at 15.7 million km, and arrived on May 17 (one day after its partner). The entry capsule was released at a distance of 25,000 km and it entered the atmosphere at 06:05 UT. It descended over 5°S 23°E on the night side and transmitted for 51 minutes. Signals ceased at an altitude of about 18 km and a pressure of 27 bar, thereby confirming the results from its partner. It was 4:18 Venus solar time and the solar zenith angle was 115 degrees.

The Venera 5 and 6 carrier spacecraft both provided measurements on the upper atmosphere and ionosphere prior to breaking up in the atmosphere.

Results:

The Venera 5 and 6 entry capsules transmitted in excess of 70 temperature readings and 50 pressure readings during the descent from about 55 km altitude to their crush depth. Atmospheric density was derived from the temperature and pressure data by using the hydrostatic equation and verified against parachute descent characteristics inferred from the radio altimeter. Doppler data provided altitude profiles of wind speed and direction, both horizontal and vertical. While there was initially confusion about the altimetry, these instruments did work properly and the temperatures and pressures measured near the radio altimeter marks were a good match to the current engineering model of the planet's atmosphere:

	Venera 5			Venera 6	
altitude (km)	36	25	18	34	22
pressure (bar)	6.6	14.8	27.5	6.8	19.8
temperature (°C)	177	266	327	188	294

Each capsule made two readings of atmospheric composition: at 0.6 and 5 bar for

Venera 5, and at 2 and 10 bar for Venera 6. Their readings were consistent, and in good agreement with the Venera 4 data:

	Venera 5	Venera 6
carbon dioxide	$97 \pm 4\%$	greater than 56%
nitrogen and other noble gases	less than 3.5%	less than 2.5%
molecular oxygen	less than 0.1%	less than 0.1%
water vapor	about 1.1% (11 mg/l)	about 0.6% (6 mg/l)

Neither photometer registered anything except darkness, although the photometer on Venera 5 did report a large reading just before termination. This could have been a flash of lightning but, given the timing, it may merely have been an electrical transient caused by the imminent breakup.

The Venera 5 and 6 carrier spacecraft returned measurements on the solar wind in the vicinity of Venus and its interaction with the planet.

THE YE-8 LUNAR ROVER SERIES: 1969–1973

Campaign objectives:

The Ye-8 series was developed to support the Soviet lunar cosmonaut program. By the time that the Soviets entered the Moon race in mid-1964 Russian engineers at OKB-1 had already developed plans for a lunar rover. This, along with all the other lunar robotic programs, was transferred to NPO-Lavochkin in 1965. In early 1966 an automated lunar surface rover entered the mission plan for supporting a cosmonaut on the lunar surface. The function of the rover was to precede the cosmonaut to the landing site, to survey and certify the site as safe for landing, to act as a radio beacon to guide the manned lander in, to inspect this lander after touchdown and certify it as safe for ascent, and, if it were not so, to transport the cosmonaut to a backup ascent vehicle that was already in place.

When the robotic lunar exploration program was transferred to NPO-Lavochkin, Georgi Babakin set to work on a design for a spacecraft to meet these requirements. The availability of the powerful four-stage Proton launch vehicle using the Block D translunar injection stage enabled the resulting Ye-8 to be much heavier and more complex than its Ye-6 predecessor. The multi-purpose, in-line module design of the Ye-6 series was abandoned for a spacecraft design suited principally for soft landing a rover, and eventually other types of payload.

Spacecraft:

The spacecraft comprised three main components; a lander stage on the bottom, the rover that was carried on top, and a pair of side-mounted 'backpacks', each of which had avionics and two cylindrical propellant tanks.

Spacecraft launched

First spacecraft:	Ye-8 No.201
Mission Type:	Lunar Lander and Rover
Country/Builder:	USSR/NPO-Lavochkin
Launch Vehicle:	Proton-K
Launch Date/Time:	February 19, 1969 at 06:48:15 UT (Baikonur)
Outcome:	Shroud failure, vehicle disintegrated.

Second spacecraft:	Luna 17 (Ye-8 No.203)
Mission Type:	Lunar Lander and Rover
Country/Builder:	USSR/NPO-Lavochkin
Launch Vehicle:	Proton-K
Launch Date/Time:	November 10, 1970 at 14:44:01 UT (Baikonur)
Lunar Orbit Insertion:	November 15, 1970
Lunar Landing:	November 17, 1970 at 03:46:50 UT
Mission End:	September 14, 1971 at 13:05 UT
Outcome:	Success.

Third spacecraft:	Luna 21 (Ye-8 No.204)
Mission Type:	Lunar Lander and Rover
Country/Builder:	USSR/NPO-Lavochkin
Launch Vehicle:	Proton-K
Launch Date/Time:	January 8, 1973 at 06:55:38 UT (Baikonur)
Lunar Orbit Insertion:	January 12, 1973
Lunar Landing:	January 15, 1973 at 22:35 UT
Mission End:	June 3, 1973
Outcome:	Success.

Cruise and lander stages:

The lander stage was based on a quartet of 88 cm diameter spherical propellant tanks arranged in a square 4 meters on a side and connected using cylindrical inter-tank sections. These tanks fed a single engine whose thrust could be varied over the range 7.4 to 18.8 kN and a set of six vernier engines, two of which were mounted next to the main engine and were for use during the final descent to the surface. The other verniers were positioned around the periphery to provide stabilization. The landing system, engine, and radar altimeter were located between the tanks on the underside of the square tank assembly. Each of the tanks supported a shock absorbing landing leg. Attitude control thrusters were located at various places around the lander. The avionics and attitude control sensors to control the translunar trajectory, lunar orbit insertion, orbital maneuvers, and landing, were housed in the inter-tank cylindrical sections. Water cooling was used for thermal control. Communications at 922 MHz and 768 MHz were by way of a cone-shaped antenna mounted on a boom. Uplink was at 115 MHz.

Figure 11.3 Luna 17 spacecraft diagram (from Ball et al.) and during test at Lavochkin.

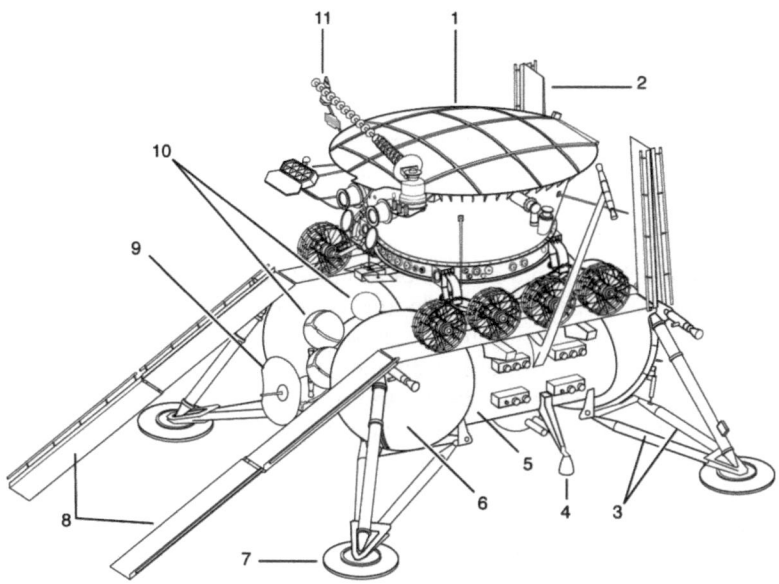

Figure 11.4 Luna 17 lander (by James Garry): 1. Lunokhod rover; 2. Folded exit ramp; 3. Shock absorbers; 4. Steering rockets; 5. Service module and avionics; 6. Propulsion tank; 7. Landing foot; 8. Extended exit ramps; 9. Radio altimeter; 10. Attitude control gas tanks; 11. Conical low-gain antenna and steerable directional helical antenna.

The two detachable 'backpacks' were mounted vertically on opposite sides of the square tank assembly and were for cruise and orbital operations. Each consisted of a pair of 88 cm diameter cylindrical tanks, between which were avionics and battery modules. The tanks contained propellants to feed the main engine. On top of each of these tanks was a smaller spherical tank of nitrogen for the cold gas attitude control system.

The Isayev design bureau built the new throttleable KTDU-417 main engine. Its purpose was to conduct midcourse maneuvers during the translunar coast, lunar orbit insertion, orbital maneuvers, and key portions of the descent. Once the operational orbit at about 100 km altitude had been achieved, descent to the surface began with a burn of about 20 m/s to lower the perilune to about 15 km directly over the landing site. The backpacks were jettisoned and, with perilune looming, the main engine was ignited for a 1,700 m/s 'dead stop' burn lasting 270 seconds designed to completely eliminate its horizontal velocity. After the spacecraft had free fallen to an altitude of about 600 meters and accelerated in the weak lunar gravity to a descent rate of about 250 m/s the main engine was reignited. This was shut down at 20 meters and the landing verniers ignited until a contact switch cut them off at a height of 2 meters. If all had gone to plan, the vehicle would then touch down at a velocity not exceeding 2.5 m/s. Unlike the Ye-6 soft landers, whose targets were constrained by the need to make a vertical descent from the translunar coast, the new spacecraft, by first going into orbit, could land anywhere.

For a rover mission, two sets of folding ramps were mounted on top of the upper side of the lander fore and aft of the rover, whose wheels were on the middle of the ramps. The ends of the ramps were carried folded up against the rover, and once on the Moon they were unfolded and lowered to provide the rover with two options for driving off the lander down onto the surface.

Lunokhod rover:

The body of the Lunokhod rover was a tub-like pressurized magnesium alloy shell for avionics, instruments and environmental controls covered by a large hinged lid. In daylight on the lunar surface the convex lid would be opened over the rear of the rover to expose solar cells on the inside surface of the lid to generate power and also to expose radiators in the top of the 'tub' for thermal control. In darkness the lid was closed. It was a very simple and effective design. The solar cells (Si on Lunokhod 1 and GaAs on Lunokhod 2) gave 1 kW of power to recharge the internal batteries. The body was mounted on a carriage of eight wheels, 51 cm in diameter and made of wire mesh with titanium blade treads. This design was in response to the data on lunar soil provided by Luna 9; the thin dust layer and firm soil that this found led to the abandonment of a caterpillar track design. Each wheel had its own suspension system using a special fluoride based lubricant to operate in vacuum, a pressurized independent DC motor and an independent brake.

The rover was controlled entirely from Earth by a five-person team, there was no automated mode, and steering required independently changing the speed settings on the wheels. It could move with only two operational wheels on each side, and any of

Figure 11.5 Lunokhod 1.

the axles could be severed to shed a wheel if it became locked. The smallest turning radius was 80 cm. Internal gyroscopes indicated its orientation. It was designed to drive over obstacles 40 cm high or 60 cm wide, to climb slopes of 20 degrees, and to maneuver on slopes as steep as 45 degrees. There were fail-safe devices to prevent movement over excessive slopes. Lunokhod 1 had only one driving speed, 800 m/hr, traveling either forward or in reverse, but Lunokhod 2 was capable of 800 and 2,000 m/hr in either direction.

The control team operated and navigated the vehicle by viewing through a pair of television cameras mounted on the front of the rover. These returned low resolution images at a rate of 20 seconds/frame for Lunokhod 1 and at the much improved rate of up to 3.2 seconds/frame for Lunokhod 2. The signal time delay to the Moon and back to Earth was 5 seconds, which had an effect on operations. Four other scanning photometer imagers of the type used on Luna 9 were mounted on the chassis. A pod on each side held a vertically mounted imager to give a 180 degree view at a 15 degree down angle, jointly providing a full panoramic view around the rover. A second imager was set above the first, nearer the top of the 'tub', and was mounted horizontally. These would jointly provide a full vertical panorama that included the sky and stars for navigation at the zenith and a vehicle level indicator at the nadir.

The rover was designed to survive three lunar nights, each lasting a fortnight over a period of 3 months. In darkness it was the kept alive by a small radioisotope heater with 11 kg of polonium-210 and by a radiator on top of the closed lid. Thermal control was by circulating internal air and by open-cycle water cooling. The rover was equipped with a conical low gain antenna, a steerable directional helical high gain antenna, television cameras, and extendable devices to impact the surface for soil density and mechanical property tests. Lunokhod 1 was 135 cm high, 170 cm long, 215 cm wide at the top, 160 cm wide at the wheels, had a wheelbase of 2.22 × 1.6 meters, and a mass of 756 kg.

Lunokhod 2 was an improvement based on experience with Lunokhod 1. It had an additional camera on the front at adult height for easier navigation. Its images could be transmitted at rates of 3.2, 5.7, 10.9 or 21.1 seconds/frame, with the fastest rate being instrumental in improving driving operations. The 8-wheel drive system was improved, and Lunokhod 2 was twice as fast for twice the range. Additional science instruments were carried.

Luna 17 launch mass: 5,660 kg (*landed mass* 1,900 kg; *rover* 756 kg)
Luna 21 launch mass: 5,700 kg (*landed mass* 1,836 kg; *rover* 836 kg)

Figure 11.6 Lunokhod 2.

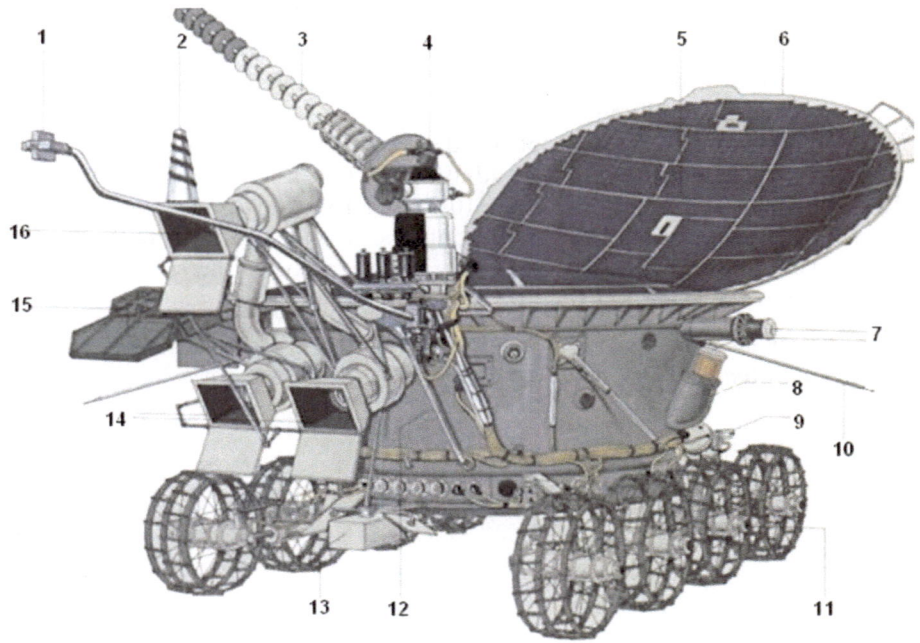

Figure 11.7 Lunokhod 2 diagram: 1. Magnetometer; 2. Low gain antenna; 3. High gain antenna; 4. Antenna pointing mechanism; 5. Solar arrays; 6. Deployed lid; 7. Imagers for horizontal and vertical panoramas; 8. Radioisotope heater with reflector and ninth wheel odometer at the rear; 9. Sampler (not deployed); 10. Boom antenna; 11. Motorized wheel; 12. Pressurized instrument compartment; 13. Soil X-ray spectrometer (not deployed); 14. Stereo TV cameras with dust-protective covers; 15. Laser reflector; 16. Human-height TV camera with dust-protective cover.

Payload:

Lunokhod 1:

1. Two television cameras for stereo images in the direction of travel
2. Four panoramic imagers
3. PrOP odometer/speedometer and soil mechanics penetrometer
4. Soil x-ray fluorescence spectrometer
5. Cosmic ray detectors
6. X-ray telescope for solar and extragalactic observations
7. Laser retro-reflector (France)
8. Radiometer

Two television cameras with a resolution of 250 horizontal lines were mounted viewing forward to provide a 50 degree stereo view of the travel direction. The other four imagers were facsimile cameras of the type flown on Luna 9, with improved sensitivity and gain control and mounted two to a side. One camera in each pair was mounted for 180 degree horizontal scanning and the other for vertical scanning from surface to sky. Each 180 degree panorama consisted of 500 × 3,000 pixels. Between them, each pair of cameras provided a 360 degree panorama. The horizontal ones provided context for the forward cameras and the vertical ones assisted navigation in the driving process.

A ninth spiked wheel trailed behind the rover with an odometer to measure distance and speed. The surface penetrometer was mounted on a pantograph. The French laser retro-reflector weighed 3.5 kg and consisted of fourteen 10 cm silica glass prisms. It was designed for 25 cm accuracy. Due to Soviet secrecy, the French were given only a drawing for how the device would be mounted and were not told in advance what kind of lunar vehicle would carry it.

Lunokhod 2:

1. Three front television cameras for stereo images in the direction of travel
2. Four panoramic imagers
3. PrOP odometer/speedometer and soil mechanics penetrometer
4. Soil x-ray fluorescence spectrometer (Rifma-M)
5. Cosmic ray detectors
6. X-ray telescope for solar and extragalactic observations
7. Laser retro-reflector (France)
8. Radiometer
9. Visible-ultraviolet photometer
10. Boom magnetometer

Based on experience with Lunokhod 1, a third forward viewing television camera was mounted higher on Lunokhod 2 to provide a better driving perspective while the lower pair of television cameras provided stereo images of potential obstacles. The visible-ultraviolet photometer was to detect Earth airglow and galactic ultraviolet sources. The magnetometer was mounted on a 1.5 meter long boom in front of the rover. A Soviet-made photocell was added to the French retro-reflector to register laser strikes and the x-ray fluorescence spectrometer was improved.

Mission description:

First attempt falls downrange

The first attempt to launch a Ye-8 with a rover failed spectacularly on February 19, 1969, when the payload stack on top of the vehicle disintegrated 51 seconds into the flight. The launcher then exploded and scattered wreckage 15 miles downrange. The investigation discovered that at the point of maximum dynamic pressure, when the loads on the vehicle were greatest, the newly designed payload shroud for the Proton failed. The radioisotope heater that was to have kept the rover warm during the lunar night was never recovered from the debris, and rumors persist that the soldiers who actually found it decided to use it to heat their barracks during that year's very cold winter.

Luna 17

A second attempt to launch a lunar rover was not made until 20 months later. After the loss of the first mission, the Ye-8 program had focused on attempts at automated lunar sample return in an effort to upstage Apollo. After Luna 16 succeeded with a returned sample in October 1970 it was decided to launch a rover next. The back to back successes of an automated sample return mission and a rover one month apart were impressive milestone achievements for the Soviet robotic lunar program.

Luna 17 was launched on November 10, 1970. After midcourse corrections on the 12th and 14th, it entered an 85 × 141 km lunar orbit on the 15th inclined at 141 degrees with a period of 115 minutes. It lowered its perilune to 19 km on the 16th and then at 03:46:50 UT on the 17th successfully touched down at a speed of about 2 m/s in the Sea of Rains at 38.25°N 325.00°E.

The Soviets announced their fourth lunar soft landing, and Westerners expected another sample return like Luna 16. However, about 3 hours after landing, at 06:28 UT, the ramps were lowered, the camera covers released, pictures of the ends of the

Figure 11.8 Lunokhod 1 operations crew.

ramps were taken to ensure that there were no obstructions, and Lunokhod 1 rolled down the front ramp and 20 meters across the surface. The next day it remained in place and recharged its batteries, then it traveled 90 and 100 meters on the next two days. On the fifth day, 197 meters from its lander, it closed its lid and shut down for the forthcoming lunar night.

The public reaction across the world was astonishing. Somehow, people resonated strongly with the idea of a robotic rover driving around on another world, even if the experience was only a virtual one. The Luna 17 rover was a triumph heralded by the Soviet and Western press alike, whereas the Luna 16 sample return had only gained fleeting admiration. The appeal of the Lunokhod may have been partly derived from its physical form. Its antics were followed ardently in the press for the first few days, and the coverage would probably have continued were it not for the requirement to

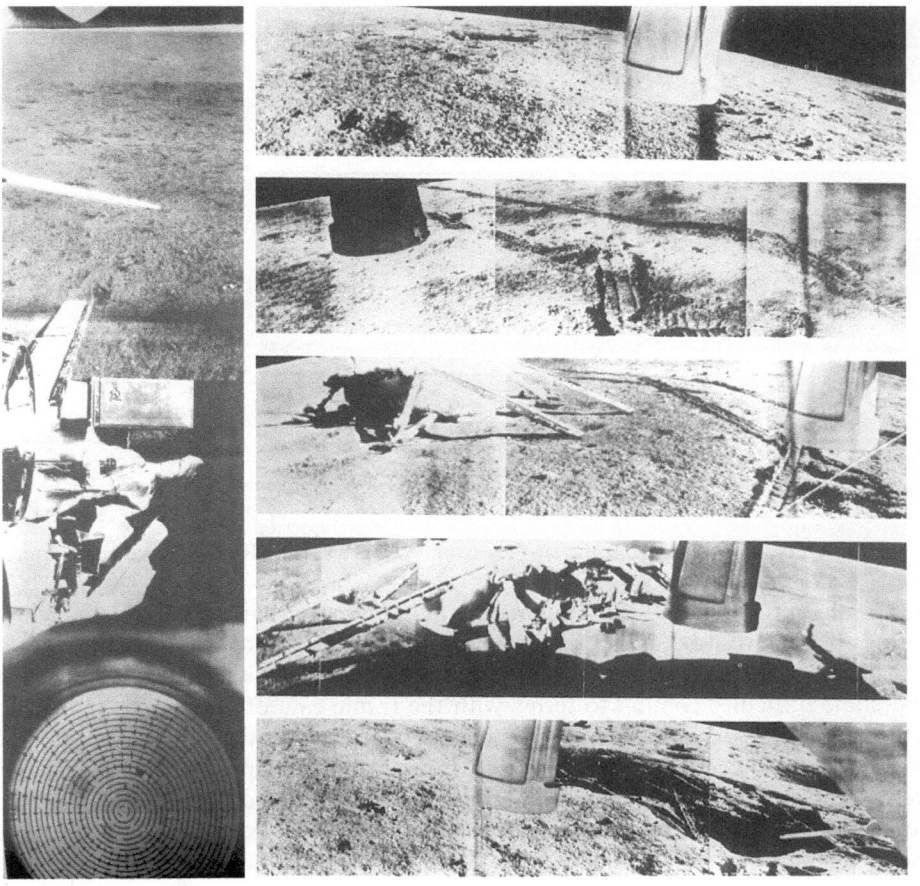

Figure 11.9 A set of Lunokhod 1 horizontal panoramas taken during its return to the landing site and vertical panorama from nadir to horizon while still mounted on the lander before deployment.

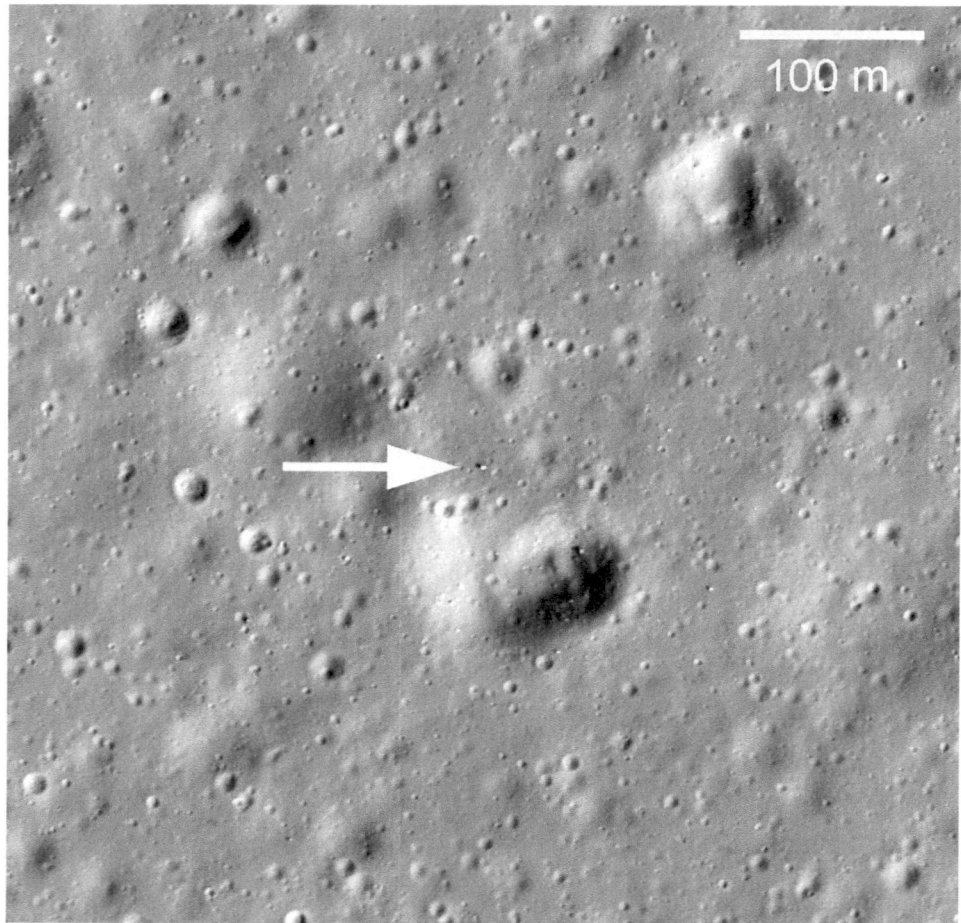

Figure 11.10 Lunokhod 1 on the surface from NASA's Lunar Reconnaissance Orbiter.

shut it down for the long lunar night. It would be more than a quarter of a century before the US would recreate the excitement of a robotic rover on another world.

Lunokhod 1 survived its first lunar night and continued its activities. The drivers had some difficulty coming to terms with the frame rate of 20 seconds, and it was realized that the driving cameras had been set too low on the vehicle because their perspective was more like sitting on a chair than standing upright. And their images were so overexposed that the contrast in the scene was poor, especially near lunar noon. Initially excluded from the control room, the scientists had difficulty in having the rover pause at interesting rocks. This was because the engineers' measure of success was distance covered. However, as the mission wore on it became easier for the scientists to achieve their objectives.

The operators drove the rover over 197 meters on the first lunar day, and as far as

2 km on the fifth lunar day. To test its navigational system, on one early excursion it returned to the lander stage. Over a period of 10 months it traversed rough hills and valleys and crossed many craters. It survived the -150°C cold of the lunar night and 100°C heat of lunar noon. It twice became stuck in craters, but after some effort was able to extract itself. The drivers had difficulty navigating because of the low mount of the cameras which meant they often did not spot a crater until the last moment. At noon the lack of shadows reduced the contrast to zero, making steering impossible. The rover survived a solar flare that might have been fatal for cosmonauts and an eclipse during which it was temporarily plunged into darkness. On the tenth lunar day it was spotted from orbit by the Apollo 15 astronauts.

The last successful communications session with Lunokhod 1 ended at 13:05 UT on September 14, 1971, after the internal pressure suddenly dropped. Officially, the mission concluded on October 4, 1971, the 14th anniversary of Sputnik. Fortunately, during its last communication cycle it had been parked with the laser retro-reflector in a position where it could continue to be used. Lunokhod 1 exceeded its expected lifetime of three lunar days by functioning for eleven lunar days. It traveled a total of 10,540 meters and transmitted more than 20,000 individual pictures, 206 panoramas, 25 x-ray elemental soil analyses, and more than 500 soil penetrometer tests. It was a spectacular success.

Luna 21

The next rover was modified to take account of the lessons from Lunokhod 1, and on January 8, 1973, Luna 21 was launched carrying Lunokhod 2. It performed a midcourse maneuver the next day, and on January 12 entered a 90 × 110 km lunar orbit inclined at 62 degrees with a period of 118 minutes. It lowered its perilune to 16 km the following day and then on January 15 fired its main engine at perilune to de-orbit itself. At an altitude of 750 meters the main engine ignited again to slow the rate of descent. At 22 meters this engine was shut down. The verniers took over to a height of 1.5 meters and were cut off. After falling the remaining distance the 7 m/s shock was absorbed by the legs. Luna 21 landed at 23:35 UT at 26.92°N 30.45°E in Le Monnier bay, an eroded and lava flooded crater cut into the Taurus Mountains on the eastern shore of the Sea of Serenity.

The Lunokhod 2 rover immediately took TV images of the surrounding area from its perch atop the lander. After rolling down onto the surface at 01:14 UT on January 16 it took pictures of the lander and the landing site. It remained in place for 2 days until its batteries were charged, then took some more pictures and began its traverse. During its first full lunar day it covered a greater distance than its predecessor had in eleven lunar days. In one day, it traveled as much as 1,148 meters. It climbed a hill 400 meters high and photographed the peaks of the Taurus mountains poking over the horizon with Earth in the sky above. In late January 1973 an American scientist attending an international conference on planetary exploration in Moscow gave a set of Apollo 17 photos of the area where Luna 21 landed to Russian scientists at the meeting. These highly detailed photographs were used to navigate Lunokhod 2 to a rille some distance east of its landing site.

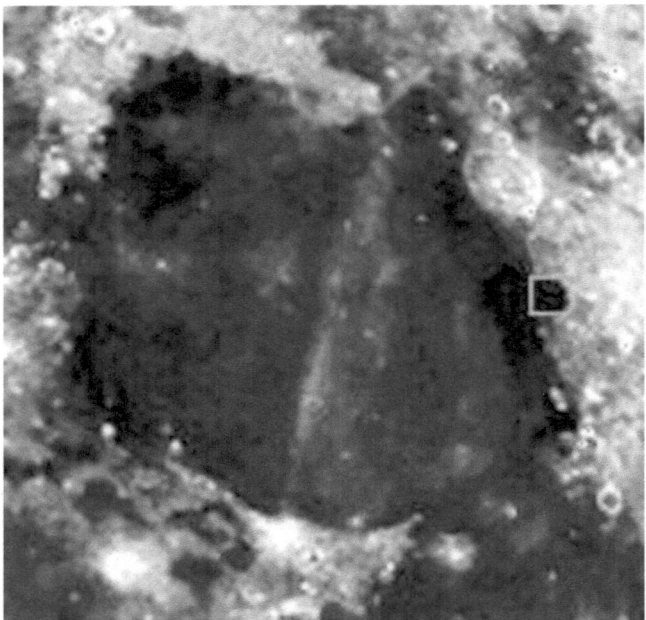

Figure 11.11 Lunokhod 2 site in the Sea of Serenity.

Figure 11.12 Picture of lander from Lunokhod 2.

Rover operations were conducted during the lunar day, stopping occasionally to recharge the battery using its solar panels. It would hibernate during the lunar night, using the radioisotope heater to maintain thermal control.

Lunokhod 2 operated for about 4 months, drove 37 km over terrain including hilly upland areas and rilles, more than four times the area of its predecessor, and returned over 80,000 individual images and 86 panoramas. It made hundreds of elemental analyses and mechanical tests of the soil, as well as being used for laser

Figure 11.13 Lunokhod 2 panorama around the landing area with Taurus Mountains in the distance.

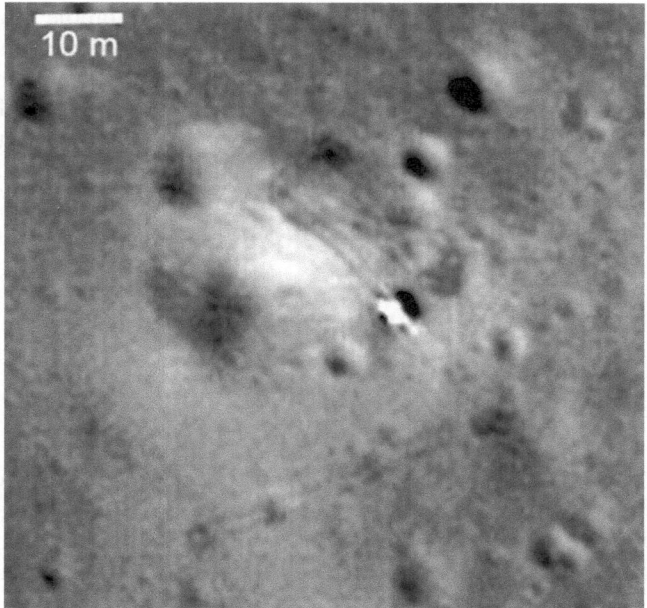

Figure 11.14 Luna 21 lander on the surface from NASA's Lunar Reconnaissance Orbiter showing rover tracks.

ranging and other experiments. On May 9, 1973, it accidentally rolled into a small 5 meter crater whose depth had been concealed by a shadow. As the rover was backing itself out, it scraped its lid on the crater wall, causing a spray of dust to cover the solar panel. When the lid was closed for the lunar night, this soil was dumped onto the radiators. On opening the lid for the next lunar day, the resulting thermal and power problems led to the vehicle's demise, which was announced on June 3.

Results:

Significant scientific results derived from analyzing the pictures of rocks and soil, wheel tracks, craters and other geological features observed by the two Lunokhods in more than 20,000 single frame images and 200 panoramas. There were many soil mechanics measurements by the penetrometer and chemical analysis results from the x-ray fluorescence spectrometer. The Sea of Rains and floor of Le Monnier proved to be a typical mare basalt, but the uplands around Le Monnier (the surviving part of the rim of the eroded crater) turned out to have higher concentrations of iron, silicon, aluminum and potassium.

Lasers fired by the French from the Pic du Midi observatory and by the Russians from the Semeis observatory in the Crimea used the retro-reflectors to determine the distance to the Moon to within 3 meters for Lunokhod 1, and 40 cm for Lunokhod 2. In the long term, such observations established the periodic and secular dynamics of the Moon. The cosmic ray instruments recorded the radiation on the Moon, and the x-ray telescope observed the Sun and the galaxy. The magnetometer on Lunokhod 2 measured a very weak magnetic field with variations due to currents induced by the interplanetary magnetic field. The photometer made some surprising observations of the brightness of the lunar sky. In particular, it determined that the day-time lunar sky was contaminated with some dust, and in Earthlight the night-time lunar sky was 15 times brighter than the sky on Earth at full Moon; findings which did not bode well for one day establishing astronomical observatories on the lunar surface.

THE N-1 LUNAR MISSION SERIES: 1969–1972

Campaign objectives:

Apollo 8 subjected the Soviets to the same anxiety felt by the Americans after the successes of the USSR in earlier years. A re-evaluation of Soviet plans resulted in a new resolution on January 8, 1969. The human circumlunar program would continue despite the clear recognition that relative to Apollo 8 it would appear both late and inferior to the world. Also, the human landing program would proceed even though it was evident that if they were not delayed by serious problems the Americans were likely to be first. Once the lunar programs were accomplished, the Americans would be upstaged in the late 1970s by using the N-1 for Korolev's originally envisioned destination, Mars. The space station program and robotic flights to the Moon, Mars and Venus were to be accelerated and represented in the press as the main thrust of the Soviet program.

The N-1 rocket, the Soviet counterpart of the Saturn V, was a key element in this strategy since it would launch the spacecraft that would see cosmonauts land on the Moon. N-1 development began in 1962 and the first launch was initially expected in 1965, but there were major delays due to organizational infighting, budgetary issues, and a total redesign of the vehicle in 1964. In particular, engine development was a key technological and organizational factor. As a result, development of the N-1 fell behind relative to the Saturn V.

In early 1969 the N-1 was deemed ready for testing, hurried by the advance of the American manned lunar landing program. These launches carried orbital versions

Spacecraft launched

First spacecraft:	7K-L1S No.3S
Mission Type:	Lunar Orbit and Return Test Flight
Country/Builder:	USSR/TsKBEM
Launch Vehicle:	N-1-3L
Launch Date/Time:	February 21, 1969 at 09:18:07 UT (Baikonur)
Outcome:	First stage failed in flight.

Second spacecraft:	7K-L1S No.5L
Mission Type:	Lunar Orbit and Return Test Flight
Country/Builder:	USSR/TsKBEM
Launch Vehicle:	N-1-5L
Launch Date/Time:	July 3, 1969 at 20:18:32 UT (Baikonur)
Outcome:	First stage exploded at liftoff.

Third spacecraft:	7K-LOK No.6A
Mission Type:	Lunar Orbit and Return Test Flight
Country/Builder:	USSR/TsKBEM
Launch Vehicle:	N-1-7L
Launch Date/Time:	November 23, 1972 at 06:11:55 UT (Baikonur)
Outcome:	First stage exploded in flight.

of the circumlunar Zond spacecraft. All the launch tests failed, in every case due to problems in the first stage. The first test was in February and the second in July, just as the US was preparing to launch Apollo 11. Although the race to Moon was now lost, the Soviets, still hoping one day to land cosmonauts on the surface, persisted in their automated testing of the circumlunar spacecraft with back to back successes by Zond 7 and 8 in August 1969 and October 1970 respectively. Another N-1 test failed in June 1971, as did the fourth in November 1972, a month before the final Apollo lunar landing. At this point the N-1 was abandoned, ignominiously ending the Soviet manned lunar program.

Spacecraft:

The 7K-LOK lunar orbital Soyuz differed significantly from the circumlunar Zond. It had a wider skirt on the service module, which now had two engines. One was the standard Soyuz engine, which on an operational mission would perform lunar orbital rendezvous. The new engine was larger, and was to boost the LOK out of lunar orbit and back to Earth. An orbital module was included just like the Earth-orbital Soyuz, but it differed in some ways. It had more ports for lunar observation and an attitude control system and docking system on the front for rendezvous and docking with the returning lunar lander. It also had a large hatch. In contrast to Apollo, the cosmonaut was to transfer between the orbital module and the lunar

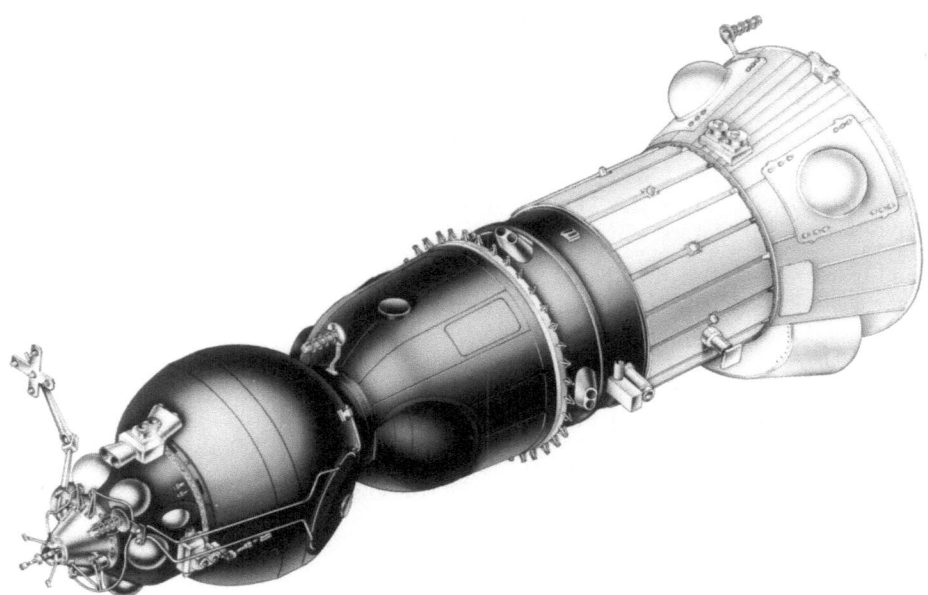

Figure 11.15 The Soyuz 7K-LOK lunar orbiter (courtesy Energiya Corp).

lander by spacewalking. As the 7K-LOK was the first Soviet spacecraft to have fuel cells for electrical power, it did not need solar panels.

On an operational mission the lunar orbital Soyuz would be carried above the LK lander, which would be on the Block D propulsion system, here serving as the fifth stage of the N-1. This whole stack would be accelerated out of parking orbit towards the Moon by the fourth stage, which would then be discarded. The Block D was to perform midcourse maneuvers, lunar orbit insertion, and preliminary maneuvers in orbit, before flying the lunar lander down to an altitude of about 1,500 meters, where it would be released and the lander would ignite its own engine for the final phase of the descent. This was quite different from the Apollo system.

The payloads for the first two N-1 test launches were the Block D, a dummy LK and instead of the 7K-LOK, which was not ready, a 7K-L1S Zond circumlunar test spacecraft fitted with an attitude control block for operations in lunar orbit. The plan was for these missions to insert the modified Zond into lunar orbit and then return it to Earth. However, as noted, both launches failed. The third N-1 was basically just a test of the launcher itself, and it carried only spacecraft mockups. However, for the fourth test it was decided to launch a complete 7K-LOK lunar Soyuz and a dummy LK lunar lander. The plan called for spending 3.7 days in lunar orbit, during which the spacecraft would image future landing sites before returning to Earth.

N-1 test launches:

N-1 no.1:	7K-L1S No.3S circumlunar vehicle and dummy LK lunar lander.
N-1 no.2:	7K-L1S No.5L circumlunar vehicle and dummy LK lunar lander.
N-1 no.3:	Mockup LOK and LK spacecraft.
N-1 no.4:	7K-LOK No.6A lunar Soyuz and dummy LK lunar lander.

7K-L1S launch mass:	6,900 kg
7K-LOK launch mass:	9,500 kg

Mission description:

The first test of the N-1 failed on February 21, 1969. Two of the thirty engines in the new first stage booster shut down just before liftoff, but the rocket was designed to handle this situation by burning the remaining engines for longer. At 5 seconds a gas pressure line broke and at 23 seconds an oxidizer line broke, resulting in a fire in the array of engine nozzles that burned through engine control system cables and caused the booster to shut down at the 70 second mark. The automatic escape system drew the Zond clear and the capsule landed safely 35 km from the launch site. The debris from the exploding vehicle fell 50 km away. The blast shattered windows across a wide area, including those in assembly buildings and a local hotel.

The second N-1 test failed spectacularly on July 3, 1969, just thirteen days before the launch of Apollo 11. After ignition, the massive rocket rose to a height of about 200 meters and then fell back onto the pad in a huge explosion that totally destroyed the pad and severely damage buildings across a wide area. A US spy satellite caught an image of the aftermath. At the 0.25 second mark the oxidizer pump in one of the first stage engines had been damaged by ingesting a foreign object through its feed lines and it exploded. The resulting fire quickly engulfed the engine compartment of the booster. At the 10 second mark the control system shut down most of the engines and the vehicle fell and exploded. The 7K-L1S escape system worked perfectly and the capsule was recovered 2 km away. This disaster banished the hope of using this rocket to compete with the Americans as they prepared for their first manned lunar landing.

N-1 launches did not resume until long after the race to the Moon was settled. The third test failed on June 28, 1971. Booster roll control was lost 48 seconds into the flight and the vehicle broke up. The fourth and final test failed on November 23, 1972, when the first stage exploded at 107 seconds into the flight, just a few seconds before it was to have shut down and handed over to the second stage. Ironically, the cause may have been excessive shock to the engine array arising from the sequenced shutdown of the central engines. Any remaining hopes the Soviets had for sending cosmonauts to the Moon were lost in this final failure.

Figure 11.16 Third N-1 launch attempt just before clearing the tower.

Results:

None.

A BOLD, NEW PROGRAM FOR MARS: 1969

Campaign objectives:

Since their origins in 1960 the Soviet Mars and Venus programs had been strongly intertwined, using slightly different versions of the same spacecraft. When NPO-Lavochkin took over the planetary program it set out to transform OKB-1's 3MV-3 design into a 1,000 kg spacecraft to be launched on an upgraded Molniya-M at the 1967 flight opportunity to Mars. But this approach was soon abandoned. The Mars program had been a disaster. Seven attempts in the period 1960 through to 1964 had failed, including one test mission. Then the Zond 2 Mars flyby spacecraft created an embarrassment by failing as Mariner 4, launched by the US at almost the same time, went on to make a successful flyby in July 1965. In that same month Zond 3, after operating successfully at the Moon, failed its Mars deep space test flight objectives. Aware that the US was turning away from Venus in favor of Mars, starting with dual flybys planned in 1969 and with orbiters and landers to follow, perhaps as early as 1973, the Soviets decided to perfect a Mars lander that would outdo the American flyby missions.

Spacecraft launched

First spacecraft:	M-69 No.521
Mission Type:	Mars Orbiter
Country/Builder:	USSR/NPO-Lavochkin
Launch Vehicle:	Proton-K
Launch Date/Time:	March 27, 1969 at 10:40:45 UT (Baikonur)
Outcome:	Launch failure, 3rd stage explosion.
Second spacecraft:	M-69 No.522
Mission Type:	Mars Orbiter
Country/Builder:	USSR/NPO-Lavochkin
Launch Vehicle:	Proton-K
Launch Date/Time:	April 2, 1969 at 10:33:00 UT (Baikonur)
Outcome:	Launch failure, booster explosion.

The entry vehicle for the 3MV Mars spacecraft had been designed in the early 1960s on the presumption that the atmospheric pressure at the surface was between 80 and 300 millibars. The Mariner 4 flyby in July 1965 showed it to be a mere 4 to 7 millibars. The design of the 3MV entry probe was therefore fatally flawed. A new technique would be required to perform entry, descent and landing in such a rarefied atmosphere. In October 1965 NPO-Lavochkin abandoned the 3MV for Mars, but retained it for Venus because it was suitable for that dense atmosphere. The Soviets skipped the 1967 Mars launch opportunity to develop a more capable spacecraft for the 1969 opportunity.

The powerful Proton launch vehicle made its debut in 1965. It doubled the mass that could be delivered to low Earth orbit compared to the three-stage Molniya, and when augmented by the Block D fourth stage (as the Proton-K) it facilitated a whole new generation of heavier, more capable and complex lunar and planetary spacecraft than the Molniya-launched 3MV. Capable of dispatching over 4 metric tons onto an interplanetary trajectory, the Proton-K became the standard launcher for lunar and Mars missions after 1966, and for Venus missions after 1972.

The engineering requirements for new Mars and Venus missions during the time period 1969–73 were defined in March 1966 by the head of NPO-Lavochkin, Georgi Babakin:

1.   Use of the Proton-K to achieve parking orbit and escape onto an interplanetary trajectory
2.   Use of a "universal" multi-purpose, modular on board propulsion system for trajectory correction while coasting and then insertion into an orbit around the target with a pericenter about 2,000 km and apocenter not exceeding 40,000 km
3.   Use of descent-from-flyby and descent-from-orbit mission designs for soft landers to place instruments on the surface
4.   Use of the main spacecraft as either a flyby vehicle or an orbiter to relay information from the lander at about 100 bits/s to the Earth
5.   Use of a telemetry system capable of transmission from the main spacecraft of about 4,000 bits/s.

It was decided that in addition to trajectory correction maneuvers, entry vehicle targeting and planetary orbit insertion and trim maneuvers, the universal propulsion system should also participate in establishing the desired interplanetary trajectory by firing after the spent Block D stage was jettisoned.

These requirements were not applied to Venus until the successful Venera type of the 3MV had fulfilled all of the objectives for that planet in 1972, but they were applied immediately to Mars for the 1969 opportunity. Also, it was decided that for the initial Mars mission the descent module would be an atmospheric probe to obtain the data required for designing a landing system for that rarefied atmosphere. Another key objective was to improve the ephemeris for Mars for future missions. The science objectives for Mars missions using this new spacecraft system were:

1.   To measure the temperature, pressure, wind speed and direction on the surface, and to measure the chemical composition of the atmosphere around the planet
2.   To achieve soft landings at chosen sites and take pictures of the surface to study the terrain and vegetation
3.   To measure the composition, bearing strength and properties of the soil
4.   To measure the radiation levels and magnetic field at the surface
5.   To detect any traces of micro-organisms in the soil
6.   To study the upper atmosphere
7.   To compile a detailed thermal radiation map from orbit
8.   To fly past Phobos and Deimos while in Mars orbit and take pictures to define their shape, size and albedo
9.   To photograph Mars from orbit in order to understand the nature of the 'seas' and 'canals' and to acquire information on seasonal changes.

These were extraordinarily demanding objectives for a program that had endured six failed missions since 1960 and had yet to achieve anything at all at Mars. In one bold leap, compelled by competition with the US and enabled by the Proton-K launcher, the Soviets would attempt the first Mars orbiters and landers at a launch opportunity that was only 33 months away, an incredibly short period of time in which to try to develop a spacecraft of such an unprecedented complexity. And by devoting part of this time to modifying the 3MV to score a success at Venus in 1967 they left themselves with only 20 months to develop the new spacecraft. Then problems with the design left them with only 13 months. Given the intense pressure to outdo the US at Mars, the risks taken were enormous.

The workload was intense during the last years of the 1960s as the Soviets tried to compete with Apollo. NPO-Lavochkin was overloaded developing the Luna rover and sample return missions, continuing to milk the successful Venus missions, and making a valiant effort on M-69. This was a brand new spacecraft like none built before, and the rushed development showed. Nothing went smoothly. The spacecraft suffered from the same development problems as OKB-1's early rushed designs and engineers were not terribly optimistic about its chances. The winter of 1968–69 was exceedingly harsh, pipes burst and heating systems failed, creating near-impossible working conditions. Control and telemetry systems were plagued with troubles and the design of the spacecraft actually prevented easy access for servicing. The entry probe had to be deleted very late in the process due to insufficient time and system mass growth, and was replaced by a compartment for additional orbital instruments.

The Soviets were to fail in their first attempt with this new spacecraft in 1969, but the engineering and science requirements for the M-69 program set a precedent for all of the Mars mission designs that were to follow. At that time almost nothing was known of these missions in the West, and 30 years would elapse before they were described in any detail.

Spacecraft:

The initial design:

As Babakin's engineers worked with their OKB-1 colleagues in 1966–67 to prepare a 3MV spacecraft for what would become the successful Venera 4 mission, others at NPO-Lavochkin were working on a new spacecraft for the Luna series that would be launched by the Proton-K instead of the Molniya. Unlike the previous 2MV, 3MV and Luna series spacecraft where the avionics compartment was the main structural element, this time a quartet of spherical propellant tanks connected together in the shape of a square using cylindrical inter-tank sections became the element on which everything else was mounted.

Given the short period of time available for the development of a Proton-launched Mars spacecraft, it was decided to exploit this work. The initial M-69 design had the entry probe attached to the tank assembly where the lunar rover would otherwise be carried, and the remaining systems attached to the 'underside'. The two solar panels were spread out from opposite sides of the square, and the antenna and engine were opposite each other on the remaining sides. This design could meet the schedule, but was not easily reconfigured and failed to satisfy some of the requirements. Also, the designers struggled with a number of engineering

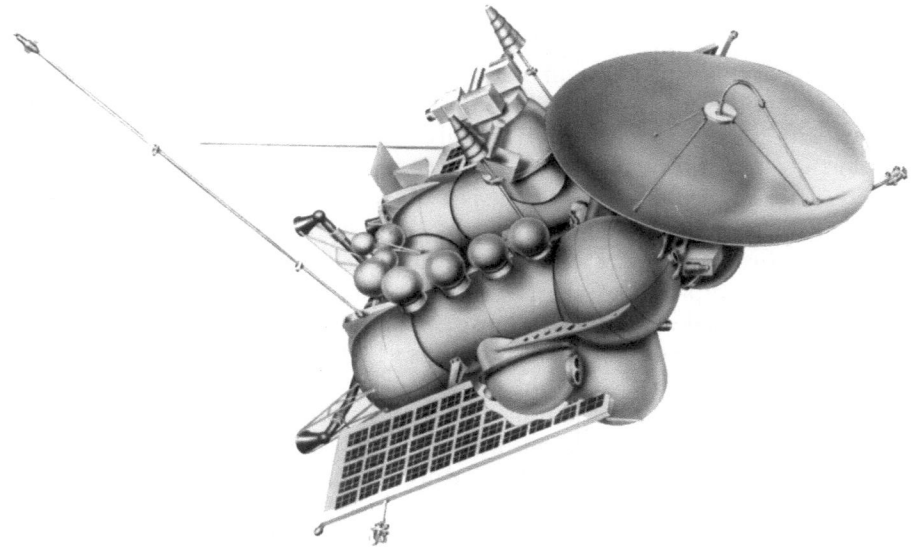

Figure 11.17 Drawing of the original Mars-69 concept.

problems in trying to adapt a lunar spacecraft for Mars exploration. The main issues centered on the fundamental tank design, and ultimately it was abandoned, forcing a total redesign 13 months before the launch date.

The final design:

The new design used a single large spherical tank at the center of the spacecraft as the main structural element. The tank had an internal baffle to separate the UDMH fuel from the nitrogen tetroxide oxidizer. The Isayev engine was attached to the base of the tank. A cylindrical interstage with a pressurized container for electronics was attached to the top of the tank, and the entry vehicle was installed above that. Two hermetically sealed cylindrical modules were attached on opposite sides of the tank, one for communication, navigation systems and optical orientation sensors, and the other for science instruments including the cameras. There were also science sensors attached to the outside of the spacecraft.

The antenna system, including both a large high gain and small conical antennas, was affixed to the cylindrical interstage. The two 3.5 square meter solar panels were mounted outboard of the instrument modules. The panels were supplemented with a NiCad battery that delivered power at 12 amps with a 110 amp-hour capacity. Both passive insulation and active thermal control were employed. The active system operated in the pressurized compartments and consisted of a ventilation and air circulation system to route air between two radiators, one exposed to sunlight and the other to shadow. The thermal control radiators were inboard of the solar panels, between the modules across the main tank. The avionics of the M-69 spacecraft were

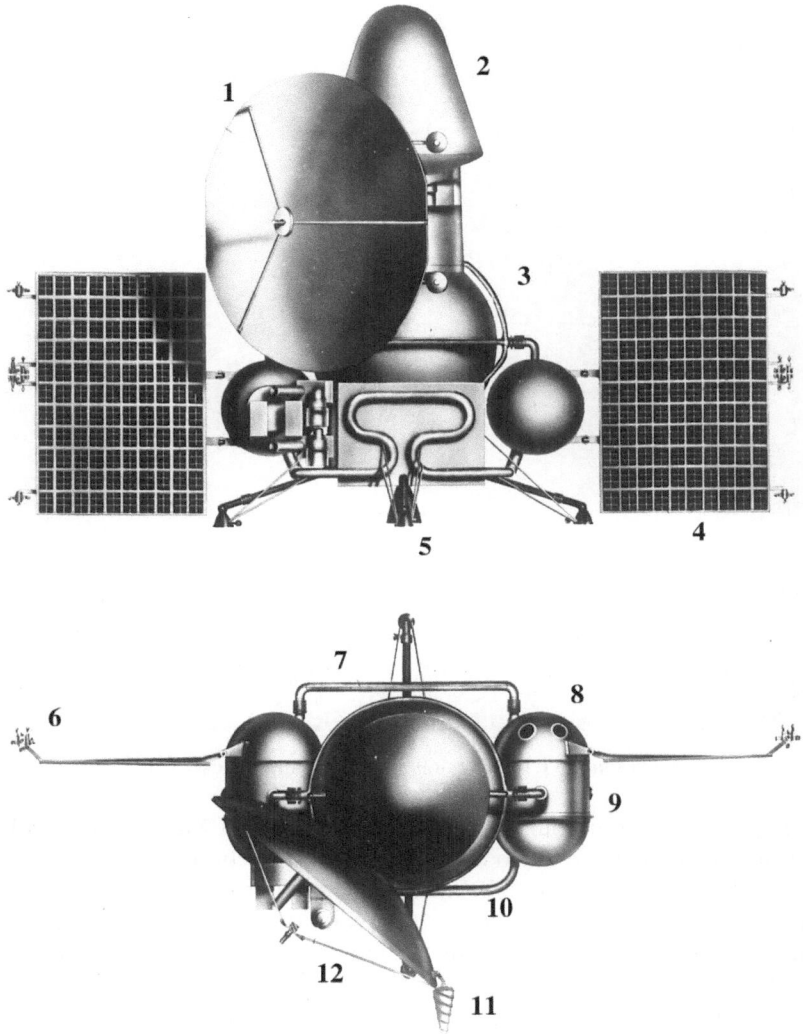

Figure 11.18 Final Mars-69 spacecraft design: 1. Parabolic high-gain antenna; 2. Entry system (not flown); 3. Fuel tank; 4. Solar Panels; 5. Propulsion system; 6. Attitude control; 7. Thermal control-cooling side nozzles; 8. Camera viewports; 9. Instrument compartment; 10. Thermal control-heating side; 11. Omni antenna; 12. Navigation system.

much improved over the 3MV series. It was the first Soviet planetary spacecraft to carry a computer. An advanced data processing system weighing only 11 kg was provided that could program the instruments and acquire, process and compress the data from both engineering and science systems for transmission to Earth.

A new telemetry system was provided that consisted of a transponder-receiver for

Figure 11.19 Mars-69 spacecraft under test.

non-imaging data and an impulse transmitter for images, a 2.8 meter parabolic high gain directional antenna and a trio of low gain semi-directional conical antennas for decimeter and centimeter bands. The arrangement of the conical antennas was such that when the solar panels were pointed at the Sun, they would be pointing at Earth. The transponder-receiver had two transmitters and three receivers in the decimeter band at 790 to 940 MHz with 100 W of power, and facilitated Doppler tracking at a transmitted data rate of 128 bits/s with 500 data channels. These transmitters and receivers could use either the conical antennas or the high gain. One receiver was always on and connected to one of the conical antennas for continuous reception. The remaining receivers and the transmitters were cycled through these antennas by timers in order to ensure the reliability of the system. As part of the payload, a new film camera system with facsimile processing was developed. The imaging system had a 5 cm impulse 50 W transmitter for a data rate of 6 kbits/s using short pulses at 25 kW.

For the attitude control system, new Sun and star sensor systems and new nitrogen gas micro-engines were developed. There were two Sun sensors, two star sensors, two Earth sensors, and two Mars sensors. Nine helium-pressurized tanks provided nitrogen gas stored in ten separate tanks to eight attitude control thrusters, two each for pitch and yaw and the other four for roll. The nitrogen tank pressure of 350 bar was regulated to 6 bar for maneuvering and 2 bar for attitude maintenance. During cruise and routine operations the vehicle used one set of sensors to maintain itself in a rough attitude that faced the solar panels towards the Sun. For high gain antenna operations, midcourse maneuvers, and orbital mapping, it used a more accurate set of sensors for precise 3-axis stabilization. Both optical sensors and gyroscope control were provided for the attitude control system.

The entry system was a prototype of that which would be used in 1971, and was to have been deployed while 2 days from Mars. But it was ultimately deleted from the 1969 mission due to mass growth of the spacecraft and insufficient time to test the parachute descent system in balloon drops. The entry probe was designed around a large spherical tank with three attached pressurized compartments. No other details are available.

Launch mass: 4,850 kg (fueled but without probe)
Orbiter mass: 3,574 kg
Probe mass: 260 kg

Payload:

Orbiter:

1. Facsimile imaging system (FTU)
2. Infrared Fourier spectrometer (UTV1) for atmosphere and surface studies
3. Infrared radiometer (RA69) for surface temperature
4. Ultraviolet spectrometer (USZ) for reflected radiation
5. Water vapor detector (IV1)
6. Mass spectrometer for ionosphere composition and hydrogen, helium detection (UMR2M)
7. Multi-channel gamma-ray spectrometer (GSZ)
8. Low-energy ion spectrometer (RIB803)
9. Charged particle detector (KM69) for solar electrons and protons
10. Magnetometer
11. Micrometeoroid detector
12. Low frequency radiation detector
13. Cosmic ray and radiation belt detector
14. X-ray radiometer
15. Gamma-ray burst detector

Total mass: 85 kg.

The new FTU was an advanced film facsimile imaging system consisting of three cameras, each with red, green and blue color filters. The image format was 1,024 × 1,024 pixels. One camera had a 35 mm lens, a second had a 50 mm lens and a field of view of 1,500 × 1,500 km, and the third had a 250 mm lens and a field of view of 100 × 100 km with a best resolution of 200 to 500 meters. The film was processed on board, encoded digitally and supplied to the impulse transmitter. The film was to be chemically activated upon arrival at Mars in order to avoid damage by radiation in cruise. Each camera had sufficient film for 160 images.

Atmosphere probe (deleted):

1.   Pressure sensors
2.   Temperature sensors
3.   Accelerometers for atmospheric density
4.   Chemical gas analyzer

Total mass: 15 kg.

Mission Description:

The plan was to use the first three stages of the Proton and the Block D upper stage to achieve parking orbit. After one orbit, the Block D would be reignited for the first part of the escape sequence under the control of the spacecraft. After burnout of the Block D and separation, the spacecraft would fire its main engine for the final boost onto the interplanetary trajectory. This would be the first time that this new scheme was used, adding more risk to an already challenging project. The spacecraft engine would also be used for two trajectory corrections during the 6 month cruise to Mars, one 40 days out from Earth and the other 10 to 15 days prior to arrival. The fourth burn of the engine would be made at the closest point of approach to Mars in order to enter a 1,700 × 34,000 km orbit inclined at 40 degrees to the equator with a period of 24 hours. No immediate trim burns were planned, despite the expectation that the errors would be considerable. After some photography and other science from this initial orbit over several weeks, the periapsis would be lowered to about 600 km for an additional 3 months of imaging and data collection. At that point the mission was expected to be concluded.

Unfortunately, neither spacecraft even reached Earth orbit. M-69A was lost to a third-stage explosion when a rotor bearing malfunction caused a turbopump to fail and catch fire. The engine shut down at the 438 second mark and the stage exploded. M-69B was lost when one of the six first stage engines exploded just at launch. The vehicle continued to climb on the five remaining engines until the 25 second mark, at which time it tipped over to the horizontal at an altitude of 1 km. The remaining engines shut down and 41 seconds into the flight the vehicle fell to the ground 3 km from the pad and exploded. Remarkably, the second stage landed intact.

The failure of the Soviets to exploit the 1969 opportunity for Mars passed largely unnoticed in the West, mainly because the two attempted launches failed so early in flight. But the Protons may have saved the Soviets from the larger embarrassment of another Mars mission failing due to the spacecraft being rushed too hard through its design and development. As one of its designers remarked, M-69 was an example of how *not* to build a spacecraft.

Results:

None.

The Proton was experiencing its worst period in development at this time, with a very high failure rate. It was responsible for the loss of many spacecraft including a large number of lunar missions. The failure of the M-69 launches was a bitter pill for the spacecraft team to swallow after all the difficult and frantic work that had gone into the preparation. To rub salt into the wound, soon thereafter the US achieved the Apollo 11 lunar landing and the successful Mariner 6 and 7 flybys of Mars.

THE YE-8-5 LUNAR SAMPLE RETURN SERIES: 1969–1976

Campaign objectives:

In late 1968 and early 1969 it became apparent to the Soviet Union that American astronauts might very well reach the Moon before Russian cosmonauts. Anxious to ensure that a Soviet mission was first to return lunar soil to Earth, NPO-Lavochkin hurriedly modified the Ye-8 spacecraft for a sample return mission. The lunar rover variant of this spacecraft was well advanced in design by the end of 1968, and could readily be modified simply by replacing the payload of the lander. Even although it would have scientific merit, the sample return mission had a far greater significance than being just another task for the Ye-8. The robotic sample return mission became the means to upstage Apollo by returning a sample to Earth before the Americans could do so. The fact that these complex spacecraft could be designed and built so readily, and ultimately work so well, is amazing in hindsight. It would seem to be a Russian characteristic to "just do it", to dismiss the hardship, use whatever you have at hand, and fix things up on the fly during and after build.

The modification of the Ye-8 lunar rover spacecraft to the Ye-8-5 for the sample return mission faced daunting problems, not the least of which were mass limitations on the return vehicle, lifting off from the Moon and navigating back to Earth. It was originally believed that the return vehicle would require the same complex avionics as any interplanetary spacecraft, to enable its position to be determined and to make midcourse correction maneuvers. The avionics necessary to meet these requirements far exceeded the available mass. However, D. Ye. Okhotsimskiy, a scientist at the Institute of Applied Mathematics, found a small set of flight trajectories for launches from the surface of the Moon that did not require midcourse corrections. In essence, the large gravitational influence of the Earth at lunar distance could, under certain conditions, assure an Earth return. These trajectories were limited to specific points on the Moon, varying within a general locus with the time of year, and required the lander to set down within 10 km of its target and the lunar liftoff for a direct ascent to occur at a precise moment. Accurate knowledge of the lunar gravitation field was also required, but this information had already been determined by the Luna 10, 11, 12 and 14 orbiter missions.

Spacecraft launched

First spacecraft: Ye-8-5 No.402
Mission Type: Lunar Sample Return
Country/Builder: USSR/NPO-Lavochkin
Launch Vehicle: Proton-K
Launch Date/Time: June 14, 1969 at 04:00:47 UT (Baikonur)
Outcome: Fourth stage failed to ignite.

Second spacecraft: Luna 15 (Ye-8-5 No.401)
Mission Type: Lunar Sample Return
Country/Builder: USSR/NPO-Lavochkin
Launch Vehicle: Proton-K
Launch Date/Time: July 13, 1969 at 02:54:42 UT (Baikonur)
Lunar Orbit Insertion: July 17, 1969 at 10:00 UT
Lunar Landing: July 21, 1969 at 15:51 UT
Outcome: Crashed.

Third spacecraft: Ye-8-5 No.403 (Cosmos 300)
Mission Type: Lunar Sample Return
Country/Builder: USSR/NPO-Lavochkin
Launch Vehicle: Proton-K
Launch Date/Time: September 23, 1969 at 14:07:36 UT (Baikonur)
Outcome: Fourth stage failure, stranded in Earth orbit.

Fourth spacecraft: Ye-8-5 No.404 (Cosmos 305)
Mission Type: Lunar Sample Return
Country/Builder: USSR/NPO-Lavochkin
Launch Vehicle: Proton-K
Launch Date/Time: October 22, 1969 at 14:09:59 UT (Baikonur)
Outcome: Fourth stage misfire, stranded in Earth orbit.

Fifth spacecraft: Ye-8-5 No.405
Mission Type: Lunar Sample Return
Country/Builder: USSR/NPO-Lavochkin
Launch Vehicle: Proton-K
Launch Date/Time: February 6, 1970 at 04:16:06 UT (Baikonur)
Outcome: Second stage premature shutdown.

Sixth spacecraft: Luna 16 (Ye-8-5 No.406)
Mission Type: Lunar Sample Return
Country/Builder: USSR/NPO-Lavochkin
Launch Vehicle: Proton-K
Launch Date/Time: September 12, 1970 at 13:25:53 UT (Baikonur)
Lunar Orbit Insertion: September 17, 1970
Lunar Landing: September 20, 1970 at 05:18 UT
Ascent Stage Liftoff: September 21, 1970 at 07:43 UT
Earth Return: September 24, 1970 at 03:26 UT
Outcome: Success.

Seventh spacecraft: Luna 18 (Ye-8-5 No.407)
Mission Type: Lunar Sample Return
Country/Builder: USSR/NPO-Lavochkin
Launch Vehicle: Proton-K
Launch Date/Time: September 2, 1971 at 13:40:40 UT (Baikonur)
Lunar Orbit Insertion: September 7, 1971
Lunar Landing: September 11, 1971 at 07:48 UT
Outcome: Failure at landing.

Eighth spacecraft: Luna 20 (Ye-8-5 No.408)
Mission Type: Lunar Sample Return
Country/Builder: USSR/NPO-Lavochkin
Launch Vehicle: Proton-K
Launch Date/Time: February 14, 1972 at 03:27:59 UT (Baikonur)
Lunar Orbit Insertion: February 18, 1972
Lunar Landing: February 21, 1972 at 19:19 UT
Ascent Stage Liftoff: February 22, 1972 at 22:58 UT
Earth Return: February 25, 1972 at 19:19 UT
Outcome: Success.

Ninth spacecraft: Luna 23 (Ye-8-5M No.410)
Mission Type: Lunar Sample Return
Country/Builder: USSR/NPO-Lavochkin
Launch Vehicle: Proton-K
Launch Date/Time: October 28, 1974 at 14:30:32 UT (Baikonur)
Lunar Orbit Insertion: November 2, 1974
Lunar Landing: November 6, 1974
Mission End: November 9, 1974
Outcome: Damaged on landing, no return attempted.

Tenth spacecraft: Ye-8-5M No.412
Mission Type: Lunar Sample Return
Country/Builder: USSR/NPO-Lavochkin
Launch Vehicle: Proton-K
Launch Date/Time: October 16, 1975 at 04:04:56 UT (Baikonur)
Outcome: Fourth stage failure.

Eleventh spacecraft: Luna 24 (Ye-8-5M No.413)
Mission Type: Lunar Sample Return
Country/Builder: USSR/NPO-Lavochkin
Launch Vehicle: Proton-K
Launch Date/Time: August 9, 1976 at 15:04:12 UT (Baikonur)
Lunar Orbit Insertion: August 14, 1976
Lunar Landing: August 18, 1976 at 06:36 UT
Ascent Stage Liftoff: August 19, 1976 at 05:25 UT
Earth Return: August 22, 1976 at 17:35 UT
Outcome: Success.

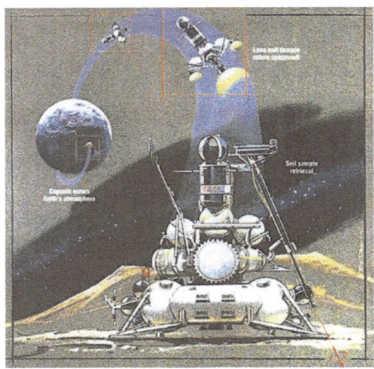 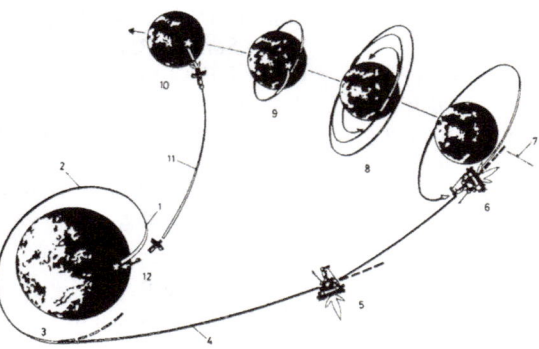

Figure 11.20 Luna sample return sequence (courtesy NPO-Lavochkin and *Space Travel Encyclopedia*): 1. Launch; 2. Parking; 3. Translunar injection burn orbit; 4. Translunar flight; 5. Trajectory correction maneuver; 6. Lunar orbit injection burn; 7. Lunar orbit; 8. Maneuvers to final orbit; 9. Descent sequence; 10. Ascent from the lunar surface; 11. Free-return trajectory to Earth; 12. Separation from return vehicle and entry.

These passive return trajectories simplified the ascent vehicle enormously. Only a single burn of the ascent vehicle was required. No active navigation was necessary, and no midcourse maneuvers were required. The only problem with a passive return was the very large error ellipse on arrival at Earth, which would make recovering the small capsule impractically difficult. This problem was solved by using a low-mass meter wave radio beacon on the ascent vehicle so that radio tracking would be able to determine its actual trajectory, supplemented by optical observations from Earth during the latter half of its flight. In addition, the return capsule would have its own radio beacon to assist in recovery operations.

Even with these ingenious solutions, the engineers could not trim the design mass of the Ye-8-5 below 5,880 kg. At that time the most that the Proton-K could send to the Moon was 5,550 kg. However, Babakin managed to cajole the Proton maker into providing sufficient additional mass capability to launch his sample return spacecraft to the Moon. This was accomplished without major changes to the launch vehicle.

Spacecraft:

Lander stage:

The lander stage was essentially the same as designed for the rover mission and its mission profile through to lunar landing was identical. The only differences were the attachment of a surface sampling system and, for the first eight spacecraft, a pair of television cameras for stereo imaging of the sampling site and floodlights for night landings. The rover and ramps were replaced by a toroidal pressurized compartment which held the instruments and avionics for surface operations. The ascent stage was mounted on top, with the entire lander and toroidal compartment acting as its launch pad.

Figure 11.21 Luna 16 spacecraft diagram (from Ball et al.) and during test at Lavochkin.

The ascent stage was powered by a silver-zinc 14 amp-hour battery, and the return capsule by a 4.8 amp-hour battery. Lander communications were provided at 922 and 768 MHz, with backups at 115 and 183 MHz. The ascent stage communicated at 101.965 and 183.537 MHz. The return capsule had beacons at 121.5 and 114.167 MHz for radio tracking.

The sampling system for the Ye-8-5 consisted of an upright 90 cm long boom arm capable of two degrees of freedom, with a drill at its end for surface sampling. Three movements were required to place the drill on the surface through a 100 degree arc of swing, and then another three to transfer the sample to the ascent stage. From the stowed position it first swung itself vertical, then rotated in azimuth to line up on the selected sample site before swinging down onto the surface. A movement in azimuth with the head on the ground might be used to clear a small area to improve drilling. This sequence was reversed to transfer the sample to the return capsule of the ascent stage. Mounted at the end of the boom was a cylindrical container 90 mm diameter and 290 mm long for a hollow rotary/percussion drill. The drill bit had a diameter of 26 mm and was 417 mm long. Its cutter was a crown with sharp teeth. The drill was equipped with different coring mechanisms for hard coring and for loose coring. At a speed of

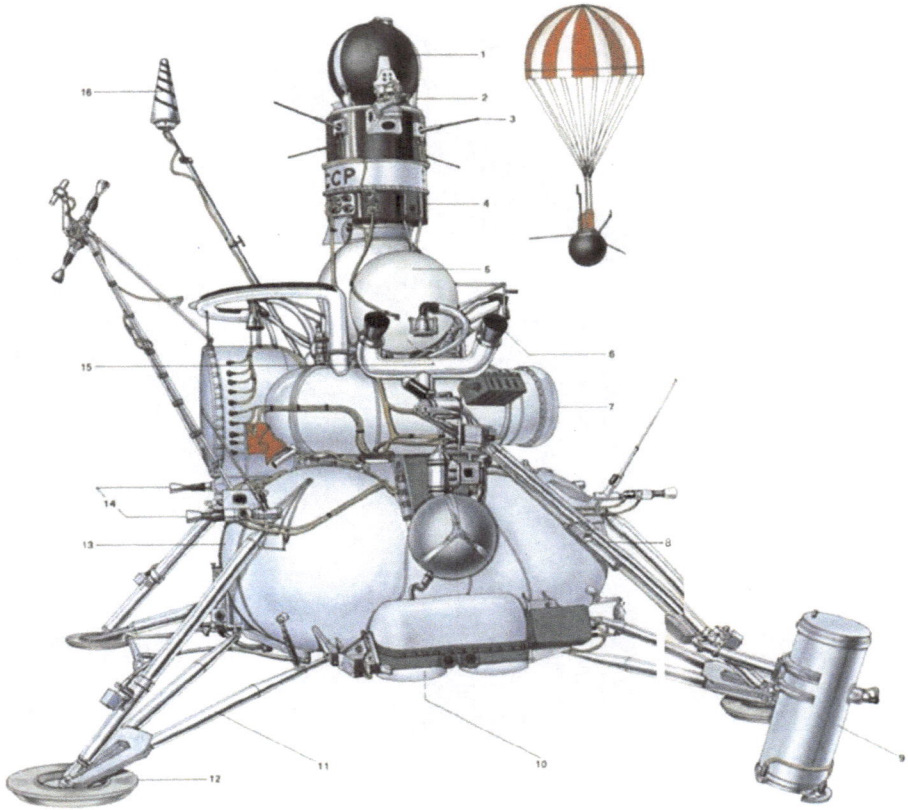

Figure 11.22 Luna 16 and Luna 20 spacecraft: 1. Return vehicle; 2. Earth entry system straps; 3. Return vehicle antennas; 4. Return vehicle instrument compartment; 5. Return vehicle fuel tanks; 6. Imaging system; 7. Lander instrument compartment; 8. Soil sampler boom; 9. Soil sampler; 10. Lander propulsion system; 11. Landing legs; 12. Footpad; 13. Lander fuel tanks; 14. Attitude control jets; 15. Return system engines; 16. Low-gain antenna.

500 rpm, it required 30 minutes to fill the entire core length of 38 cm. The drill was both insulated and hermetically sealed, and to enable the mechanism to be lubricated using oil vapor it was not opened until just before use. Some parts used a lubricant designed to reduce friction in a vacuum. A standby motor was provided as a contingency to overcome obstacles encountered during drilling. The whole device weighed 13.6 kg.

An improved drill system was provided for the Ye-8-5M version, which had a rail mounted deployment mechanism. This drill was capable of penetrating to a depth of 2.5 meters and preserving the stratigraphy, but it could not be articulated to select a sampling site. It used an elevator mechanism rather than the articulated boom arm to transfer the sample to the return capsule.

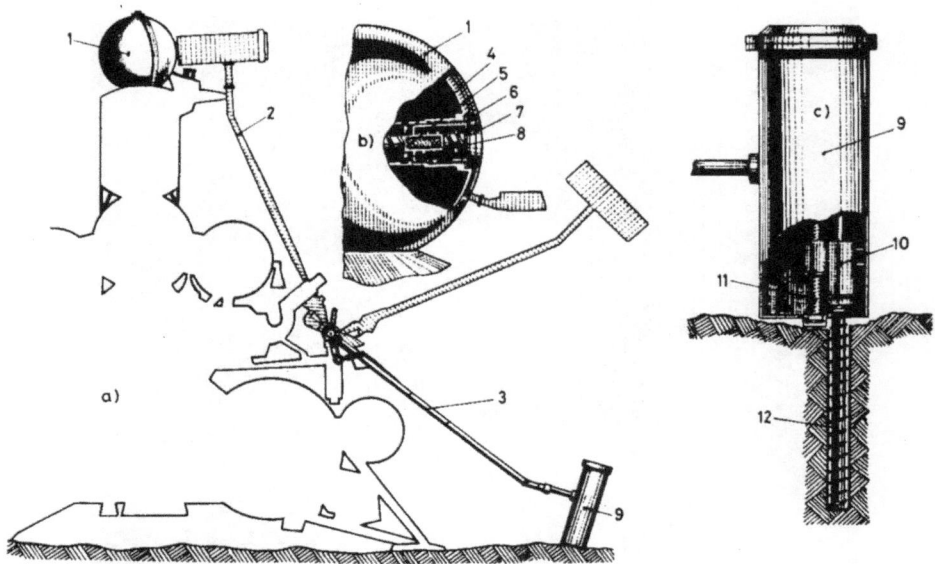

Figure 11.23 Luna 16 and Luna 20 sampling system (from *Space Travel Encyclopedia*): 1. Entry capsule; 2. Stowed position of the drill arm; 3. Deployed position of the drill arm; 4. Soil container; 5. Soil sample with drill bit; 6. Locking cover; 7. Hermetically sealing sample container cover; 8. Spring; 9. Drill unit container; 10. Drill motor; 11. Drill motor transmission; 12. Drill head.

Ascent stage:

The ascent stage was a smaller, vertically mounted open structure composed of a pressurized cylindrical avionics compartment above three spherical propellant tanks and the rocket engine. This was the same engine as used on the lander, but was not throttled. Four vernier engines were attached outboard of the propellant tanks. There were perpendicular antennas mounted radially at 90 degree intervals near the top of the avionics compartment. The spherical return capsule was held in place on top by deployable straps. Including the return capsule, the ascent stage was 2 meters tall. It weighed 245 kg dry and 520 kg with propellant. The KRD-61 Isayev engine burned nitric acid and UDMH and produced a thrust of 18.8 kN for 53 seconds to impart a velocity of 2.6 to 2.7 km/s, which was enough to escape from the Moon on a direct ascent trajectory.

Return capsule:

The return capsule was a 50 cm sphere covered with ablative material for entry at a speed of about 11 km/s and a peak deceleration load of 315 G. It had three internal sections. The upper section contained the parachutes (a 1.5 square meter drogue and a 10 square meter main) and beacon antennas, the middle section contained the lunar sample, and the base had the heavy equipment including batteries and transmitters.

Figure 11.24 Luna 16 ascent stage.

On the Moon, the sample was inserted into the capsule through a hatch in the side. The capsule weighed 39 kg, and the distribution of mass was designed to stabilize it on entry.

Luna 15 launch mass: 5,667 kg
Luna 16 launch mass: 5,727 kg
Luna 18 launch mass: 5,750 kg
Luna 20 launch mass: 5,750 kg
Luna 23 launch mass: 5,795 kg
Luna 24 launch mass: 5,795 kg

On-orbit dry mass:	4,800 kg (Luna 24)
Landed mass:	1,880 kg
Ascent stage mass:	520 kg (515 kg for Luna 23 and 24)
Capsule entry mass:	35 kg (34 kg for Luna 23 and 24)

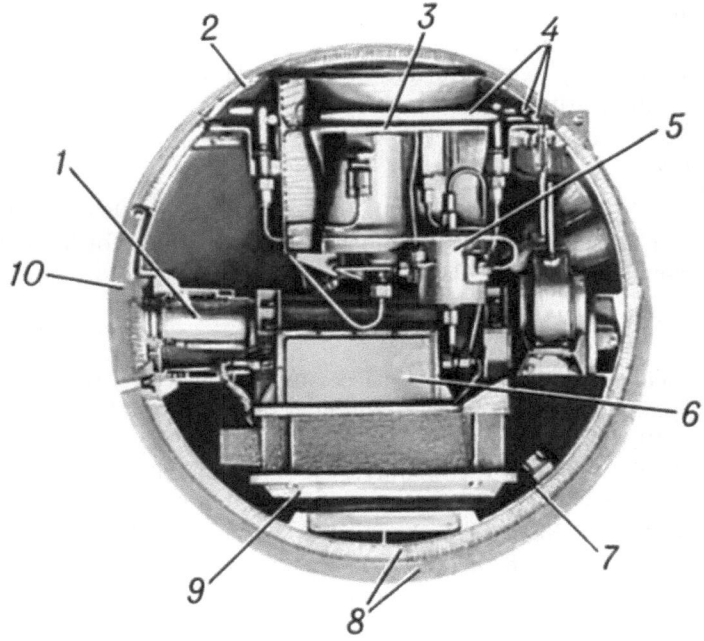

Figure 11.25 Luna 16 and Luna 20 return capsule: 1. Soil sample container; 2 Parachute container cover; 3. Parachute container; 4. Antennae; 5. Antenna release; 6. Transmitter; 7. Entry capsule interior wall; 8. Heat insulation material; 9. Battery; 10. Soil sample container cover.

Payload:

1. Stereo panoramic imaging system with lamps (deleted on Luna 23 and 24)
2. Remote arm for sample collection (improved drill on Luna 23 and 24)
3. Radiation detectors
4. Temperature sensor inside capsule

The stereo imaging system had two 300 × 6,000 panoramic scan cameras of the type used on the earlier Ye-6 landers and Lunokhod rovers. Mounted on the lander just below the level of the ascent stage on the same side as the sampling system, they were spaced 50 cm apart, angled at 50 degrees to the vertical, and gave a field of view of 30 degrees. The orientation of the lander was determined by measuring the position of Earth in a panoramic image. Stereo images were taken of the surface between the two landing legs to select the position to be sampled. They also imaged sampling and drilling operations. For the Luna 23 and 24 sample return missions the cameras and lamps were deleted.

Mission description:

Only six of the eleven spacecraft in this series were launched successfully. Of these, three succeeded in returning lunar samples to Earth.

The first attempt

The first launch (Ye-8-5 No.402) was attempted on June 14, 1969, one month prior to the Apollo 11 launch date, but the Block D failed to ignite for its first burn and the payload re-entered over the Pacific Ocean.

Luna 15

The second spacecraft in this series was successfully launched on July 13, 1969, just 3 days before Apollo 11, and the Soviets announced that Luna 15 was to land on the Moon on July 19, one day ahead of the Americans, with the objective of returning something to Earth. At 10:00 UT on July 17 it entered a 240 × 870 km lunar orbit inclined at 126 degrees. This orbit was much higher than intended, so the next day it was trimmed to 94 × 220 km. Another trim a day later yielded an orbit 85 × 221 km. Ideally the orbit should have been near-circular at about 100 km, but the Soviets had underestimated the effect of the lunar mascons and they were also suffering attitude control problems. Meanwhile, Apollo 11 had arrived and entered an equatorial orbit The drama was palpable. In Russia its nature was clear, but in America the ultimate purpose of Luna 15 was mysterious and opinions ranged from the suspicious to the sublime to the ridiculous. Apollo 8 astronaut Frank Borman, just back from a visit to the Soviet Union, appealed for information and the Academy of Sciences supplied orbit data, operational frequencies, and assurances that Luna 15 would not endanger the Apollo 11 mission.

On July 20, after several more orbit changes, Luna 15 began its descent sequence during its 39th orbit by lowering its perilune to 16 km above the landing site in the Sea of Crises. The intention was to land just 2 hours before Apollo 11 landed further west in the Sea of Tranquility. But when controllers saw the radar data from the first perilune pass they became concerned. The one and only target appeared uneven and potentially dangerous. It must have been with the utmost reluctance and dismay that the decision was taken to postpone the landing to test the radar and perform further observations. As a result of this delay, not only was there now no chance of landing ahead of the Americans, the nature of the return trajectory would make it impossible to get a sample back first. All of this was unknown to an anxious world, wondering what stunt Luna 15 was going to pull in order to upstage Apollo 11. Eighteen hours later, on its 52nd orbit, Luna 15 was commanded to land at 15:46:43 UT on July 21, after Armstrong and Aldrin had already walked on the Moon. The descent maneuver failed and, for reasons still to be explained, the transmission ceased 4 minutes after the de-orbit burn started. It crashed at 17°N 60°E, about 800 km east of Tranquility Base. Jodrell Bank flashed notification to the Americans that Luna 15 had impacted at a velocity of 480 m/s just as Apollo 11's lunar module was preparing to leave the Moon. The Soviets reported that Luna 15 had "reached the lunar surface in a preset area" but remained silent on its true mission and there was no propaganda victory.

Three duds and a launcher review

The next three sample return attempts were lost to launcher failures. In September Ye-8-5 No.403 was stranded in parking orbit when the Block D failed to restart. An oxygen fuel valve had stuck open after the first firing and allowed all the oxidizer to escape. It was designated Cosmos 300 by the Soviets, and re-entered after 4 days. In October a programming error caused the Block D to misfire and spacecraft No.404, designated Cosmos 305, re-entered during its first orbit. In February 1970 spacecraft No.405 was lost when a pressure sensor command error shut down the second stage after 127 seconds of flight and the vehicle was destroyed. This precipitated a review of the Proton launch vehicle, which had a miserable record with many more failures than successes in the Ye-8 and lunar Zond series at that time. Changes were made as a result of this review, and in August 1970 a successful suborbital diagnostic flight was flown. The following month another sample return attempt was made and, after five successive failures in a period of 15 months, success was finally achieved.

Luna 16

Luna 16 was launched at 13:25:53 UT on September 12, 1970. Seventy minutes after entering parking orbit, the Block D reignited and performed the translunar injection burn. After a course correction on September 13, the spacecraft entered a nearly circular 110 × 119 km orbit at 70 degrees inclination on September 17. After gravity data had been acquired in this orbit, two orbital adjustments were made on September 18 and 19, the first into an elliptical orbit with its perilune 15.1 km above the landing site and an apolune of 106 km, and the second to adjust the orbit plane to 71 degrees inclination. As the spacecraft approached perilune on September 20, the extra tanks were jettisoned. The engine was ignited at perilune, 05:12 UT, and fired for 270 seconds to cancel the orbital velocity and initiate the free fall. Triggered by the radar altimeter, the engine was restarted at an altitude of 600 meters and velocity of 700 km/hr. It was shut down at an altitude of 20 meters and a velocity of 2 m/s. The primary verniers were ignited for the terminal phase, and cut off at a height of 2 meters, then the spacecraft dropped to impact at about 4.8 m/s. Touchdown occurred at 05:18 UT in the Sea of Fertility at 0.68°S 56.30°E, only 1.5 km from the planned point.

Because Luna 16 touched down 60 hours after local sunset and was not fitted with floodlights it did not provide any images. The drill was deployed after an hour, and in 7 minutes of operations it penetrated to a depth of 35 cm before encountering an obstacle. The boom then lifted the core sample from the surface and swung it up to

Figure 11.26 Luna 16 lander with canister sampler deployed.

the return capsule atop the ascent stage, inserting it through an open hatch that then closed. Some soil was lost from the sampler during this operation. At 07:43 UT on September 21, after 26 hours and 25 minutes on the Moon, the ascent stage lifted off and escaped at 2.7 km/s. The lower stage remained on the surface and continued to transmit lunar temperature and radiation data. The ascent stage returned directly to Earth. On September 24, at a distance of 48,000 km, the straps released the return capsule. Four hours later it hit the atmosphere traveling at 11 km/s on a trajectory at 30 degrees to the vertical on a ballistic entry with a peak deceleration of 350 G. At an altitude of 14.5 km the top of the capsule was ejected and the drogue parachute deployed. At 11 km the main chute was unfurled and the beacon antennas deployed. The capsule landed at 03:26 UT on September 24, approximately 80 km southeast of the city of Dzhezkazgan in Kazakhstan.

Luna 16 proved to contain 101 grams of lunar material. It was a triumph, and the Soviet press made the best of it, hyping the use of robots over manned missions. For the Americans, Luna 16 confirmed what they suspected about Luna 15, that it was a sample return attempt intended to upstage Apollo 11.

Luna 18

The next attempt at sample return was made one year later. Luna 18 was launched on September 2, 1971, made midcourse corrections on the 4th and 6th, and entered a 101 km circular lunar orbit inclined at 35 degrees the following day. After lowering its perilune to 18 km it was commanded to land on September 11, but the signals cut off abruptly at 07:48 UT at an altitude of about 100 meters. The main engine had run out of fuel due to excessive consumption in earlier operations, and the wreckage lies at 3.57°N 56.50°E in rugged highland terrain.

Luna 20

Luna 20 was the second attempt after Luna 18 to obtain a sample from in the lunar highlands. Launched on February 14, 1972, it was tracked by telescopes to compute an accurate trajectory, made a midcourse correction the following day and entered a 100 km circular lunar orbit at 65 degrees on February 18. It lowered its perilune to 21 km the following day. And finally, at 19:19 UT on February 21, it touched down in the Apollonius highlands near the Sea of Fertility at 3.53°N 56.55°E, only 1.8 km from where Luna 18 crashed. The Sun was 60 degrees above the horizon. Images of the surface were transmitted prior to the sampling operations. The drill encountered resistance, and had to be paused three times to prevent overheating. It was ultimately able to achieve a depth of only 25 cm. The recovered sample core was transferred to the return capsule.

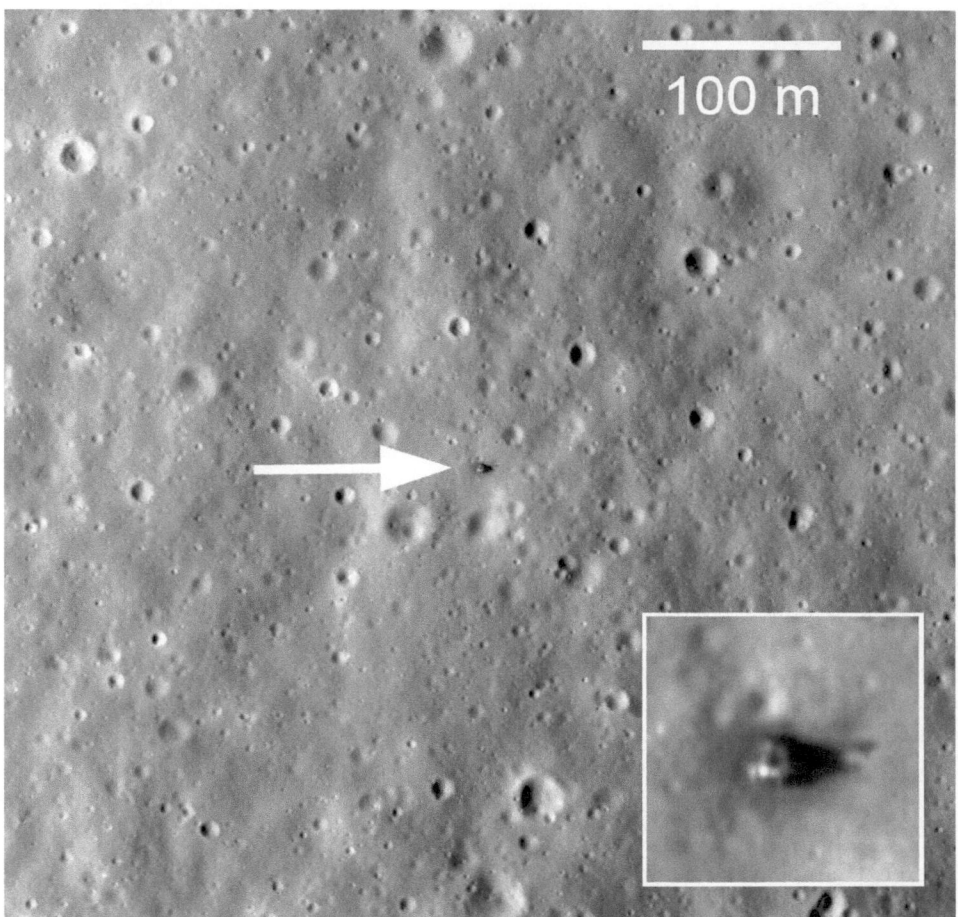

Figure 11.27 Luna 20 on the Moon and closeup from NASA's Lunar Reconnaissance Orbiter.

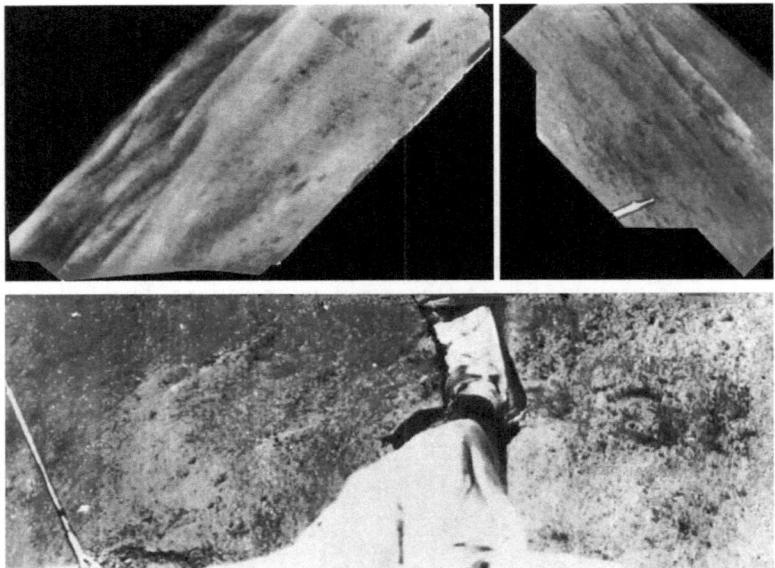

Figure 11.28 Pictures from the Luna 20 imaging system showing the sampler on the surface and (above) views to the lunar horizon at left and right.

The ascent stage lifted off at 22:58 UT on February 22, and then upon its return to Earth fell into a strong snow storm. Although it came down over the Karkingir river 40 km north of Dzhezkazgan, when it landed at 19:19 UT on February 25 it touched down on an island. The ice, wind, and snow presented the recovery team with severe difficulties. It was recovered the following day, and when opened proved to contain only 55 grams of lunar soil.

Luna 23

The first improved Ye-8-5M sample return vehicle, No.410, became Luna 23 with a successful launch on October 28, 1974. After a midcourse correction on October 31, it entered an almost circular 94 × 104 km orbit at 138 degrees on November 2. Upon lowering its perilune to 17 km it was commanded to land on November 6. Although Luna 23 landed on target in the southern part of Sea of Crises at 12.68°N 62.28°E, it did so at 11 m/s and the impact shock wrecked the sample collection apparatus and caused other damage. No samples were obtained and the ascent stage was not fired. The lander continued to transmit until contact was lost on November 9.

One more failure and one last sample return, Luna 24

The next sample return attempt with spacecraft Ye-8-5M No.412, the tenth in the series, was lost when the Block D failed during its first burn and was unable to reach parking orbit. The final attempt, with No.413, was successful. Luna 24 was launched on August 9, 1976, made a midcourse correction on the 11th, and entered a circular lunar orbit at 115 km with an inclination of 120 degrees on August 14. It adjusted its orbit to 120 × 12 km on the 17th and at 02:00:00 UT on the 18th it landed in darkness at 12.75°N 62.20°E in the Sea of Crises, only about 2,400 meters from the Luna 23 lander and near the Luna 15 target. The focus of the Ye-8-5M series was to obtain a deep core from the surface of the lunar mascon in Mare Crisium. The drill was able to reach the planned depth of 2.25 meters at a slightly inclined angle that equated to a vertical penetration of 2 meters. The sample was transferred to the return capsule and then at 05:25 UT on the 19th the ascent stage lifted off. The capsule landed 200 km southeast of Surgut in Siberia at 17:53 UT on August 22. It proved to contain a sample of 170.1 grams.

End of the Moon, but the beginning for a propulsion stage

Luna 24 was the last of the Luna series and the final Soviet lunar mission. A third rover was built and another sample return spacecraft was prepared, but in 1977 the

Figure 11.29 Luna 24 lander with rail-mounted drill.

Figure 11.30 Luna 16 recovery and soil sample (below). The balloons were inflated post-landing to right the capsule for antenna exposure.

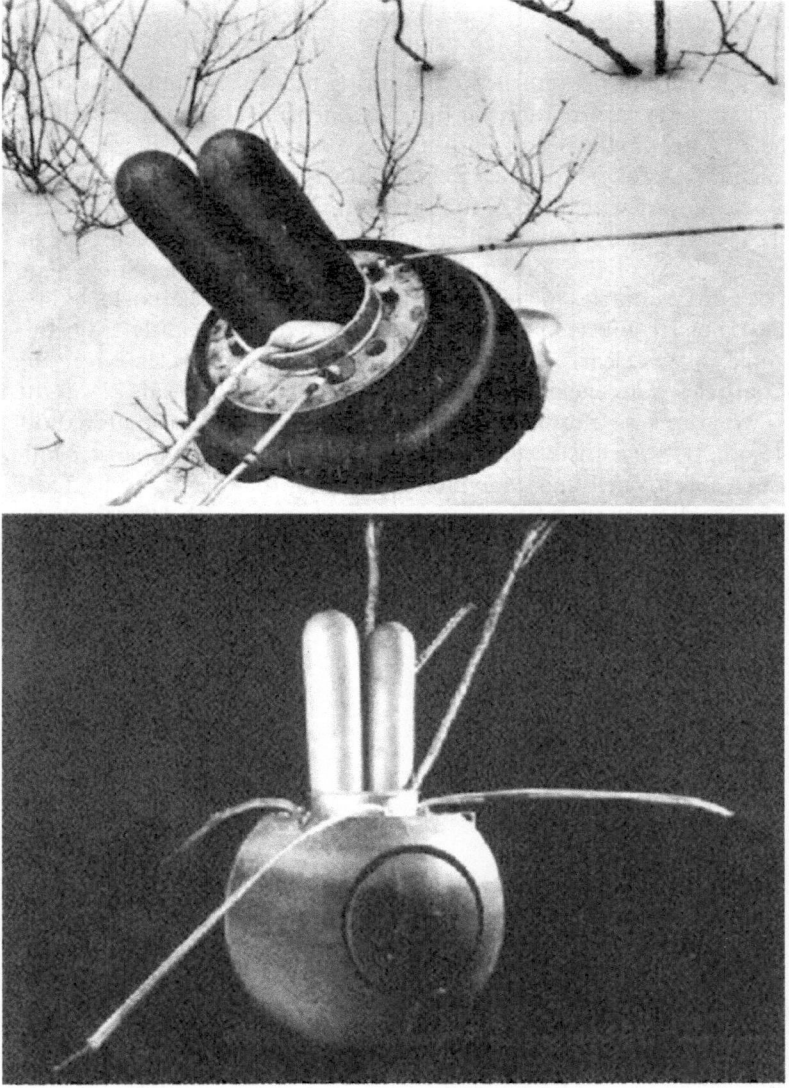

Figure 11.31 Luna 20 recovery (top) and Luna 24 return capsule (bottom).

launcher for the rover was requisitioned for a communications satellite and attention swung towards the ill-fated 5M Mars sample return project. Later all lunar and Mars plans were canceled in favor of continuing the successful Venus missions.

However, the reliable Ye-8 cruise stage was modified to produce the autonomous propulsion stage for the 1988 Phobos spacecraft, and went on to become the Fregat upper stage that is currently used by Proton-K and Soyuz launchers.

Results:

The altimeter on Luna 15 returned data during the descent until loss of signal, and this provided useful information on the mean density of the lunar soil. The sample that Luna 16 got from the Sea of Fertility was a dark, powdery mare basalt similar to that obtained by Apollo 12 in the Ocean of Storms. In 1971 three grams of Luna 16 soil were exchanged with NASA for three grams of Apollo 11 samples and three grams of Apollo 12 samples. The Vernadsky Institute of Geochemistry in Moscow conducted analyses of the soils returned from these missions. Some samples were donated to other countries including France, Austria and Czechoslovakia. Although small, the Luna 20 soil from the highlands differed significantly from the Luna 16 mare sample. It was clearly lighter in color, with larger particles. More than half of the rock particles were ancient anorthosite compared to less than 2% in the Luna 16 sample. Two grams were exchanged for one gram of highlands material obtained by Apollo 15, and US scientists were able to date the Luna 20 sample at 3 billion years. The 2 meter core from Luna 24 exhibited layering that clearly indicated successive deposition. Small portions of this core were exchanged with US scientists.

12

Landing on the Moon, Venus, and Mars

TIMELINE: AUG 1970–FEB 1972

The Soviets reached the zenith of their success at the Moon with robotic missions in 1970 and 1971. In September 1970 the Luna 16 mission successfully returned a sample of the Moon to Earth; an impressive achievement still unmatched by the US. In November the Luna 17 mission successfully deployed the first robotic rover on the Moon, Lunokhod 1; another achievement unmatched by the US. An attempt at sample return in September 1971 failed when communications were lost as Luna 18 was landing. It was followed immediately by Luna 19, a successful orbiter version of the spacecraft. Luna 20 became the second successful sample return mission in February 1972.

The Soviets also finally achieved a landing on Venus after eleven attempts since February 1961. Venera 7 was launched on August 17, 1970, with a descent capsule modified to withstand the massive surface pressure on Venus, and this succeeded in descending through the entire atmosphere and gently impacting the surface where it continued to operate for 23 minutes before succumbing to the high temperature. The Soviets finally had some success at Mars in 1971 after eight attempts since October 1960. The 1971 opportunity was not as energetically favorable as in 1969, requiring the landers to be released in the approach rather than after entering orbit around the planet. This and several engineering problems with the Mars-69 spacecraft forced a complete redesign. The 1971 Mars spacecraft became the basic design reference for all Soviet Proton-launched planetary spacecraft thereafter.

The Soviet plan in 1971 was to start with an orbiter to Mars which would provide precise information on the position of the planet to the spacecraft that followed, to enable these to deploy their landers on the necessarily very precise entry trajectories before themselves entering into orbit. This plan was foiled when the launch of the leading orbiter failed on May 10. Fortunately, the Soviets had a backup plan in which the approaching carrier spacecraft would use on board optical navigation to determine the position of Mars and autonomously update their navigation system so that they could properly deploy their landers. This complex and sophisticated system was far in advance of its time, but very risky. The Mars 2 system worked, but due to a software error it dispatched its lander on an entry angle which was too steep and resulted in a crash. It worked perfectly for Mars 3 whose entry system placed the first successful lander on Mars, but after sending 20 seconds of uninterpretable data it fell silent. Both spacecraft entered orbit around the planet and transmitted images of its surface and data on its atmosphere, surface and plasma environment.

The US also had a major success at Mars in 1971. Mariner 9 was the first mission from this launch opportunity to arrive and became the first spacecraft to enter orbit. With more sophisticated cameras and systems, and an excellent instrument suite, its accomplishments completely eclipsed those of the much heavier Soviet orbiters.

REACHING THE SURFACE OF VENUS: 1970

Campaign objectives:

The Venera 4, 5 and 6 missions were major successes for the Soviets, sending back detailed information while descending on their parachutes, as they were designed to, but they did not survive all the way to the surface. The controversy about the surface conditions in the wake of the Venera 4 and Mariner 5 missions was not resolved in time to permit modification of the Venera 5 and 6 probes to survive the temperature and pressure at the surface. It was decided to launch them anyway in order to obtain more information on atmospheric conditions, which they did well.

Spacecraft launched

First spacecraft:	Venera 7 (3V No.630)
Mission Type:	Venus Atmosphere/Surface Probe
Country/Builder:	USSR/NPO-Lavochkin
Launch Vehicle:	Molniya-M
Launch Date/Time:	August 17, 1970 at 05:38:22 UT (Baikonur)
Encounter Date/Time:	December 15, 1970
Outcome:	Successful, transmitted from surface.

Second spacecraft:	Cosmos 359 (3V No.631)
Mission Type:	Venus Atmosphere/Surface Probe
Country/Builder:	USSR/NPO-Lavochkin
Launch Vehicle:	Molniya-M
Launch Date/Time:	August 22, 1970 at 05:06:09 UT (Baikonur)
Outcome:	Failed to depart Earth orbit.

For the 1970 Venus opportunity the Soviets were determined to reach the surface. They now had unequivocal scientific proof that the pressure at the surface was about 100 bar and the temperature exceeded 450°C. But after years of controversy among scientists, the engineers were wary and they designed the new capsule to withstand 180 bar and a temperature of 540°C for 90 minutes. The additional mass required a significant reduction in the number of science instruments.

One mission was dispatched successfully but the other was a launcher failure. The entry capsule of Venera 7 was the first to reach the surface in an operational state and became the first successful planetary lander. Over the decade 1960–70 the Soviets had seventeen unsuccessful attempts at Venus missions, seven of which were intended to reach the surface. Now persistence had paid off. At the same time, their first Lunokhod rover was traversing the surface of the Moon.

Spacecraft:

Carrier spacecraft:

The carrier vehicle was unchanged from Venera 4, 5, and 6, but fewer instruments were carried to accommodate the larger mass of the entry system.

Entry vehicle:

The descent capsule was significantly modified as described above. In particular the entry system was made more egg-shaped to accommodate extra thermal insulation and a new shock absorber. A new spherical pressure vessel was used instead of the previous flat-capped hemisphere. The pressure vessel was made of titanium, and to minimize weak points it had the fewest possible number of feed-throughs and welds. The temperature and pressure sensors were on the exterior of the shell, under the top hatch. The design was verified in a new test chamber at 150 bar and 540°C.

The goal was to reach the surface as fast as possible without losing the capsule in order to maximize its lifetime on the surface. A single parachute was used. On first opening, it was reefed by a cord wrapped around the shroud lines to limit its area to 1.8 square meters. Being smaller than the main parachute of previous missions, this would produce a faster descent. But the reefing cord was designed to melt at 200°C, deep in the atmosphere, opening the parachute to its full 2.5 square meters in order to achieve a soft landing. The parachute was built to survive the high temperatures at the surface. The capsule was pre-cooled to -8°C prior to being released by the carrier spacecraft to maximize its survival time. The total design lifetime of the capsule was 90 minutes.

Launch mass: 1,180 kg (*entry capsule* 490 kg)

Figure 12.1 Venera 7 spacecraft.

Payload:

Carrier spacecraft:

1.   Solar wind charged particle detector

Descent/landing capsule:

1. Temperature, pressure and density sensors
2. Radio altimeter
3. Gamma-ray spectrometer
4. Doppler wind experiment

Due to the additional mass required to accommodate the very high pressure limit, the spacecraft had only the solar wind charged-particle detector and it was necessary to delete the atmospheric composition experiment and airglow photometer from the descent capsule. The aneroid barometer could measure pressures of 0.5 to 150 bar, and the resistance thermometer had a range of 25 to 540°C. Density was measured by an accelerometer during the entry phase. A gamma-ray instrument was added for measurement of surface rock type.

Mission description:

Venera 7 was dispatched successfully on August 17, 1970, and conducted midcourse maneuvers on October 2 and November 17. The second launch on August 22 failed

Figure 12.2 Venera 7 descent capsule.

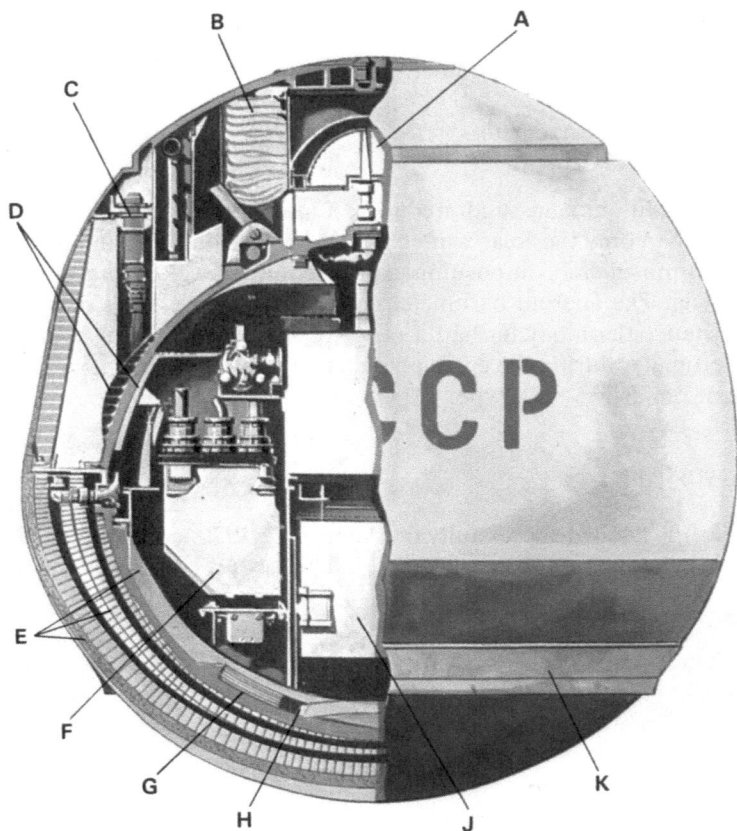

Figure 12.3 Venera 7 entry vehicle (from *Robot Explorers*): A. Antenna; B. Parachute; C. Top hatch release bolt; D. Internal heat shield; E. Insulating layers; F. Instrument commutator; G. Pressure shell; H. Shock damper; J. Transmitter; K. Spacecraft adaptor.

to depart low Earth orbit when the fourth stage misfired. As a result of a sequencer problem and a power system failure the engine ignited late and shut down after only 25 seconds. This vehicle was designated Cosmos 359 by the Soviets and re-entered on November 6. Venera 7 initiated its planetary encounter activities on December 12 when the capsule batteries were charged by the solar panels. The internal equipment compartment was activated and chilled down to -8°C. The capsule was released at 04:58:44 UT on December 15. It struck the atmosphere at an altitude of 135 km at 11.5 km/s. By the time it was down to 54 km it had been slowed to 200 m/s, and a pressure reading of 0.7 bar triggered the deployment of the parachute just above the cloud layer. The capsule transmitted for 35 minutes during its descent in darkness. It survived the impact at 05:34:10 UT and a weak signal was received for another 23 minutes. The landing site was at 5°S 351°E, where it was 4:42 Venus solar time and the solar zenith angle was 117 degrees.

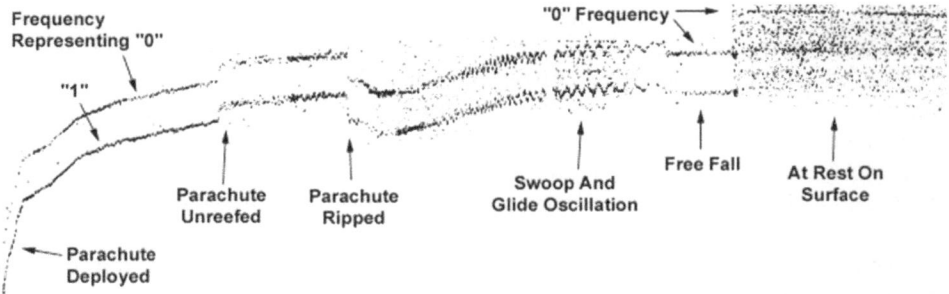

Figure 12.4 Doppler frequency plot for the Venera 7 descent capsule (from Don Mitchell).

This success was not immediately obvious, and it was originally thought that the capsule had *failed* to reach the surface. The 35 minute descent to the surface turned out to be a wild ride. After the first 13 minutes the reefing cord melted away and the parachute opened fully, just as it was meant to do. Six minutes later the parachute ripped, and over the next several minutes the descent rate increased and the capsule oscillated wildly as the rip extended. A few minutes before reaching the surface the parachute failed and the capsule fell freely. All of this was evident from the Doppler shift in the transmitter carrier frequency. The capsule hit the surface at 16.5 m/s, the signal disappeared into noise, and it was concluded that the capsule must have been destroyed. There was no immediate announcement of the success of Venera 7. After the New Year holiday, an expert in signal processing re-ran the data tapes and in all the noise found the barely perceptible signal from the capsule on surface. The signal strength had reduced to 3% at impact, returned to full strength for one second, then dropped back to 3% for the next 23 minutes before terminating. The capsule would seem to have bounced on impact and come to rest tilted at about 50 degrees to the vertical, aiming the radiation pattern of its antenna well off Earth and resulting in a very low received power. The team members were elated by this discovery. It may not have been a graceful touchdown, but it was another first for the Soviet planetary program.

The Venera 7 carrier spacecraft returned measurements on the upper atmosphere and ionosphere prior to breaking up in the atmosphere.

Results:

The Venera 7 probe measured the temperature of the atmosphere from an altitude of 55 km to the surface, but a commutator failure resulted in no pressure or altimetry information being transmitted. Initially it was thought the temperature measurements returned were internal but fortunately they did turn out to be atmospheric, and when combined with Doppler data and with thermodynamic and aerodynamic modeling it was possible to construct a profile of temperature and pressure down to the surface. Altitude profiles of horizontal wind speed and direction were also obtained from the Doppler data and aerodynamic modeling. There was a fast wind at high altitudes in the same retrograde direction as the axial rotation. This confirmed astronomical evidence from ultraviolet cloud flow that the upper atmosphere was 'super-rotating'. Wind speeds of less than 2.5 m/s were measured at the surface. The probe temperature sensor oscillated between binary readings of 457°C and 474°C on the surface. The computed surface pressure was 92 bar. Doppler data at the moment of touchdown, plus the fact that the capsule survived the high speed impact, implied that the surface was harder than sand but no harder than pumice. No surface composition measurements were returned because of the stuck commutator.

THE FIRST LANDER ON MARS: 1971

Campaign objectives:

The Soviets had a strong desire to follow their original long-term plan for the 1971 campaign and build a new entry vehicle containing a soft lander, but the M-69 losses meant NPO-Lavochkin lacked both the detailed ephemeris for the planet and the atmospheric data which was required to design a soft lander. One option was to repeat the atmospheric probe mission with a hard lander in 1971 to obtain this data, and postpone the soft lander to 1973. But the 1973 opportunity would require more energy, and so would require separate rather than combined launches for an orbiter and lander. This would mean launching at least four vehicles, two orbiters and two flyby spacecraft carrying landers, and redesigning the entry vehicle to accommodate entry from the initial approach rather than from orbit. This scenario was deemed too expensive at the time, but it is exactly what the Soviets ended up doing in 1973. An alternative was to get the data from the US. The Mariner 4, 6 and 7 flyby missions in 1965 and 1969 had studied the atmosphere and estimates of the surface pressure had been published, but the crucial ephemeris had not been published and the Americans were unwilling to supply it to the Soviets since the antagonism of the Cold War was rife at the time.

Ultimately, the Soviets settled upon a clever but risky approach to implementing a soft lander which facilitated the launch of combined orbiter/landers in 1971 without requiring pre-launch data on the planet's ephemeris. This involved sending another spacecraft ahead of the two orbiter/landers to enter orbit around Mars and serve as a radio beacon that the other spacecraft would use to achieve the desired navigational accuracy. On this orbiter the mass which would normally have been allocated to the entry system facilitated the larger propellant load required to achieve a high energy, fast trajectory and increased the scientific payload. Optical tracking during approach and radio tracking in orbit would enable the ephemeris to be derived in sufficient time for trajectory corrections to be sent to the orbiter/landers. Once in orbit, the leading spacecraft would act as a radio beacon to assist the entry vehicles navigate

Spacecraft launched

First spacecraft:	M-71S (M-71 No.170 and Cosmos 419)
Mission Type:	Mars Orbiter
Country/Builder:	USSR/NPO-Lavochkin
Launch Vehicle:	Proton-K
Launch Date/Time:	May 10, 1971 at 16:58:42 UT (Baikonur)
Outcome:	Stranded in orbit, fourth stage failed to reignite.

Second spacecraft:	Mars 2 (M-71 No.171)
Mission Type:	Mars Orbiter/Lander
Country/Builder:	USSR/NPO-Lavochkin
Launch Vehicle:	Proton-K
Launch Date/Time:	May 19, 1971 at 16:22:44 UT (Baikonur)
Encounter Date/Time:	November 27, 1971
Mission End:	August 22, 1972
Outcome:	Orbiter successful, lander crashed

Third spacecraft:	Mars 3 (M-71 No.172)
Mission Type:	Mars Orbiter/Lander
Country/Builder:	USSR/NPO-Lavochkin
Launch Vehicle:	Proton-K
Launch Date/Time:	May 28, 1971 at 15:26:30 UT (Baikonur)
Encounter Date/Time:	December 2, 1971
Mission End:	August 22, 1972
Outcome:	Orbiter successful, lander failed on the surface

their approach following release by their carriers. The Americans were planning to send two Mariner spacecraft to enter orbit around Mars at this launch opportunity. Sending a spacecraft on ahead offered the Soviets the propaganda advantage of being first to insert a spacecraft into orbit around the planet.

The scientific objectives of all the Soviet orbiters were to image the surface of the planet and its clouds, study the topography, composition and physical properties of the surface, measure properties of the atmosphere, make temperature measurements, and study the solar wind and interplanetary and planetary magnetic fields. The two carrier vehicles were also to relay back to Earth the transmissions from their landers. The entry system was to make atmospheric measurements during entry and deliver the lander to the surface. The objectives of the lander were to return images from the surface, obtain data on meteorological conditions and atmospheric composition, and deploy a small rover that would measure the mechanical and chemical properties of the soil.

These Soviet missions and the US Mariner 9 orbiter in 1971 had the potential to transcend the pervasive competition between the two spacefaring powers with the first cooperation by a telephone 'hot line' that was set up between the Jet Propulsion Laboratory in Pasadena and the Soviet space center in Yevpatoriya, Crimea, for the exchange of results.

Spacecraft:

Orbiters:

Designated M-71S (S for Sputnik, or orbiter), the lead orbiter would require much larger tanks than the M-69 spacecraft to enable it to fly the higher energy trajectory required to arrive at Mars ahead of the orbiter/landers. In conjunction with a number of engineering problems with the multiple instrument modules of the M-69 design, this prompted the Soviets once again to redesign the entire spacecraft. Instead of the propellant tank being the main structural element, this function was assigned to the KTDU-425A propulsion system. The fuel and oxidizer tanks formed a 3 meter long cylinder on top of the propulsion system. The avionics and science instruments were in a hermetically sealed module at the base of the cylinder, forming a toroid around the propulsion system. The gimbaled engine nozzle attached at the base of the tank protruded through the center of the instrument module. Instruments could be reached during testing simply by detaching the lower half of the toroidal cover.

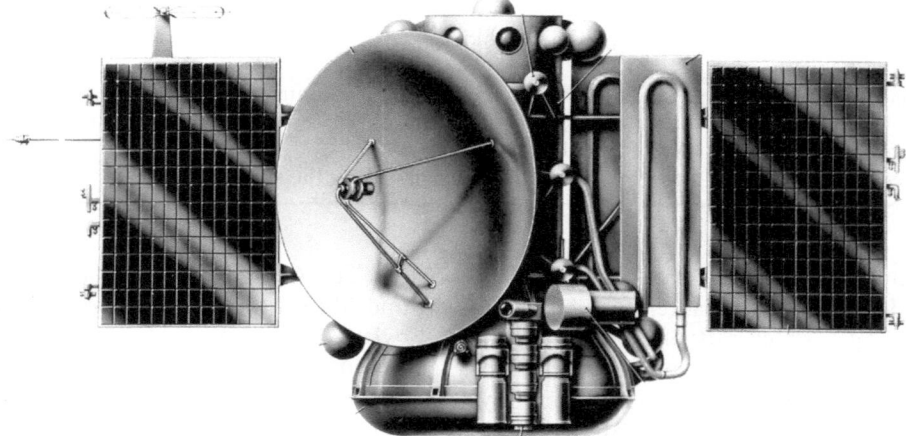

Figure 12.5 Mars-71S orbiter spacecraft.

Two 2.3 × 1.4 meter solar arrays extended from opposite sides of the cylindrical tank. Attached to the solar arrays were cold gas attitude control jets, an antenna for relaying the lander's transmission, and the magnetometer booms. A parabolic high-gain antenna 2.5 meters in diameter was mounted on the side to support redundant transmitters for 5 and 32.5 cm (5.8 GHz and 928.4 MHz). Three omnidirectional spiral antennas were installed near the high-gain antenna. The thermal control radiators and tanks of attitude control propellant were on the side of the cylinder. Navigational optics were on the outside of the instrument module – a pair of star sensors pointing downward in terms of the vehicle's structure, three Sun sensors in a vertical stack, all pointing radially out, an Earth sensor that was aligned with the parabolic antenna, and a Mars sensor aimed horizontally off to one side.

M-71S launch mass: 4,549 kg (*dry mass* 2,164 kg)

The orbiter/landers to follow the M-71S were designated M-71P (P for Posadka, or lander). They had shorter tanks with less propellant, the mass being used for the entry system carried on top of the tank, but otherwise they were almost identical to the M-71S and they were almost identical to each other. With its lander the M-71P was 4.1 meters high with a base diameter of 2 meters. The span across the deployed solar panels was 5.9 meters. They incorporated a new digital guidance and control computer based on the prototype for the Block D stage of the N-1 rocket. This was capable of significantly greater navigational accuracy, but with a mass of 167 kg and a power rating of 800 W it was rather demanding. The extra mass was compensated by deleting the control system from the Block D and instead using the spacecraft to control the stack. This is an interface design that would never have been considered in the US.

Figure 12.6 Mars 3 spacecraft.

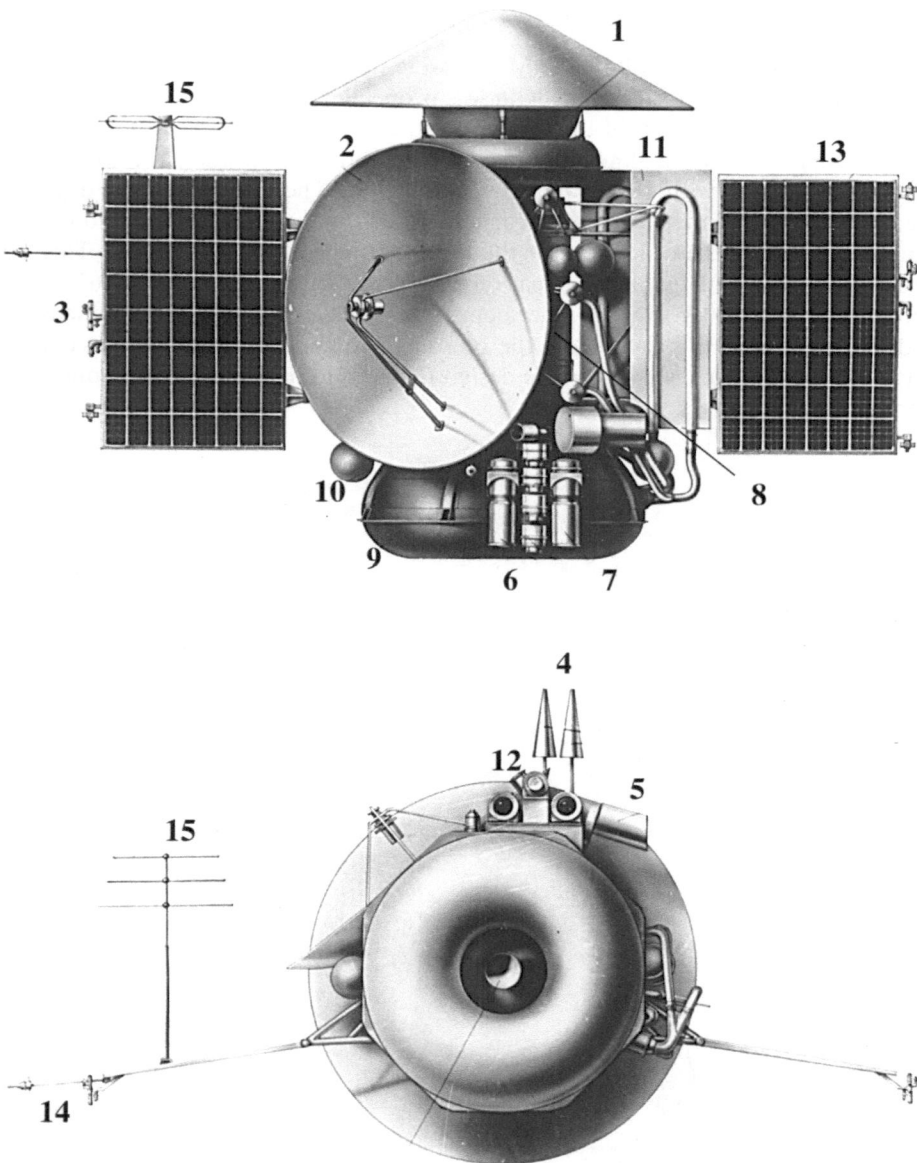

Figure 12.7 Mars-71 orbiter/lander spacecraft: 1. Lander; 2. Parabolic antenna; 3. Attitude control jets; 4. Spiral antenna; 5. Mars sensor; 6. Star sensor; 7. Star sensor; 8. Propulsion system; 9. Instrument compartment; 10. Attitude control gas tanks; 11. Thermal radiators; 12. Earth sensor; 13. Solar panels; 14. Magnetometer; 15. 'STEREO' experiment antenna.

Mars 2 and 3 launch mass: 3,440 kg (orbiter; *dry mass* 2,265 kg)
1,210 kg (entry vehicle)
635 kg (lander system on descent)
358 kg (lander)
4,650 kg (total)

The 1971 spacecraft were much easier to work on in testing operations, and were more readily modified for various planetary missions by changing instruments in the module, attaching various modules to the top of the tank, and changing the length of the tank itself. The 1971 design formed the basis for all subsequent Mars spacecraft, and all Venera spacecraft beginning with Venera 9 through the Vega spacecraft, and for astrophysics spacecraft in Earth orbit.

Entry system:

A new entry system was required to slow the spacecraft rapidly in the thin Martian atmosphere for a soft landing. The steep cone angle of the entry vehicle designed for the (unflown) 1969 atmospheric probe would not be adequate. For a soft landing in 1971 a much larger entry shell 3.2 meters in diameter and with an open vertex angle of 120 degrees was devised to maximize the altitude at which the parachute opened. Furthermore, the parachute would have to open at a supersonic velocity of Mach 3.5, a feat that had never been done before. This engineering and test challenge was met by a program of drop tests using balloons at an altitude of 35 km and meteorological rockets at 130 km. Due to the lack of data on the Martian atmosphere, the aerobrake for the M-71 system was designed for an uncontrolled ballistic descent instead of the controlled descent to be used by the Viking entry vehicles that the Americans were designing.

The entry system comprised four stacked assemblies: the aerobrake at the forward end, the egg-shaped lander nested in the aerobrake, the toroidal parachute container above the lander, and the propulsion assembly at the rear – with the latter including a structural ring. The stack was held together by four crossbars linking the rim of the aerobrake to the ring at the rear. Unlike US designs, there was no monolithic back shell. The role of the solid rocket in the center of the propulsion ring assembly was to separate the entry system from the orbiter after release and to transfer from the flyby trajectory to the desired entry trajectory. The carrier would remain on the flyby trajectory until firing its own engine for orbit insertion. For attitude control, tanks mounted on the interior of the propulsion ring assembly provided nitrogen to the cold gas micro-engines located on the crossbars near the rim. Small solid rocket micro-engines were affixed to the aerobrake rim in order to spin the vehicle prior to entry and to de-spin it following entry in readiness for deploying the parachute. The vehicle was actively 3-axis controlled from its release to the spin-up for entry, passively aerodynamically controlled during entry, and passively controlled for parachute descent. The toroidal section holding the parachutes, deployment devices, and terminal rocket engines was attached to the lander. The aerobrake was connected to the parachute container by metal bands on the underside. The avionics to control the sequence of entry, descent and landing were contained in a small

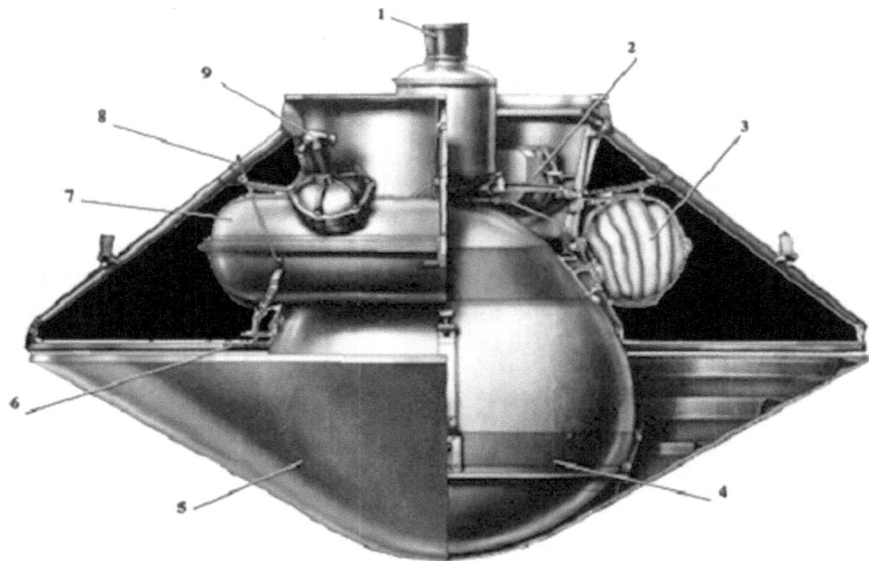

Figure 12.8 Mars 3 entry system diagram: 1. Main solid rocket; 2. Avionics; 3. Main parachute; 4. Lander surface station; 5. Aeroshell; 6. Altimeter antenna; 7. Parachute container; 8. Relay antennas; 9. Drogue parachute pyro.

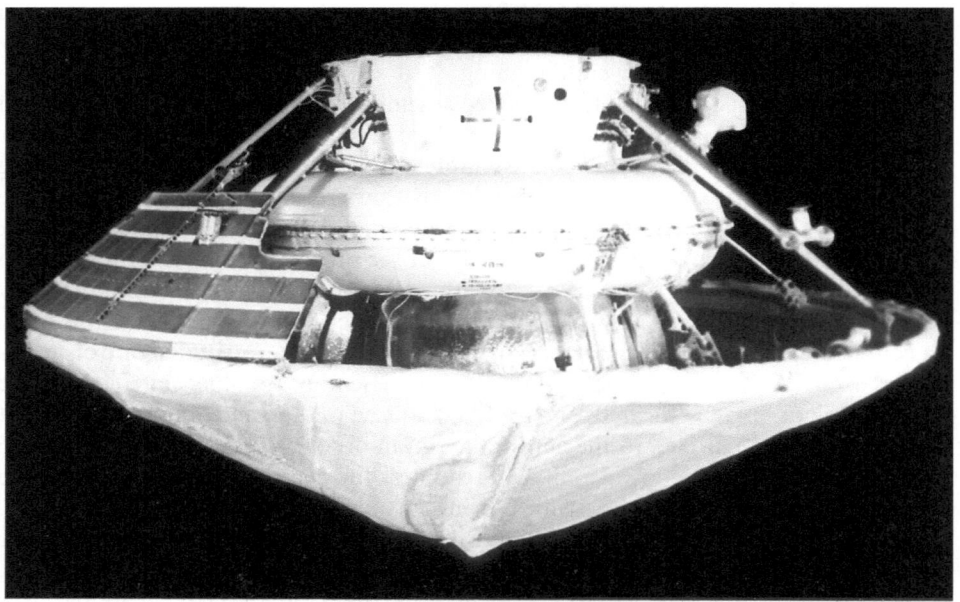

Figure 12.9 Mars-71 entry system.

cylinder attached to the underside of the toroid, which was itself designed to separate into two halves. A solid rocket device with four small nozzles affixed to the side of the upper half dragged the 13 square meter drogue parachute from the toroid. The upper half of the toroid was separated and carried away by the drogue, which in turn pulled out the 140 square meter main parachute whose lines were connected to the bottom half. The solid terminal rockets were deployed in a container part way up the

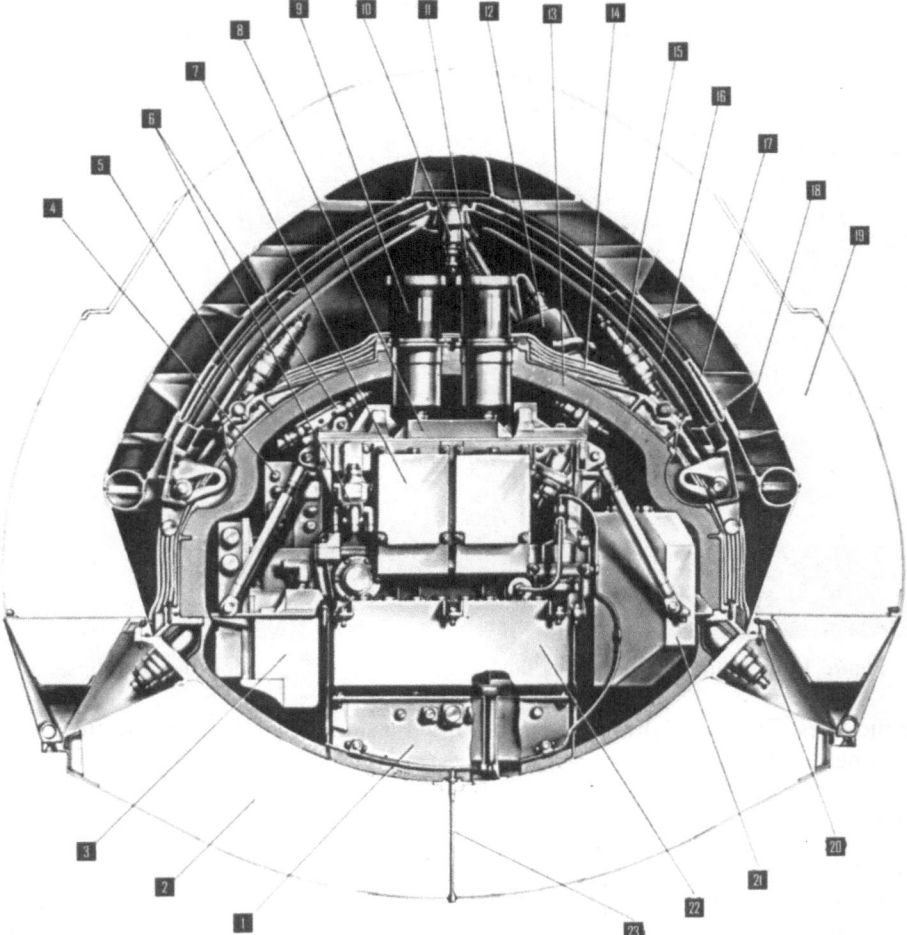

Figure 12.10 Mars-71 lander diagram: 1. Radar altimeters; 2. Shock absorber; 3. Telemetry units; 4. Automatic radio system; 5. Antennae; 6. Radio; 7. Radio system units; 8. Science instrument module; 9. Imaging system; 10. Petal locking pin; 11. Instrument deployment system; 12. Science sensors; 13. Internal thermal insulation; 14. External thermal insulation; 15. Petal deployment mechanisms; 16. Petals; 17. Aeroshell cap displacement balloon; 18. Aeroshell cap; 19. Aeroshell cap shock absorber; 20. Gas cartridge for displacement balloon; 21. Control system; 22. Batteries; 23. Pressure sensor.

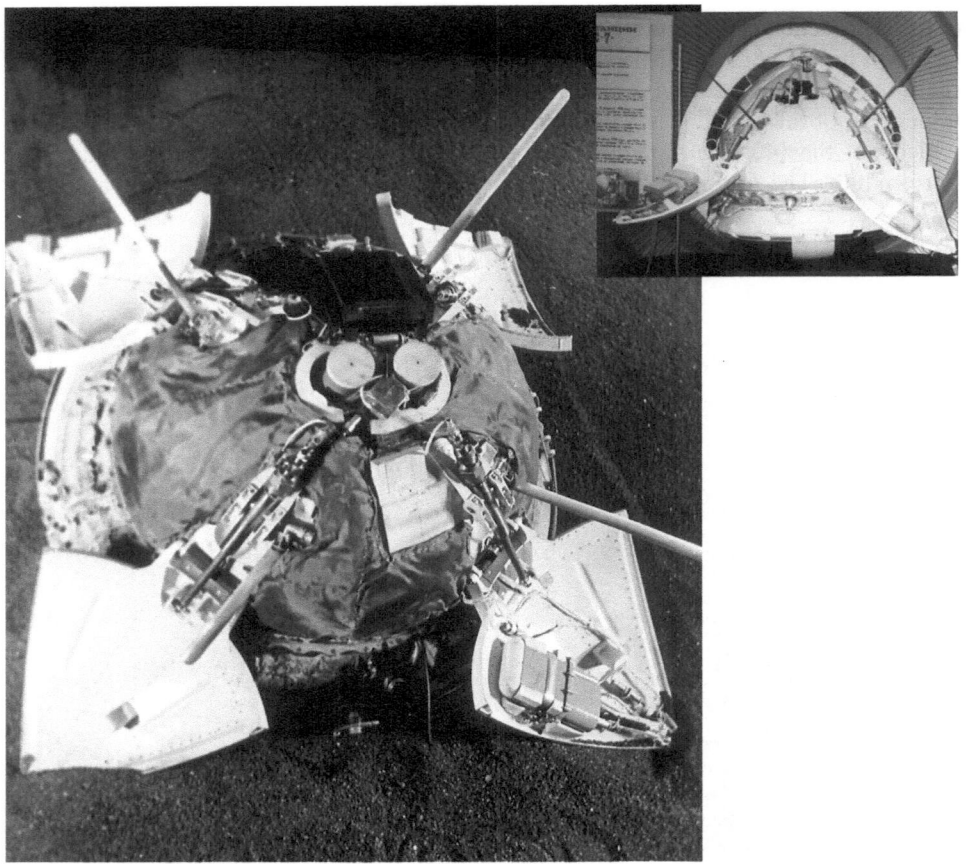

Figure 12.11 Mars-71 engineering lander in test bed and a sectioned model (insert).

shroud lines. The radar altimeter was mounted inside the lander at the bottom of the instrument compartment.

Lander:

The lander was an egg-shaped capsule 1.2 meters in diameter across the middle that was entirely covered with a 20 cm thick layer of foam. The foam was in two pieces, one an aeroshell cover in the form of an ejectable cap over the larger top portion of the lander capsule and fitting onto a small skirt encircling the bottom of the capsule; and the other a lens-shape which was permanently mounted on the bottom, under the encircling skirt, in order to absorb the shock of landing. The foam aeroshell cap was ejected after landing by inflating a balloon to allow the petals to open, in the process righting the lander and exposing its internal instruments. Two camera ports and four deployable elastic aerials protruded from the top of the sphere for communicating with the orbiter. The tethered rover was mounted on a deployable

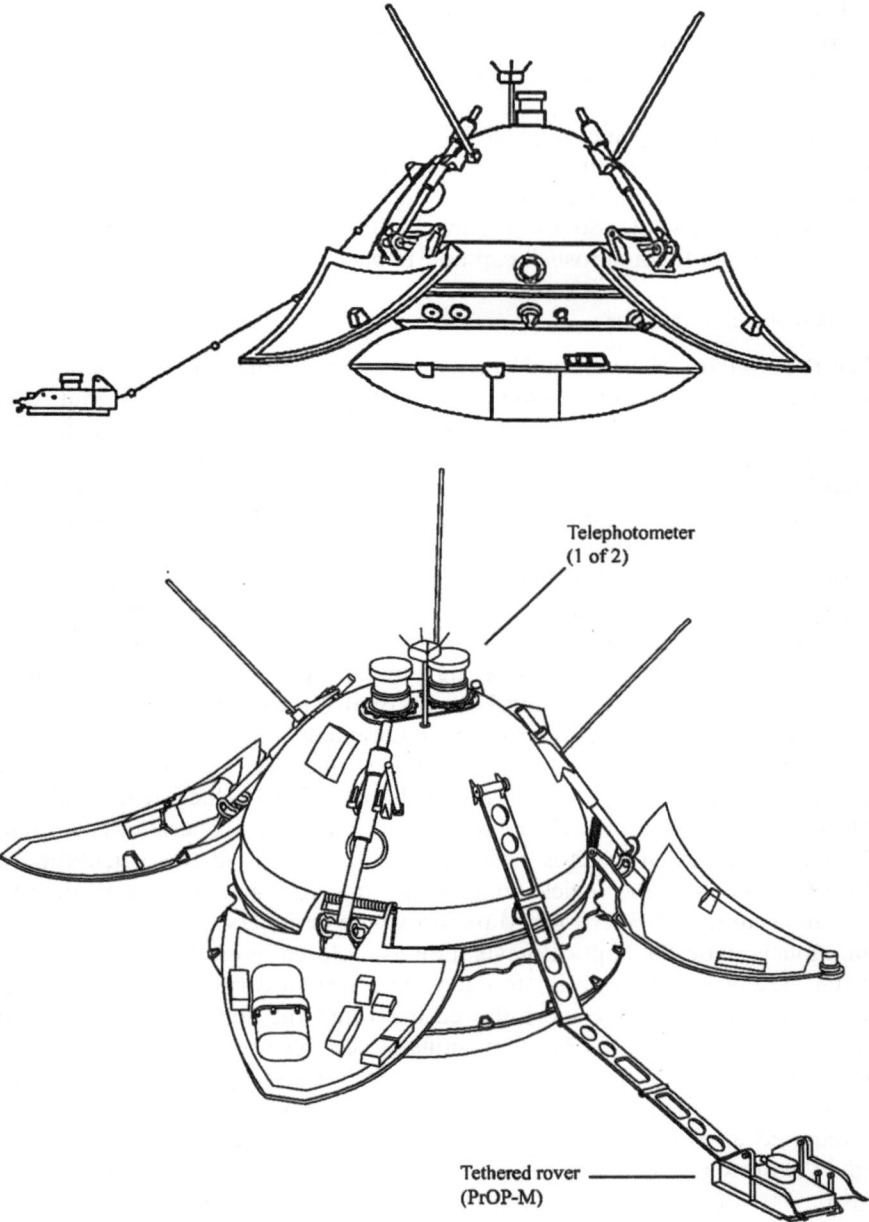

Figure 12.12 Lander diagrams showing surface deployments and impact shock absorber (from Ball et al.).

arm. The lander was powered with batteries that would be charged by the orbiter prior to separation. Temperature control was by thermal insulation covering the exposed portions and a system of radiators. It was designed to survive the chill of the Martian night.

The entire lander capsule weighed 358 kg and was sterilized prior to launch by germicidal lamps to prevent contamination of the Martian environment. It was tested using catapults and rated for horizontal speeds of 28.5 m/s, vertical speeds of 12 m/s and impacts of 180 G. Figure 12.10 shows it with petals closed and encapsulated in its foam aeroshell cap and impact shock absorber.

Entry, descent and landing:

Rather than having the inbound carrier spacecraft target the atmospheric entry point, release a passive entry system, and then perform a deflection maneuver to reach the position where it would perform orbit insertion, the Soviet mission design targeted the carrier at its insertion point and required a more complex entry system that had a propulsion system with which to maneuver for the requisite atmospheric entry point and angle of attack. The difference in entry strategies for Venus and Mars was due to the nature of their atmospheres. The atmosphere of Venus is so thick that a simple spherical shell with an offset center of mass for attitude alignment is readily able to reduce the entry velocity to subsonic far above the surface. The atmosphere of Mars is rarefied and requires a large conical aeroshell to slow the velocity rapidly enough and high enough in the atmosphere for parachutes and terminal rockets to be able to cancel the residual velocity prior to surface contact. The Martian atmosphere levied stringent requirements on the entry angle: if it were too steep then the vehicle would reach the surface before the various velocity reduction steps could be completed; too shallow, and the vehicle would skip out of the atmosphere. Furthermore the conical shield had to be properly oriented relative to the incoming velocity vector and spin stabilized to hold this orientation. The requirement to deliver the vehicle on a precise trajectory and entry angle despite the lack of an accurate ephemeris for Mars, drove the designers to enable the carrier to autonomously undertake optical navigation as it closed in on the planet and release the entry system just hours prior to entry. The Venera carriers released their entry systems 2 days before entry and followed them into the atmosphere and destroyed themselves. But for the 1971 Mars missions the carrier was to enter orbit. To have maneuvered the entire spacecraft to the trajectory for atmospheric entry, released the entry system, and then performed a deflection maneuver so near the planet would have required a prohibitive amount of propellant. The tradeoff in mass therefore favored the orbiter by complicating the entry system with a maneuvering engine and active 3-axis attitude control capability.

Figures 12.13 to 12.17 illustrate the approach, separation, trajectory correction, entry, descent and landing sequences for the Mars-71 entry system. All events after the entry system separates from the orbiter occur automatically, without command from Earth. The entry mission begins with the pyrotechnic separation of the entry system from the orbiter at a distance from Mars of about 46,000 km. At this time the

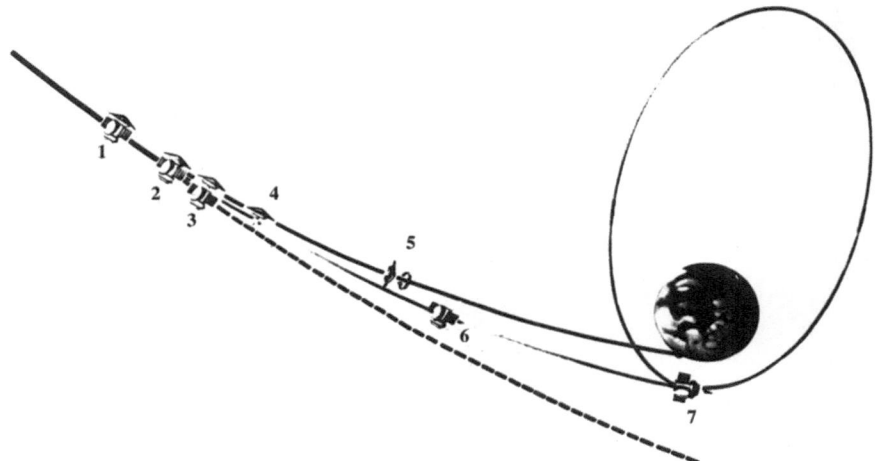

Figure 12.13 Mars-71 approach and targeting sequence: 1. First optical navigation measurement at ~70,000 km range to update orbiter and entry vehicle trajectory parameters; 2. Trajectory correction maneuver (the third since leaving Earth) to target the orbiter, with a velocity change of less than 100 m/s changing the periapsis from ~2,350 ± 1,000 km to 1,500 ± 200 km; 3. Entry vehicle separation about 6 hours before entry; 4. Entry vehicle trajectory correction maneuver to target entry vehicle. Entry angle accuracy ~5 deg, velocity change ~100 m/s, propulsion system ejected post maneuver; 5. Entry vehicle reorientation to entry attitude and spin-up; 6. Second optical navigation measurement at ~20,000 km range to update orbit insertion parameters; 7. Mars orbit insertion maneuver. Velocity change ~1,190 m/s, orbital period accuracy ~2 hrs.

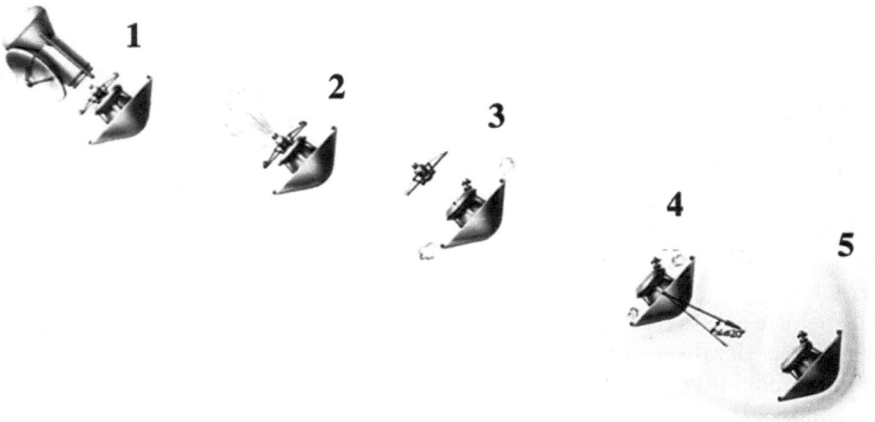

Figure 12.14 Mars-71 entry sequence: 1. Entry system separation 6 hours from entry; 2. Solid rocket ignition to retarget from flyby to entry trajectory; 3. Separation of the propulsion system and spin-up; 4. Spin-down after peak deceleration; 5. Aerobraking.

Figure 12.15 Mars-71 pilot parachute braking sequence: 1. Accelerometer initiates descent program timer at t = 0, auxiliary parachute cover is severed and extraction rocket is ignited; 2. Drogue parachute and cover is extracted from its container; 3. Drogue parachute shroud line is extracted from the container and tension built up in suspension lines; 4. Drogue parachute is released from the extraction mechanism and opened at t = 0.7 sec; 5. Top half of the toroidal main parachute cover is severed and drawn away; 6. Main parachute is extracted with shroud lines attached to the bottom half of the toroidal compartment; 7. Main parachute is deployed, but reefed by a ripcord to prevent overload. Descent science instruments activated at t = 3.1 sec.

entry system is under 3-axis attitude control. After 900 seconds (by now hopefully a safe distance from the orbiter) the main solid rocket is fired to provide an impulse of 120 m/s and adopt the required entry trajectory. 100 seconds later, the vehicle rotates to the proper entry attitude. After another 50 seconds, a set of solid micro-engines on the aerobrake rim are ignited, each delivering 0.5 kN for 0.3 second to spin up the vehicle to 10 rpm. Then the propulsion ring assembly is jettisoned, taking with it the attitude control system and the mounting bars. The spin-stabilized vehicle coasts to its target.

The vehicle enters the atmosphere at about 5.8 km/s. When the load drops to 2 G after peak deceleration, spin stabilization is no longer required and the second set of solid micro-engines on the aerobrake rim are fired to de-spin the vehicle. After about 100 seconds, at a preset G equivalent to about Mach 3.5, an accelerometer triggers the start of the descent program timer at t = 0 and deploys the 13 square meter drogue parachute. The toroidal section is bisected at t = 2.1 seconds and its top half is pulled away by the drogue, drawing out the main parachute. The drogue is then released. The 140 square meter main parachute is reefed to prevent over stressing it

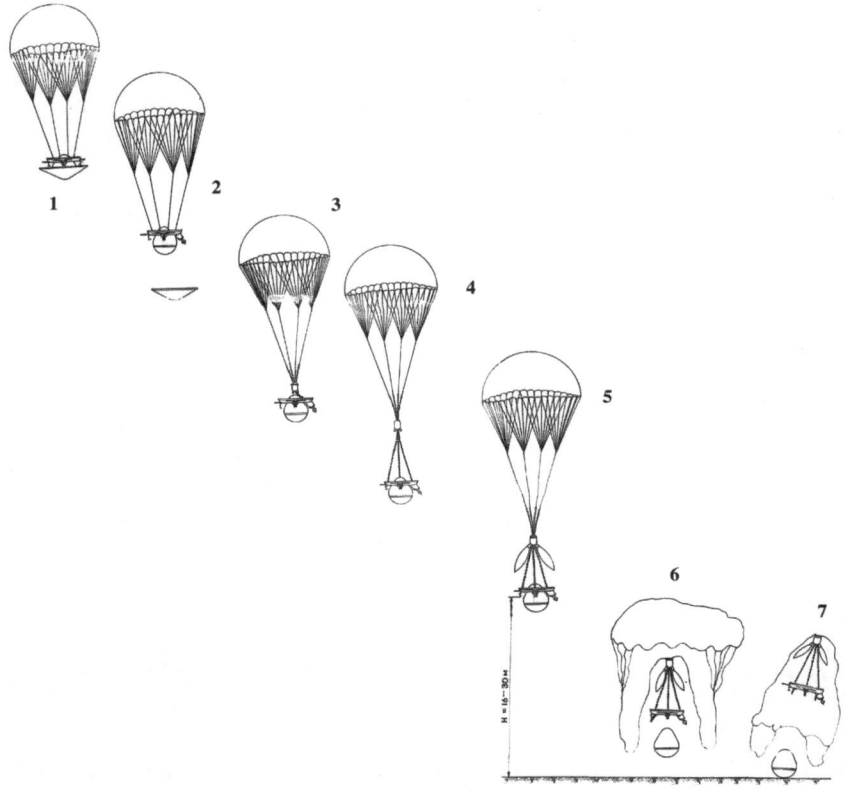

Figure 12.16 Mars-71 main parachute descent sequence: 1. Ripcord cut at 12.1 sec to fully open the main parachute; 2. Heat shield separated at t = 14 sec. At t = 19 sec the high altitude radar altimeter is activated; 3. At t = 25 sec, pyros are fired to release the terminal rocket; 4. The main parachute extracts the rocket on a new set of shroud lines. At t = 27 sec the low altitude radar is activated; 5. After 30 to 200 seconds on the parachute, at a height of 16 to 30 meters the low altitude radar turns off the descent science instruments and ignites the terminal landing rockets; 6. The parachute is carried away by another rocket and the lander is dropped; 7. The lander free falls to the surface.

at such a high speed. The descent science instruments are activated at t = 3.1 seconds. At t = 12.1 seconds, after the speed has become subsonic, the reef lines are cut and the canopy opens fully. The aerobrake is jettisoned at t = 14 seconds.

The high altitude radar is activated at t = 19 seconds and a descent rate of about 65 m/s. At t = 25 seconds the lower shroud lines are withdrawn from the toroid with the terminal solid rocket system at their top, and at t = 27 seconds the low altitude radar is activated. After 30 to 200 seconds on the parachute, at a height of 16 to 30 meters the radar triggers the landing sequence in which, in rapid succession, a second timer is initiated, the descent science instruments are turned off, the lander terminal

Figure 12.17 Mars-71 landing sequence: 1. The terminal rockets are ignited and another rocket carries the parachute away; 2. The lander is dropped and comes to rest on the surface; 3. The displacement balloon inflates to separate the top cover of the lander (at right); 4. Petals open on the upper hemisphere to stabilize the lander, the antennas and booms are deployed, and the science package is activated.

solid rocket is ignited to deliver 56 kN for 1.1 seconds, and the parachute is carried away by a second rocket that fires for 1 second and delivers a thrust of 9 kN. After terminal rocket firing, the lander is released to fall to the surface and two small rockets on the side of the terminal rocket container deliver a horizontal impulse of 1 kN for 4 seconds in order to prevent it from falling onto the lander. Meanwhile, the lander should impact at a vertical velocity no greater than 12 m/s.

Fifteen seconds after the lander makes physical contact with the surface, a timer commands the ejection of the foam cap covering the petals and initiates the lander's sequence. This deploys the four petals, antennas, and booms, and starts to transmit to the main spacecraft at a rate of 72 kbits/s on two independent VHF channels. This communication session lasting about 20 minutes has to occur before the spacecraft makes its insertion maneuver. It includes a panoramic image of 500 × 6,000 pixels. The lander is then powered down, as it will be between all communications sessions. The sessions are initiated by timer and may be as short as 1 minute depending on the location of the site, the nature of the terrain, and the mutual orbiter/lander positions. The lander was designed to operate for several local days.

The entire descent sequence was tested by fifteen M-100B sounding rocket flights using scale models dropped from 130 km.

Payloads:

M-71S orbiter:

The scientific payloads of the orbiters were almost identical, except that the M-71S and Mars 3 spacecraft both had the French STEREO instrument to measure solar outbursts. This was the first time a Soviet spacecraft carried a Western instrument. However, the Soviets still guarded their secrecy and the French simply handed over the equipment "at the border". They were not involved in its integration and testing. In fact, they were not shown any drawings, and were not told where and on which spacecraft the instruments would be mounted. The loss of the M-71S orbiter left this experiment with only one instrument, compromising the stereoscopic aspect of the project.

Mars 2 and 3 orbiters:

Most of the orbiter scientific instruments were mounted in the hermetically sealed instrument module, and were generally intended to be operated for 30 minutes near each periapsis. Others were externally mounted or had externally mounted sensors for in-situ investigation of the space environment:

1. FPU dual camera facsimile imaging system
2. Infrared radiometer (8 to 40 microns) for measurement of surface temperatures
3. Infrared narrow-band 1.38 micron photometer for measurement of water vapor content in the atmosphere
4. Infrared spectrometer in the 2.06 micron absorption band of carbon dioxide to measure atmosphere optical thickness and as an indicator of surface topography
5. Ultraviolet photometer with filters in the intervals 1,050 to 1,180, 1,050 to 1,340 and 1,225 to 1,340 angstroms to detect atomic hydrogen, oxygen, and argon
6. Lyman-alpha photometer (French-Soviet) for measurement of upper atmosphere hydrogen
7. Six channel visible photometer in range 0.35 to 0.7 microns for measurement of color and albedo of the surface and atmosphere
8. Microwave radiometer (3.4 cm) for measurement of dielectric constant and subsurface temperatures to depths of 25 to 50 cm
9. Radio science investigation to determine atmospheric structure (temperature and density profiles)

10. Cosmic ray charged particle detector consisting of a Cherenkov counter, four gas discharge detectors and seven silicon solid-state detectors
11. Solar wind plasma sensors (8) for measurement of speed, temperature and composition in the 30 eV to 10 keV energy range
12. Boom mounted three-axis fluxgate magnetometer
13. STEREO instrument on M-71S and Mars 3 to measure solar radiation outbursts at 169 MHz in conjunction with Earth-based receivers (French-Soviet).

The Mars 2 and 3 photo-television imaging system was an improvement over the M-69 system, and consisted of two bore-sighted film cameras, one with a 52 mm wide angle lens and several color filters and the other with a 350 mm narrow angle lens and an orange filter. At the planned periapsis altitude, surface resolutions of 100 to 1,000 meters were expected. There was film for 480 images, most of which were pre-programmed for the first 40 days of the orbital mission.

The science instruments on the Mars 3 orbiter weighed a total of 89.2 kg.

Mars 2 and 3 entry systems:

A radio altimeter attached to the toroid provided data during the descent. The lander payload had a mass of 16 kg and consisted of:

1. Accelerometer for atmospheric density during entry
2. Temperature and pressure sensors for descent and landing
3. Radio altimeter for providing altitudes on descent
4. Mass spectrometer for atmospheric composition on descent and landing
5. Atmospheric density and wind velocity on the surface
6. Two panoramic television cameras for stereo viewing of the surface
7. X-ray spectrometer for soil composition deployed to the surface from a petal
8. PrOP-M walking robot deployed to the surface from this same petal with onboard gamma-ray densitometer and conical penetrometer.

The cameras were similar to those of the Luna 9 lander with a single photometer and a scanning mirror that tilted to scan vertically and rotated to scan horizontally, returning a single brightness value for each scan position. A full panorama spanned 500 × 6,000 pixels. The mass spectrometer was an early form of the Bennett radio-frequency instrument being developed for Venera 9 and 10. There was no telemetry during the descent. All data obtained during this time was stored for transmission in the communication session programmed for immediately after touchdown.

The 4.5 kg PrOP-M rover was a box 250 × 250 × 40 mm with a small protrusion rising from the center of its upper surface. The body was supported by two skis, one projecting down from each side. By moving the skis in alternating fashion the rover was able to 'walk', and by moving them in opposite directions it could turn. There were obstacle-sensing bars at the front, and it was programmed to reverse in order to circumnavigate an obstacle. The rover was to be deployed by a 6-joint manipulator arm and moved into the field of view of the cameras. It was tethered by a 15 meter long cable for direct communication with the lander, and was to pause at intervals of

Figure 12.18 PrOP-M 'Marsokhodnik' rover.

1.5 meters to make measurements. It carried a dynamic penetrometer and a gamma-ray densitometer, and its tracks were to be photographed to investigate the physical properties of the surface.

Mission description:

M-71S:

The Soviets must have breathed a sigh of relief on May 8, 1971, when the launch of Mariner 8 failed. Their plan was for the M-71S spacecraft to arrive at Mars and enter orbit before the two US spacecraft arrived, and their chances of achieving this had just improved. The M-71S orbiter was launched two days later, on May 10, but the failure of the Block D to reignite due to an ignition timing error – "a most gross and unforgivable mistake" – left the spacecraft stranded in parking orbit. The timer was intended to have been set to reignite the engine 1.5 hours after the Block D achieved orbit, but the 8-bit code was erroneously specified as 150 hours by the programmer who input the command with the bits in reverse order. The coupled spacecraft and stage was named Cosmos 419 by the Soviets to hide its purpose. It re-entered 2 days later.

This failure not only cost the Soviets a chance to be the first to orbit Mars, it also threatened the success of the Mars 71 campaign because it meant there would be no radio beacon orbiting the planet to assist in refining the trajectories of the spacecraft carrying landers. The French were not informed of the loss of their first STEREO instrument. The Soviets would have to resort to the backup method of correcting the trajectory, which was less accurate and much more risky. Lacking an accurate Mars ephemeris to calculate a pre-determined release point and how to orientate the entry system in relation to Mars, each approaching spacecraft would have to use on board optical sensors to determine its position relative to the planet and then calculate for itself the release point, the trajectory correction required to reach this point, and the orientation that the entry system must adopt in readiness for atmospheric entry. This autonomous procedure using an optical navigation instrument had been developed as a back up contingency, but the M-71S failure made it the only option. It was bold, very complex, highly sophisticated, and far ahead of its time. Several decades would pass before American mission designers adopted automated optical navigation; had they known so at the time they would have been aghast at its use for Mars 2 and 3.

The mission plan for the Mars-71 campaign allowed as many as three midcourse correction maneuvers, but nominally used only two, the first soon after leaving Earth and the second on approaching Mars. Another correction now became essential, and was dedicated to the autonomous entry system targeting procedure. The first step, at about 70,000 km from Mars, would be to make the optical navigation observations required to correctly target the entry system. After a new vector had been calculated and the course corrected, the entry system would be released to pursue its standard procedures. The main spacecraft would then undertake a second optical observation about 20,000 km from Mars in order to identify any change required for the orbit insertion maneuver. All of these operations were to be performed autonomously.

Mars 2:

After a successful launch on May 19, 1971, the first trajectory correction maneuver was conducted on June 5. Almost simultaneously on June 25 communications with both Mars 2 and Mars 3 in the primary decimeter band were lost, evidently owing to problems with the transmitters. After working for a brief period the decimeter back-up transmitter also failed on Mars 2. It proved impossible to activate its centimeter band telemetry system. The primary decimeter transmitter remained unreliable, but conditions were identified in which the back-up transmitter could be made to work. The loss of the centimeter band system was never understood, but it worked reliably on subsequent missions. There were no further incidents and on November 21, with 6 days remaining to arrival, Mars 2 performed an optical navigation sequence and 7 hours later made its second trajectory correction. The third maneuver, to target the entry system, was made on November 27 but it proved to be fatally imprecise. After being released 4.5 hours before the main spacecraft was to perform its orbit insertion maneuver, the entry system ran through its standard procedures. The orbiter made a trim burn, then the 1.19 km/s insertion maneuver and settled into a 1,380 × 24,940 km orbit inclined at 48.9 degrees. The problem with the third targeting maneuver resulted in a lower apoapsis than intended, with a period of 18 instead of 24 hours.

Meanwhile, having entered the Martian atmosphere at a velocity of approximately 6.0 km/s at a steeper angle than planned, the descent system malfunctioned and the lander hit the surface before it could deploy its parachute. It fell at 44.2°S 313.2°W, delivering a coat of arms of the USSR. Post-flight analysis showed that the computer codes were not sufficiently developed owing to lack of development time to address all situations, including that faced by Mars 2 in which the trajectory prior to the third correction was fairly close to that desired and the ensuing procedure over-corrected and produced an overly steep entry angle.

Mars 3:

Mars 3 was launched on May 28, 1971, and performed its first midcourse correction on June 8. The primary decimeter band transmitter failed on 25 June, but the back-up functioned. The cruise was uneventful, and on November 14 the spacecraft made a second midcourse maneuver. On approaching Mars on December 2 it executed the autonomous final targeting. At 09:14 UT, some 4 hours 35 minutes prior to orbital insertion, the spacecraft cut loose the entry system. Fifteen minutes later, the entry system performed its separation maneuver and adopted the required orientation. At 13:47 UT it entered the Martian atmosphere at 5.7 km/s at an entry angle of less than 10 degrees. The drogue parachute was deployed. This drew out the main parachute, which remained reefed until the speed became subsonic and the canopy could fully open. The heat shield was jettisoned and the low altitude radar was activated. At a height of 20 to 30 meters, falling at 60 to 110 m/s, the parachute was discarded and a small rocket lifted it away from the lander. Simultaneously, the lander fired its own retro-rockets. After a descent lasting a little over 3 minutes, Mars 3 touched down at 13:50:35 UT at a speed of 20.7 m/s. The landing site was at 44.9°S 158.0°W, in the planned area.

The foam cover was immediately ejected and the four petals opened. At 13:52:05 UT, 90 seconds after landing, the capsule began to transmit to its parent. However, after 20 seconds the transmission ceased and no further signals were received. It was several hours before the main spacecraft, which had to devote its attention to making the orbit insertion maneuver, was able to replay to Earth the transmission that it had recorded from the lander. The partial image returned by the lander is uninterpretable, being essentially noise. The only real information was an imaging calibration signal. The cause of this loss of signal may have been related to the planet-wide dust storm that was raging at the time. This would also explain the bland image lighting. It has been suggested that the transmitter failed due to coronal discharge in the dusty, low-pressure atmosphere. In any event, because the data collected during the descent was stored on board the lander for transmission in that first communication session this was lost as well.

Meanwhile a computer programming error caused the Mars 3 orbiter to cut short the insertion burn and it ended up in a 1,530 × 190,000 km orbit that had a period of 12.79 days instead of 25 hours. As a result there were only seven opportunities for periapsis observations during its limited operating life. As in the case of Mars 2, the inclination of the orbit was 49 degrees.

In the 4 month interval between December 1971 and March 1972 the two orbiters transmitted a large amount of science data. Mars 2 had the better orbit for planetary observations but, still suffering communications problems, its telemetry was of poor quality and almost all of the planetary data were lost except radio occultations as the spacecraft crossed the planetary limb. The telemetry system on Mars 3 was working properly, although its impulse transmitter was malfunctioning. Its orbit was ill-suited for planetary observations but Mars 3 was able to return useful planetary data. After the science observations finished in March, both orbiters continued to operate until contact was lost almost simultaneously in July 1972 when their attitude control gas ran out. The missions were announced to have been completed on August 22, 1972, by which time Mars 2 had made 362 of its shorter than intended orbits and Mars 3 only 20 of its exceedingly long orbits.

These spacecraft were highly sophisticated engineering marvels. They were the first of a new generation of large, complex spacecraft designed for comprehensive and bold investigation of our planetary neighbors. Their success on this initial outing led to a whole new generation of spacecraft for exploring the planets and conducting astrophysical investigations.

Results:

Orbiters:

Imagery

The Mars 2 and 3 orbiters suffered from a combination of circumstances. First, the telemetry systems had some problems. Very little telemetry at all was received from Mars 2. The Mars 3 impulse transmitter failed, and only lower resolution 250-line images were returned using the PCM decimeter band transmitter. Then there was the dust storm that began in October and had fully engulfed the planet by the time the spacecraft arrived. Third, the imaging sequences were pre-programmed, and with all but the very tallest mountain summits obscured imaging was impractical. Lastly the cameras had been set at the wrong exposure. And once the ampoules containing the chemicals to process the film were opened, the time available for photography was limited. Nearly all of the Mars 3 imagery was returned in four batches. The first two batches taken on 10 and 12 December 1971 showed very little detail due to the dust storm. Due to control system problems the next two batches were postponed to 28 February and 12 March 1972, by which time the dust storm had abated. A total of 60 pictures were returned, including color images of volcanoes whose summits rose as high as 22 km and depressions as deep as 1.2 km, but the image quality was rather poor.

Only one picture was released during the mission, a relatively featureless view of the whole planet taken from the apoapsis of Mars 3's extremely eccentric orbit. The imaging results of the Soviet missions paled in comparison to the 7,000 pictures that Mariner 9 provided, showing about 70% of the planet in unprecedented detail. The flood of orbital data from the American spacecraft revealed a much more interesting Mars than the dry, cratered, Moon-like perception created by the Mariner 4, 6 and 7 flybys. The canyons, dry river beds, flood plains and volcanoes imaged by Mariner 9 hinted at a much wetter past and raised the prospect of there being subsurface water and maybe even life. The accomplishments of the Mars 2 and 3 orbiters were lost in the glare of Mariner 9, and the Soviets could only think about what might have been had they been blessed with a little more luck.

Dust storm

The dust storm abated in late January 1972 allowing the orbiting cameras a view of the surface, but it was many months before the very light particles of dust settled out of the atmosphere. Dust clouds were found to extend to altitudes of 10 km, but were not evenly distributed around the planet. Dust particle sizes were determined, and small micron-sized dust grains were found as high as 7 km in the atmosphere during the storm. Bright ultraviolet clouds indicated the presence of even smaller particles at higher altitudes. During the dust storm the water vapor content of the atmosphere was very low, on the order of a few precipitable microns. After the storm the water vapor content increased to 20 microns, with greater humidity at the equator than in the northern polar region. The dust diverted a significant amount of sunlight, and the surface temperatures rose by about 25°C after the atmosphere had cleared.

Upper atmosphere and ionosphere

Temperature and pressure profiles of the upper atmosphere were obtained from the frequent Mars 2 radio occultations, as was data showing there to be a neutral upper atmosphere made almost entirely of carbon dioxide with about 2% atomic oxygen. A night-side airglow was detected 200 km beyond the terminator. The base of the ionosphere was at 80 to 110 km. From 100 to 800 km, the carbon dioxide became increasingly dissociated into atomic oxygen and carbon monoxide. And the Lyman-alpha experiment detected an atomic hydrogen corona out to 20,000 km. There was less atomic hydrogen than Mariner 6 and 7 had found in 1969, presumably because the hydrogen atoms derived from water dissociation and during the dust storm there was less water present in the atmosphere. Charged particles were measured in the ionosphere, and the bow shock that defines the interaction of the solar wind with the planetary ionosphere was detected.

Lower atmosphere

The data obtained included temporal and spatial changes of temperature in the lower atmosphere, and water vapor concentrations 5,000 times weaker than in the Earth's atmosphere. Images of the limb showed the layered structure of the atmosphere and its extent out to 200 km. Clouds were observed in the lower atmosphere composed of sub-micron particles at altitudes as high as 40 km. The composition was reported to be 90% carbon dioxide, 0.027% molecular nitrogen, 0.02% molecular oxygen, 0.016% argon, and water vapor variable in the range 10 to 20 precipitable microns.

Surface topography

Altimetry based on the column density of carbon dioxide measured by the infrared photometer in the 2.06 micron absorption band was obtained along the orbiter tracks across the surface. The inferred altitudes were in general agreement with terrestrial radar observations.

Surface properties

The large diurnal variations of surface temperature indicated a low heat conductivity characteristic of a dry and dusty surface. Latitudinal surface temperature variations ranged from -110°C at the northern polar cap to +13°C near the equator. Equatorial temperatures averaged -40°C, and at 60°S latitude they were -70°C without much diurnal variation. Dark areas on the surface were 10 to 15 degrees warmer than the light areas. The surface cooled rapidly during the night in low latitudes, indicating a dry, porous soil with a low thermal conductivity. Subsurface temperatures down to a depth of 0.5 meter were no higher than -40°C. There were thermal 'hot spots' some 10°C warmer than their surroundings. Temperatures at the northern polar cap were close to the carbon dioxide condensation temperature. Surface pressures of 5.5 to 6 millibars were measured. Soil density, heat conductivity, dielectric permeability and reflectivity were derived from microwave and thermal radiometry. Soil densities of 1.2 to 1.6 g/cc were reported, with values increasing to 3.5 g/cc in some places. The surface was presumed to be covered with silicon dioxide dust to an average depth of about 1 mm. Heat flow anomalies on the surface were discovered.

Global properties

Global data on the Martian gravity and magnetic fields was acquired. No intrinsic planetary magnetic field was detected, and plasma data for the interaction of the ionosphere with the solar wind indicated a magnetic moment at least 4,000 times weaker than that of Earth. A key discovery were large local mass concentrations in the gravity field, similar to those of the Moon, which created significant changes in the orbits of the spacecraft. In addition, the polar diameter was measurably less than that at the equator.

Landers:

Although the Mars 2 lander crashed, it is significant as the first human artifact to reach the surface of Mars.

The Mars 3 lander gained the distinction of being the first successful landing on Mars, but it fell silent almost immediately. Figure 12.19 shows the data returned by

Figure 12.19 Image from the Mars 3 lander.

the scanning-photometer imager, released in recent years, which analysis indicates to be mostly noise.

THE YE-8 LUNAR ORBITER SERIES: 1971–1974

Campaign objectives:

In addition to the lunar surface rover and sample return missions, the Ye-8 modular spacecraft also flew orbital missions to support the engineering requirements of the planned manned missions. The principal requirements were to photograph the lunar surface at high resolution and conduct remote surface composition measurements to assist in selecting landing sites. A secondary objective was to acquire data on the radiation and plasma in lunar orbit in order to understand the risks to humans. Two Ye-8LS orbiters were launched successfully as Luna 19 and 22. Their tracking data continued the accurate mapping of the lunar gravity field that Luna 14 initiated.

Spacecraft launched

First spacecraft:	Luna 19 (Ye-8LS No.202)
Mission Type:	Lunar Orbiter
Country/Builder:	USSR/NPO-Lavochkin
Launch Vehicle:	Proton-K
Launch Date/Time:	September 28, 1971 at 10:00:22UT (Baikonur)
Encounter Date/Time:	October 3, 1971
Mission End:	October 3, 1972
Outcome:	Success.
Second spacecraft:	Luna 22 (Ye-8LS No.206)
Mission Type:	Lunar Orbiter
Country/Builder:	USSR/NPO-Lavochkin
Launch Vehicle:	Proton-K
Launch Date/Time:	May 29, 1974 at 08:56:51 UT (Baikonur)
Encounter Date/Time:	June 2, 1974
Mission End:	November 1975
Outcome:	Success.

Spacecraft:

The spacecraft was essentially the same as the lander stage for the lunar rover, but with a payload consisting of a pressurized module for the orbital instruments. This was a squat cylinder and, just like the Lunokhod, had a hinged lid that exposed solar panels on the underside.

Figure 12.20 Luna 19 spacecraft.

Luna 19 launch mass: 5,700 kg
Luna 22 launch mass: 5,700 kg

Payload:

1. Imaging system
2. Gravitational field experiment
3. Gamma-ray spectrometer for surface composition
4. Radiation sensors
5. Magnetometer
6. Micrometeoroid detector
7. Altimeter
8. Radio occultation experiment

For these orbiters, new linear-scan cameras were developed based on the Luna 9 and 13 panoramic imagers. Basically the motion of the spacecraft provided the long axis of the image and the photometer scanned only perpendicular to the direction of orbital motion. The field of view was 180 degrees centered on the nadir and gave a 'cylindrical fish-eye' image. At 4 lines per second from an altitude of 100 km it had a resolution of 100 meters in the direction of travel and 400 meters perpendicular to it. Luna 22 carried an additional camera and engineering tests of solid lubricants for operation in vacuum and wafer tests of surface reflection properties.

Mission description:

Luna 19 was launched on September 28, 1971, and on October 3 it entered a 2 hour circular lunar orbit at 140 km inclined by 41 degrees. Three days later the orbit was changed to an elliptical one of 127 × 385 km. Several months later the perilune was lowered to 77 km to undertake closer photography. After more than 4,000 orbits the spacecraft ceased operations on October 3, 1972.

Luna 22 was launched on May 29, 1974, and entered a 219 × 221 km lunar orbit at an inclination of 19.6 degrees on June 2. It made many orbit adjustments over its 18 month lifetime to optimize experiment operation, at times lowering its perilune to 25 km to improve photography. Sporadic contact was maintained after the supply of attitude control gas ran out on September 2, and the mission was concluded in early November 1974.

Figure 12.21 Section of a panoramic image from Luna 19 of Sinus Aestuum with the crater Eratosthenes at the right.

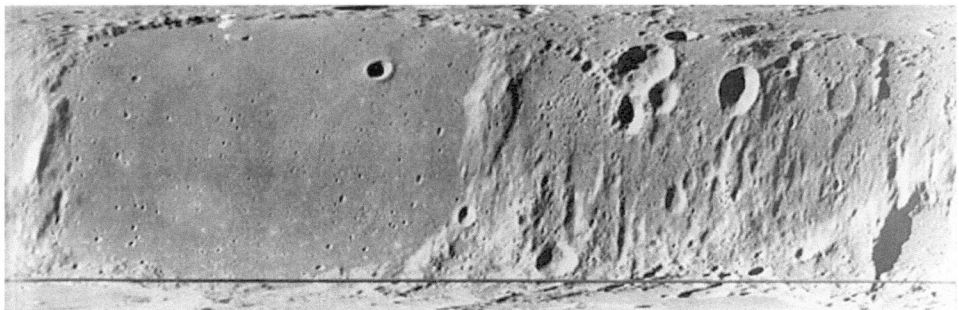

Figure 12.22 Luna 22 panorama fragment illustrating the 'cylindrical fish-eye' effect of the linear scanning photometer imager.

Results:

Luna 19 and 22 both returned images of the lunar surface from orbit, with Luna 19 apparently returning about 5 panoramas and Luna 22 ten panoramas. Both spacecraft extended the systematic study to locate mass concentrations (mascons) begun by the earlier Luna orbiters. They also remotely sensed the composition of the surface and directly measured the properties of the orbital environment including the radiation, plasma, magnetic fields and micrometeoroid flux. Altimetry measurements of lunar topography were made, and the electromagnetic properties of the regolith examined. The results must have been substantial but few results were published, particularly from Luna 22.

13

Closeouts on a Venus spacecraft, a Moon rocket, and desperation at Mars

TIMELINE: MAR 1972–DEC 1973

The year 1972 opened with the last launch of a 3MV spacecraft, Venera 8, marking the final use of the Molniya launcher for a Soviet lunar or planetary mission. Using Venera 7's measurement of the temperature at the surface, and an inferred pressure, the over-engineered descent capsule was simplified to enable Venera 8 to carry more instruments. It was dispatched on March 27 and the lander operated successfully on the surface of Venus. Two other significant closeouts were in 1972, the launch of the fourth N-1 rocket in November whose failure led to that project being canceled, and the final Apollo landing in December.

The second and final robotic lunar rover, Lunokhod 2, was delivered to the Moon by the Luna 21 mission in January 1973. Apart from this, the year 1973 witnessed perhaps the most frustrating campaign in the history of robotic planetary flight. The Soviets were acutely aware of US plans to launch sophisticated orbiter/landers to Mars in 1975. Encouraged by the near-success of the Mars 3 lander, they devised an audacious campaign for 1973. This opportunity was less favorable than that of 1971, which precluded sending orbiter/landers, so instead they launched four spacecraft in July and August: Mars 4 and 5 were orbiters, and Mars 6 and 7 carried landers to be released during flybys. During the development of these spacecraft, the Soviets had incorporated a new transistor into many spacecraft systems. These were discovered before launch to be faulty, with limited lifetimes due to technological "innovations" in their manufacture. They could not be trusted to last for the duration of the flight to Mars, yet they could not be replaced in time to achieve the launch window. It was a terrible dilemma. Rather than stand down and wait for the next Mars opportunity in 1975, when the US missions would be launched, the Soviets decided to proceed and hope that at least one of the landers would succeed. But the transistor lifetime tests proved to be accurate, and there were failures in all four spacecraft which ultimately resulted in disaster for this campaign.

Launch date

1972

2 Mar	Pioneer 10 Jupiter flyby	Successful flyby on Dec 3, 1973
27 Mar	Venera 8 entry probe	Success
31 Mar	Venera entry probe	Fourth stage failure
16 Apr	Apollo 16 lunar landing	Success
23 Nov	N-1 Moon Rocket test	Booster exploded, 4th & final test
7 Dec	Apollo 17 lunar landing	Success, last Apollo flight

1973

8 Jan	Luna 21 rover	Success, Lunokhod 2
5 Apr	Pioneer 11 Jupiter/Saturn	Successful Jupiter and Saturn flybys
21 Jul	Mars 4 orbiter	Failed to achieve orbit, flew past Mars
25 Jul	Mars 5 orbiter	Completed 22 orbits, early failure
5 Aug	Mars 6 flyby/lander	Successful descent, lost at landing
9 Aug	Mars 7 flyby/lander	Entry system failed, flew past Mars
3 Nov	Mariner 10 Venus/Mercury	Successful flybys at Venus & Mercury

Crippled by ailing systems, Mars 4 was unable to enter Mars orbit and sailed past the planet. Mars 5 achieved orbit, but fell silent after only 22 orbits. After an early failure in its telemetry system, the Mars 6 flyby spacecraft had to be commanded in the blind. It miraculously made automated midcourse maneuvers, optical navigation on approaching Mars, and dispatched its entry system on the correct trajectory. The entry system entered the atmosphere, descended by parachute, and sent back mostly garbled data on its compromised communications system. Contact was lost when the lander was released near the surface. The Mars 7 flyby spacecraft released its entry system as planned, but this failed to function and missed the planet. The lack of any real results from this massive campaign was not explained until after the collapse of the USSR.

Meanwhile the US launched Pioneer 10 on March 2, 1972, for the first mission to the outer Solar System. It went on to make a flyby of Jupiter, the largest of planets. In November 1973 Mariner 10 was launched to use a flyby of Venus for a gravity-assist to reach Mercury. And Pioneer 11, launched in April 1973, went on to use this same technique at Jupiter to reach Saturn.

SCIENCE ON THE SURFACE OF VENUS: 1972

Campaign objectives:

The Venera 7 landing on the surface of Venus was a jubilant success for the Soviets. Once the pressure and temperature at the surface were finally confirmed, the NPO-Lavochkin engineers scaled back the pressure design limit from 180 bar to 105 bar and used the saved mass for a stronger parachute and more scientific instruments. In anticipation of their next generation of larger more complex Venus landers, a photometer was added to determine the illumination at the surface. All of the previous entry probes had been targeted at the night-time hemisphere, mainly to ensure direct-to-Earth communications, but day-side light-level measurements were required in order to design imagers for future landers. So the 1972 missions were to land in early morning daylight at sites near the terminator from which it would still be possible to transmit to Earth. A redundant deployable antenna was included as a precaution against poor primary antenna pointing or obscuration of the line of sight by rough terrain.

Spacecraft launched

First spacecraft:	Venera 8 (3V No.670)
Mission Type:	Venus Atmosphere/Surface Probe
Country/Builder:	USSR/NPO-Lavochkin
Launch Vehicle:	Molniya-M
Launch Date/Time:	March 27, 1972 at 04:15:01 UT (Baikonur)
Encounter Date/Time:	July 22, 1972
Outcome:	Successful, transmitted from surface.
Second spacecraft:	Cosmos 482 (3V No.671)
Mission Type:	Venus Atmosphere/Surface Probe
Country/Builder:	USSR/NPO-Lavochkin
Launch Vehicle:	Molniya-M
Launch Date/Time:	March 31, 1972 at 04:02:33 UT (Baikonur)
Outcome:	Failed to depart Earth orbit.

Two launches were attempted, the first successfully dispatching Venera 8 and the second stranding its spacecraft in parking orbit. Venera 8 was the ultimate success of the 3MV series and, as events transpired, the last of its type. It achieved all that the Soviets had worked so hard for over so many years and so many attempts. It was the final reward for dogged persistence. During its construction, NPO-Lavochkin was working on the new generation of Luna spacecraft to undertake sample return, rover and orbiter missions, and the orbiters and landers for Mars, both of which would use the Proton launcher, so this was the final planetary campaign to employ the 8K78M Molniya.

Venera 8 supplied the data needed to design the much more sophisticated landers that would be delivered by the next generation of advanced Venera spacecraft to be launched by the Proton rocket beginning in 1975.

Spacecraft:

The carrier spacecraft for Venera 8 was essentially the same as for all missions since Venera 4, but the entry probe was modified. The pressure design limit was reduced to accommodate additional science instruments and the parachute was strengthened, although the size of the canopy was the same as for Venera 7 in order to make the same rapid descent through the atmosphere. Since the probe was to land further from the center of the planet as viewed from Earth, the antenna transmission pattern was changed from the egg-shape that was appropriate when Earth was at the zenith to a funnel-shape for when Earth was low on the horizon. In case the capsule were to come to rest on its side, a second antenna was provided that was to be ejected onto the surface and this was a flat disk with a spiral antenna on each side to enable it to work irrespective of how it settled.

A new honeycomb composite material was used as the primary insulation of the lander. Further thermal protection was provided by using lithium nitrate trihydrate, a phase-change material that absorbs heat by melting at 30°C. In addition to forming 'thermal accumulators' inside the pressure vessel, this jacketed the instruments that projected outside.

Launch mass:	1,184 kg
Entry capsule mass:	495 kg

Payload:

Carrier spacecraft:

1. Solar wind charged particle detector
2. Cosmic ray gas discharge and solid state detectors
3. Ultraviolet spectrometer for Lyman-alpha measurements

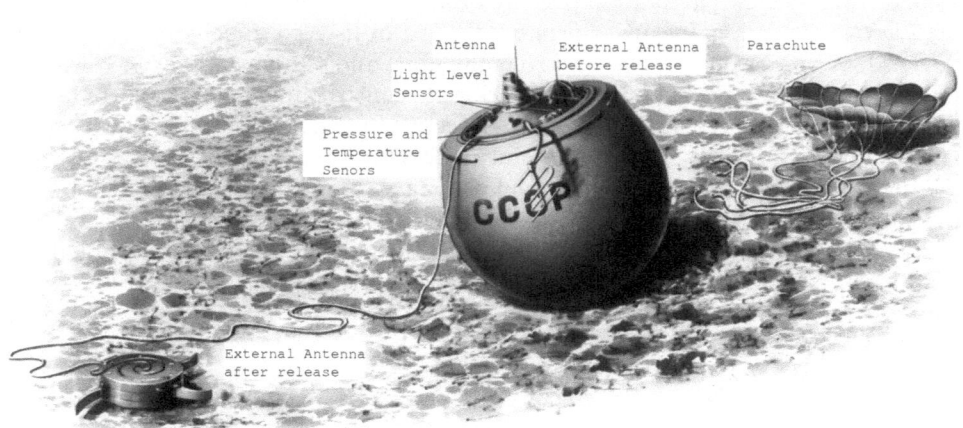

Figure 13.1 Depiction of Venera 8 deployed on the surface with ejected parachute and deployed second antenna (courtesy NPO-Lavochkin).

Figure 13.2 Venera 8 spacecraft in test at Lavochkin.

Descent/landing capsule:

1. Temperature, pressure and density sensors
2. Atmospheric chemical gas analyzer
3. Broad-band visible photometers (2)
4. Gamma-ray spectrometer
5. Radio altimeter
6. Doppler experiment

An accelerometer was to measure atmospheric density during the descent prior to parachute deployment. The altimeter had been redesigned to provide an accuracy of several hundred meters for the instruments that would operate during the parachute descent. The atmospheric composition experiment now included an ammonia litmus

Figure 13.3 Venera 8 probe. Radio altimeter deployed at left, primary antenna in the center, secondary antenna and deployment mechanism at the right. Small cylinders on the rim are the two photometers, one on each side, and the gas analyzer.

test, and the atmospheric structure experiment carried four resistance thermometers, three aneroid barometers, and a capacitance barometer. A pair of single-channel broadband cadmium sulfide photometers were carried to measure the integrated downward flux with a 60 degree field of view in the wavelength range 0.52 to 0.72 microns. The optical unit was outside the capsule, mounted on top inside a separate unit sealed against high pressure and insulated against high temperature. The light reached the electronics by a 1 meter long light guide of fiber optic. The photometers were sensitive over the range 1 to 10,000 lux and encoded logarithmically.

The gamma-ray spectrometer was mounted inside the hermetically sealed probe. It was sensitive to emissions from potassium, thorium and uranium, and had been calibrated for these elements against a suite of Earth rocks.

Mission description:

Venera 8 was launched on March 27, 1972, made its midcourse correction maneuver on April 6, and arrived at Venus on July 22. The solar panels charged the batteries of the capsule and a system in the cruise module pre-cooled the capsule by circulating air through it at -15°C. After being released 53 minutes prior to entry, the capsule hit the atmosphere at 11.6 km/s at 08:37 UT at an angle of 77 degrees on the sunlit side,

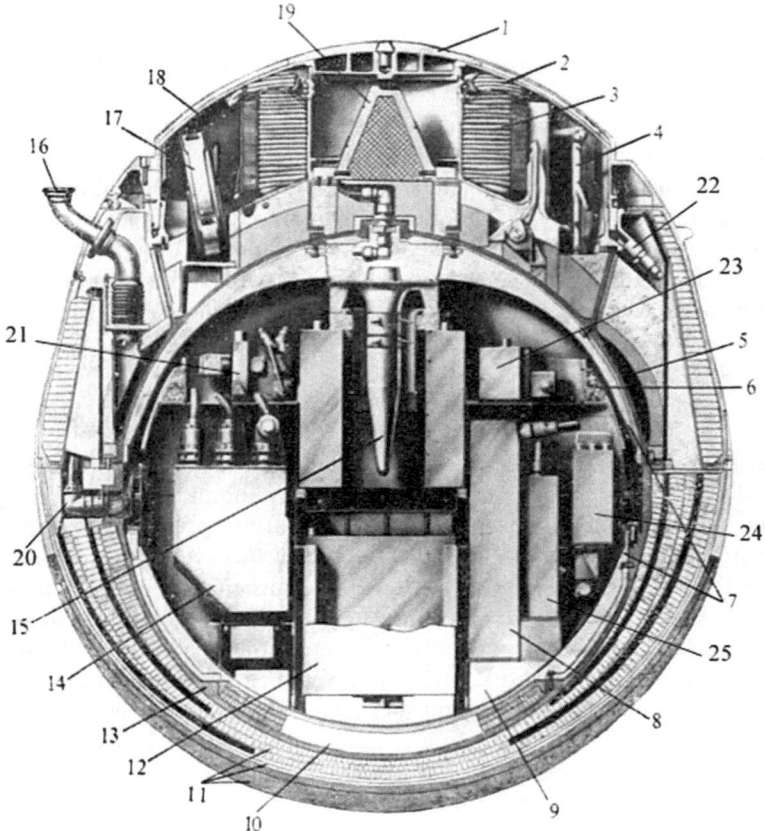

Figure 13.4 Venera 8 probe diagram: 1. Parachute housing cover; 2. Drogue parachute; 3. Main parachute; 4. Deployable radio altimeter antenna; 5. Heat exchanger; 6. Heat accumulator; 7. Internal thermal insulation; 8. Program timing unit; 9. Heat accumulator; 10. Shock absorber; 11. External thermal insulation; 12. Transmitter; 13. Pressurized sphere; 14. Commutation unit; 15. Fan; 16. Cooling conduit from carrier; 17. Deployable secondary antenna; 18. Parachute housing; 19. Primary antenna; 20. Electrical umbilical; 21. Antenna feed system; 22. Cover explosive bolts; 23. Telemetry unit; 24. Stable quartz oscillator; 25. Commutation unit.

approximately 500 km from the morning terminator. Eighteen seconds later it had slowed to 250 m/s and deployed its pilot parachute. The reefed main parachute opened at 60 km altitude and the canopy was fully opened at 30 km. The instruments were activated at 50 km and transmitted data during the 55 minute descent. There was a clear line of sight to Yevpatoria. The capsule thumped onto the surface at 10.70°S 335.25°E. It was 6:24 Venus solar time and the solar zenith angle was 84.5 degrees. The parachute was jettisoned on impact and the secondary antenna was deployed onto the surface. The capsule transmitted for another 63 minutes reporting measurements on the surface, starting with a 13 minute stream from the primary antenna, then a 20 minute stream from the secondary antenna, and finally a 30 minute stream from the primary.

The Venera 8 carrier spacecraft returned measurements on the upper atmosphere and ionosphere prior to breaking up in the atmosphere.

The second spacecraft to be launched failed to depart from low Earth orbit due to a fourth-stage misfire when the failure of a timer caused the engine to stop after only 125 seconds. It was stranded in a highly elliptical orbit and designated Cosmos 482. At the end of June a fragment separated. This was probably the entry capsule, and it remained in orbit when the main spacecraft re-entered on May 5, 1981.

Results:

The Venera 8 capsule returned a wealth of data about the atmosphere and surface. It determined atmospheric density from accelerometer data in the altitude range 100 to 65 km and directly measured atmospheric temperature, pressure, composition and down-welling light flux from 55 km down to the surface. Although imprecise, these first profiles of the solar flux versus altitude were sufficient to confirm that the high temperatures were caused by the greenhouse effect. The illumination at the surface was measured and the pattern of change in attenuation attributed to clouds. Profiles of the speed and direction of horizontal winds from 55 km down to the surface were obtained from Doppler data. The wind speed was 100 m/s above 50 km, 40 to 70 m/s in the haze layer near 45 km, surprisingly rapid at 20 to 40 m/s below this down to 20 km, and only about 1 m/s from 10 km to the surface. The wind was super-rotating coincident with the motion of the high ultraviolet clouds.

The first report from the radio altimeter was at an altitude of 45.5 km and it gave a total of 35 readings, the last at 900 meters. The capsule drifted 60 km horizontally as it descended. The altimeter produced a ground profile with two mountains 1,000 and 2,000 meters tall, a hollow 2,000 meters deep, and a gentle upward slope toward the landing site. Two echo intensity profiles were obtained from which it was possible to compute the dielectric constant and a surface density of 1.4 g/cc. The photometers made 27 measurements, and the light level declined steadily from 50 to 35 km as the probe descended through the clouds. Venera 8 was the first to distinguish three main optical regions in the atmosphere: two cloud layers with a thicker upper layer of fog from 65 to 49 km and a lower haze layer from 49 to 32 km. Then the light level was essentially constant to the surface, indicating a relatively clear atmosphere below the clouds. The illumination in this part of the atmosphere was comparable to a cloudy day on Earth at twilight. The weak surface brightness indicated that only 1% of the incident sunlight reached the surface. On the other hand, the Sun was only 5 degrees above the horizon. The important finding was that the illumination was sufficient for the next lander to operate a camera.

The gas analyzer returned a composition of 97% carbon dioxide, 2% nitrogen,

0.9% water vapor, and 0.15% oxygen. Although the ammonia test gave a positive detection at altitudes between 44 and 32 km with readings of 0.1 to 0.01%, this was compromised by sulfuric acid which also reacts positively. A significant point is that the gas analyzer confirmed the presence of sulfuric acid in the clouds. This had been offered as an explanation of why the clouds were so arid and yet were able to form cloud droplets. And the fact that such droplets would reflect sunlight so efficiently explained why the planet had such a high albedo.

On the surface, Venera 8 reported a pressure of 93 ± 1.5 bar and a temperature of $470 \pm 8°C$, confirming the measurements by Venera 7 and in good agreement with an extrapolation of the data from the Venera 4, 5 and 6 probes down to the surface using models of the adiabatic temperature lapse rate.

The gamma-ray spectrometer made measurements in the descent, and two on the surface. It reported 4% potassium, 6.5 ppm thorium, and 2.2 ppm uranium indicative of a more granitic than basaltic composition. However, this result was contested and all later Venera landers found more common basaltic compositions. Radar mapping many years later showed that Venera 8 landed in an upland volcanic region that was probably older than the lava plains that constitute most of the planet. Alternatively, a potassium-rich basalt that is relatively rare on Earth could account for this particular data.

A MASSIVE ASSAULT ON MARS FAILS: 1973

Campaign objectives:

In planning of their 1973 campaign for Mars the Soviets were aware of the US plan to send orbiter/landers to the planet in 1975. They also knew that these Vikings were considerably more capable than their own 1971 lander. Adding to this problem, the 1973 opportunity required more energy than that of 1971 and they could not repeat the M-71 orbiter/lander strategy. If they were to send landers to Mars in 1973, then the carrier spacecraft would have to fly past the planet. Their spacecraft were too massive and the Proton-K launcher did not have enough capability to combine an orbiter and a lander in 1973. Chagrined by the poor performance of their spacecraft in 1971 in comparison to the success of the Mariner 9 orbiter, yet encouraged by the near-success of the Mars 3 lander, the Soviets were determined to score a success at Mars ahead of the Viking missions. In fact the US had hoped to launch these in 1973 but financial problems forced a postponement to 1975. Knowing there would be no American competition in 1973, the Soviets decided upon four launches to send an armada consisting of two orbiters and two flyby/landers. The orbiters would be sent first so that they could act as communications relays for the landers on the surface. After releasing its lander, a carrier spacecraft would relay telemetry from the entry system to Earth in real-time during the entry and descent, and then perform its own remote-sensing observations of the planet in making its flyby.

Spacecraft launched

First spacecraft: Mars 4 (M-73 No.52S)
Mission Type: Mars Orbiter
Country/Builder: USSR/NPO-Lavochkin
Launch Vehicle: Proton-K
Launch Date/Time: July 21, 1973 at 19:30:59 UT (Baikonur)
Encounter Date/Time: February 10, 1974
Outcome: Failed orbit inserted burn, flew past planet.

Second spacecraft: Mars 5 (M-73 No.53S)
Mission Type: Mars Orbiter
Country/Builder::   USSR/NPO-Lavochkin
Launch Vehicle: Proton-K
Launch Date/Time: July 25, 1973 at 18:55:48 UT (Baikonur)
Encounter Date/Time: February 12, 1974
Mission End: February 28, 1974
Outcome: Successful, but short-lived.

Third spacecraft: Mars 6 (M-73 No.50P)
Mission Type: Mars Flyby/Lander
Country/Builder: USSR/NPO-Lavochkin
Launch Vehicle: Proton-K
Launch Date/Time: August 5, 1973 at 17:45:48 UT (Baikonur)
Encounter Date/Time: March 12, 1974
Outcome: Successful descent, but lander lost at touchdown.

Fourth spacecraft: Mars 7 (M-73 No.51P)
Mission Type: Mars Flyby/Lander
Country/Builder: USSR/NPO-Lavochkin
Launch Vehicle: Proton-K
Launch Date/Time: August 9, 1973 at 17:00:17 UT (Baikonur)
Encounter Date/Time: March 9, 1974
Outcome: Entry system failed, flew past Mars.

By 1973 the US and USSR were in a period of detente, and cooperation between the two space programs had increased with a number of joint working groups and a joint Apollo-Soyuz mission scheduled for 1975. The Soviets gave the US the data from Mars 2 and 3 and from Venera 8 in return for an accurate ephemeris for Mars, models of its atmosphere, and Mariner 9 orbital imagery of the zones chosen for the Mars 6 and 7 landers.

In order to save cost and reduce risk, Soviet engineers used the same spacecraft as in 1971 with minimum changes. But they were plagued with electronics problems in test and in flight. One crucial difference between the 1971 electronics packages and those built for 1973 were that gold leads had been replaced with aluminum on a key transistor used throughout the spacecraft. This difference was discovered very late in the integration and test program, when some of the new 2T-312 transistors suffered failures. The difficulties with the 2T-312 transistor were to be the Achilles' heel of the

1973 campaign. They were in almost every engineering subsystem and science instrument on the spacecraft. Tests showed that these transistors generally failed 1.5 to 2 years after production. This would correspond to Mars arrival, and they could not all be replaced in time to meet the launch window. An analysis estimated a 50% likelihood of a complete mission failure owing to these transistors. In a US program this would have mandated postponing the mission, but the Soviet decision makers, in their rush to beat the Americans to the surface of Mars, dismissed the concerns of the engineers and opted to take the risk and launch with the suspect transistors.

The problem of the transistors became manifest almost immediately after launch, and plagued the entire campaign. All four spacecraft reached Mars but with three in a seriously crippled state. One orbiter flew past the planet, one lander missed and the other lander was lost just before touchdown after returning a degraded set of descent data. Mars 5 was able to achieve orbit, but then failed after less than a month.

The paucity of data from this flotilla compared to the mass of data returned from the single long-lived Mariner 9 orbiter, considered together with the lack of a viable competitor to the forthcoming Viking missions, led the Soviets to abandon Mars in favor of Venus for the foreseeable future – and it would be 15 years before they tried again.

Spacecraft:

The overall design and subsystems of the spacecraft for the 1973 Mars campaign were the same as before, although the scientific instruments were slightly different. The most significant engineering difference was the installation of a new telemetry system solely to enable the flyby carrier to relay data from its entry system to Earth in real-time; this would prove crucial in the case of Mars 6. The lander's telemetry system remained the same and communications from the surface would be through the orbiters.

The M-73 orbiters were almost identical to the M-71S spacecraft. Their function was to enter into orbit around Mars, communicate with the landers, and obtain their own information on the composition, structure, and properties of the atmosphere and surface of the planet. The science payload was mounted on top of the spacecraft, in the same place as the entry systems of the flyby spacecraft. The principal function of the flyby carrier was to release the entry system on a proper entry trajectory and provide a real-time telemetry relay during the descent. It had science instruments for cruise and Mars flyby observations. Because the Mars 3 lander had made it to the surface, the entry systems were the same. However, the science payload of the landers was upgraded.

Mars 4 and 5 launch mass: 3,440 kg (orbiter; *dry mass* 2,270 kg)
Mars 6 and 7 launch mass: 3,260 kg (flyby vehicle)
                                         1,210 kg (entry vehicle)
                                             635 kg (lander system on descent)
                                           358 kg (lander)
                              4,470 kg (total)

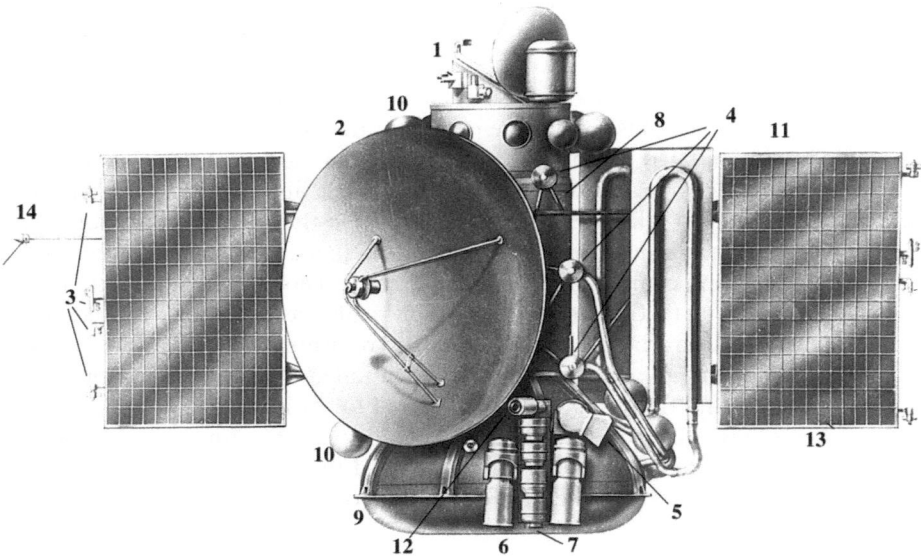

Figure 13.5 Mars 4 and Mars 5 spacecraft diagram: 1. Science instrument compartment; 2. Parabolic high-gain antenna; 3. Attitude control system; 4. Spiral antennas; 5. Mars sensor; 6. Star sensor; 7. Sun sensor; 8. Fuel tank and propulsion system; 9. Avionics compartment; 10. Attitude control gas tanks; 11. Thermal control radiators; 12. Earth sensor; 13. Solar panels; 14. Magnetometer.

Figure 13.6 Mars 4 and Mars 5.

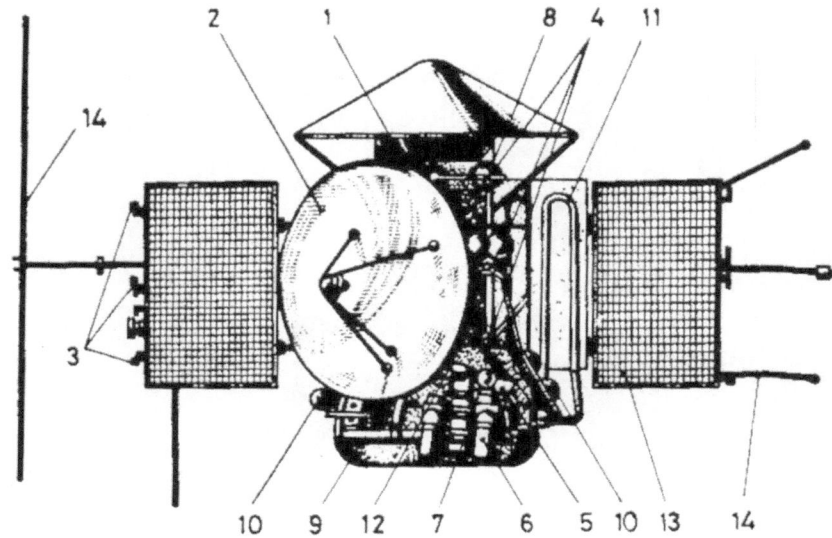

Figure 13.7 Mars 6 and Mars 7 spacecraft diagram (from *Space Travel Encyclopedia*): 1. Lander; 2. Parabolic antenna; 3. Attitude control gas jets; 4. Spiral antenna; 5. Mars sensor; 6. Star sensor; 7. Sun sensor; 8. Propulsion system; 9. Instrument compartment; 10. Attitude control gas tanks; 11. Radiators; 12. Earth sensor; 13. Solar panels; 14. 'STEREO' radio emission antennae.

Figure 13.8 Mars 6 and Mars 7.

Figure 13.9 Mars 6 entry system in test.

Payloads:

Some of the M-73 scientific instruments were redesigned forms of those carried by Mars 2 and 3, but others were new. The Mars 4 and 5 orbiters and the Mars 6 and 7 flyby spacecraft were equipped as follows:

1. FTU facsimile imaging system
2. Optical-mechanical panoramic imaging system
3. Infrared radiometer (8 to 40 microns) for measurement of surface temperature (Mars 5 only)
4. Infrared photometer in five carbon dioxide bands around 2 microns to obtain surface altitude profiles
5. Microwave polarimeter (3.5 cm) for measurement of dielectric constant and subsurface temperatures
6. Two polarimeters in ten bands from 0.32 to 0.70 microns to characterize surface texture (French-Soviet)
7. Four band visible photometer in range 0.37 to 0.6 microns for measurement of color and albedo of the surface and atmosphere (Mars 5 only)
8. Infrared narrow-band 1.38 micron photometer for measurement of water vapor content in the atmosphere
9. Ultraviolet photometer (0.260 and 0.280 microns) for measurement of ozone
10. Scanning photometer (0.3 to 0.9 microns) to study emissions in the upper atmosphere (Mars 5 only)
11. Gamma-ray spectrometer for surface elemental composition
12. Micrometeoroid sensors (Mars 6 and 7 only)
13. Lyman-alpha photometer for measurement of upper atmosphere hydrogen (French-Soviet)

14. Solar wind plasma sensors (8) for measurement of speed, temperature and composition in the 30 eV to 10 keV energy range (Mars 4 and 7 only)
15. Boom mounted three-axis fluxgate magnetometer (Mars 4 and 7 only)
16. STEREO-2 to study solar radio emissions (French-Soviet, Mars 7 only)
17. ZHEMO to study solar protons and electrons (French-Soviet, Mars 6 and 7 only)
18. Multichannel electrostatic analyzer (Mars 4 and 5 only)
19. Dual-frequency radio occultation experiment to profile ionospheric electrons and tropospheric density.

Two types of imaging system were employed. The first was a version of the M-71 FTU camera with various technical improvements and more film and faster scanning rates. The second was a single line push-broom panoramic imager that was to scan a 30 degree field of view from horizon to horizon and was sensitive in both the visible and near-infrared. It stored its data on a 90 minute analog tape recorder for replay to Earth.

The FTU optics were as before: two bore-sighted cameras, one an f/2.8 lens with a focal length of 52 mm and a 35.7 degree field of view, and the other an f/4.5 lens with a 350 mm focal length and a 5.67 degree field of view. They each weighed about 9 kg. The wide angle camera was equipped with red, green, blue and orange filters, and the narrow angle one used an orange long-pass filter. Twenty meters of 25.4 mm film was contained in a radiation-shielded magazine. This was sufficient for at most 480 frames. Exposure times alternated between 1/50th and 1/150th of a second. Each camera produced a 23 × 22.5 mm frame and the film could be scanned at up to ten resolutions, only three of which were used actually: 235 × 220, 940 × 800, and 1,880 × 1,760 pixels. The scanned images were transmitted at either 512 or 1,024 pixels per second by the dedicated impulse transmitter. At the intended operating altitude these cameras would provide resolutions of between 100 and 1,000 meters.

The push-broom panoramic camera system was first used by Luna 19 in 1971. It comprised two optical-mechanical cameras, each with a single photomultiplier tube and a rotating prism to scan a 30 degree field of view across the spacecraft's track. One camera had red and orange filters and was sensitive across the visible spectrum and the other used a red long-pass filter with a photomultiplier that was sensitive in the infrared. They scanned at 4 lines/second and produced 250 cycles/line for video recording on magnetic tape at 1,000 Hz. The readout rate was 1 line/second and the transmission was commanded at 256 or 512 pixels/line resolution.

The payload of the entry system and lander was essentially the same as flown on Mars 2 and 3, but with upgraded imagers, mass spectrometer, and temperature and pressure sensors. The Doppler experiment to measure winds during the descent was new. Most significantly, it was now possible to transmit the descent data in real-time rather than storing it for transmission after landing. They were equipped as follows:

1. Accelerometer for atmospheric density during entry
2. Doppler experiment for winds and turbulence on descent
3. Temperature and pressure sensors on descent and landing
4. Radio altimeter for providing altitudes on descent

5. Mass spectrometer for atmospheric composition on descent and landing
6. Atmospheric density and wind velocity on the surface
7. Two panoramic television cameras for stereo viewing of the surface
8. Gamma-ray spectrometer for soil composition mounted in a petal
9. X-ray spectrometer for soil composition deployed to the surface from a petal
10. PrOP-M walking robot deployed to the surface from this same petal with onboard gamma-ray densitometer and conical penetrometer.

Doppler measurements and the radio altimeter, accelerometer, temperature, and pressure sensors operated from the beginning of parachute deployment right down to the surface.

Mission descriptions:

All four spacecraft were dispatched successfully but predictably and inevitably first Mars 6, then Mars 7, then Mars 4 suffered system-wide failures within a matter of weeks owing to the faulty transistor. Only Mars 5 arrived relatively trouble free and was able to enter orbit around Mars, but it suffered a pressure leak and shortly thereafter fell silent.

Mars 4:

Mars 4 was launched on July 21, 1973. It made a midcourse correction on July 30, but the computer developed problems that prevented a second midcourse correction. It reached the planet on February 10, 1974, but the engine failed to ignite for the orbit insertion maneuver and the spacecraft flew past the planet at a range of 1,844 km.

Mars 5:

Mars 5 was launched on July 25, 1973. It made midcourse corrections on August 3 and on February 2, 1974, and followed through with the orbit insertion maneuver at 15:44:25 UT on February 12 to achieve a 1,760 km × 32,585 km orbit with a period of 24.88 hours inclined at 35.3 degrees to the equator. Some unknown event during orbit insertion triggered a slow leak in the instrument compartment. An accelerated plan of observations was prepared that focused on obtaining high resolution imagery of the surface. But when the pressure fell below operating levels in the transmitter housing on February 28 after only 22 orbits, this prematurely ended the mission and ruled out the use of this spacecraft as a relay for the landers that were scheduled to arrive in early March.

Mars 6:

Mars 6 was launched on August 4, 1973, and conducted a midcourse correction on August 13. At the end of September the science and operations downlink was lost, almost certainly due to the failure of a 2T-312 transistor. Only two channels in the telemetry system remained operational, neither of which provided any information on spacecraft status. Refusing to give up, the engineers continued to send commands to the spacecraft in the hope that the receivers might still be functioning. As it turned out the command uplink was unaffected, and Mars 6 dutifully obeyed the commands and executed its autonomous functions. Unable to report to Earth, in February 1974 it autonomously determined its position, calculated the second midcourse correction and carried this out. Upon approaching the planet on March 12, it properly executed the optical navigation and targeting, and then released its entry system at a range of 48,000 km from the planet with just 3 hours remaining to atmospheric entry.

Ground controllers first realized that Mars 6 had done its job shortly thereafter, at 08:39.07 UT, when data began to arrive through the dedicated relay channel. At that time the entry system was 4,800 km from its target. The spacecraft was able to relay data throughout entry and descent, and then continued past the planet, passing within 1,600 km of the surface. This performance was a monumental achievement in spacecraft autonomy for the Soviet program so early in the history of planetary exploration.

The entry system penetrated the atmosphere at 09:05:53 UT at a speed of 5.6 km/s and an 11.7 degree angle of attack. Loss of signal due to plasma effects occurred at 09:06:20 at an altitude of 75 km. The signal was regained at 09:07:20 at 29 km, with the start of the data transmission. The main parachute was deployed at 09:08:32 after the entry system had slowed to 600 m/s at an altitude of 20 km. The canopy opened fully at 09:08:44, and the capsule started to transmit data on altitudes, temperatures and pressures. The mass spectrometry was stored for transmission after the landing. The Doppler shifts on the signal were noted. The capsule appeared to be swaying under the parachute more than expected, impairing the transmission quality. Ignition of the rocket engines was confirmed but the transmission cut off at 09:11:05 when the lander was "in direct proximity to the surface", probably when it hit the surface. The velocity at the time of signal loss was 61 m/s, which was excessive for a safe touchdown. The transmitter was programmed to turn off after the lander had been released in order to switch over to a different set of VHF antennas on the lander. No further signals were received. The fate of the lander is unknown. It lies at 23.90°S 19.42°W in the Margaritifer Sinus region, in the vicinity of Samara Valley, where the landscape is characterized by steep slopes.

Mars 7:

Mars 7 was launched 4 days after Mars 6 but flew a faster trajectory which arrived at the planet 3 days earlier. It made only one course correction on August 16, 1973. An early failure in the communications system cost it one transmitter, but it was able to remain in contact. At Mars the targeting maneuvers to set up the entry system were executed properly and the entry system was released on March 9, 1974. However, most likely owing to a failed 2T-312 transistor, the entry system computer did not issue the command to fire the retro-rocket, with the result that it missed the planet by 1,300 km. The intended target was in the crater Galle at 51.2°S 30.9°W. The carrier spacecraft provided some data during its flyby.

Results:

Orbiters:

Imagery

After Mars 4 failed to execute its orbit insertion maneuver it flew past the planet and continued to return interplanetary data from solar orbit. During the flyby it returned one swath of twelve pictures and two panoramas in a 6 minute imaging cycle from a range of 1,900 to 2,100 km. Two dual-frequency radio occultation profiles were also obtained, one on passing behind the planet as viewed from Earth and the other upon exiting and these supplied the first indication of a night-side ionosphere.

The Mars 5 orbiter operated for only 25 days after orbit insertion, and returned atmospheric data and images of a small portion of the southern hemisphere of Mars. In all it returned 108 pictures, but the narrow angle ones were motion blurred. The useful data included 43 wide angle images and five panoramas which were returned in five imaging sessions over a 9 day period, all of roughly the same area, which was near the imaging track of Mariner 6 and showed swaths of the area south of Valles Marineris from 5°N 330°W to 20°S 130°W. Measurements by other remote sensing instruments were also made near periapsis along seven adjacent arcs in this region. High cirrus clouds and yellow clouds of fine dust were identified.

Surface properties

Data returned from orbit by the infrared radiometer on Mars 5 showed a maximum surface temperature of -1°C, -43°C near the terminator, and -73°C at night. It gave a thermal inertia for the soil that was consistent with grains 0.1 to 0.5 mm in size, and polarization data in the visible spectrum implied grain sizes smaller than 0.04 mm in aeolian deposits of variable cover. The polarization at 3.5 cm suggested a dielectric constant of 2.5 to 4 at depths of several tens of cm. The oxygen, silicon, aluminum, iron, uranium, thorium, and potassium sensed by the gamma-ray spectrometer from orbit suggested a surface similar to terrestrial mafic rocks.

Lower atmosphere

Six altitude profiles were measured by the carbon dioxide photometer along a path between 20 and 120°W in longitude and spanning 20 to 40°S in latitude, and these were in general agreement with the Mariner 9 ultraviolet spectrometer data. Surface pressures as high as 6.7 millibars were determined. Mars 3 had found only 10 to 20 precipitable microns of water vapor while the dust storm was raging in 1971. Two years later, Mars 5 found abundances as high as 100 precipitable microns south of the Tharsis region. The water vapor content was variable by a factor of 4 to 5 across the planet. An ozone layer about 7 km thick was detected at 40 km altitude in the equatorial region, not near the surface as anticipated, with a concentration of about 1/1,000th that of Earth's ozone layer. The Mariners were able to detect ozone only at the poles where it is more abundant. The existence of argon in the atmosphere was confirmed.

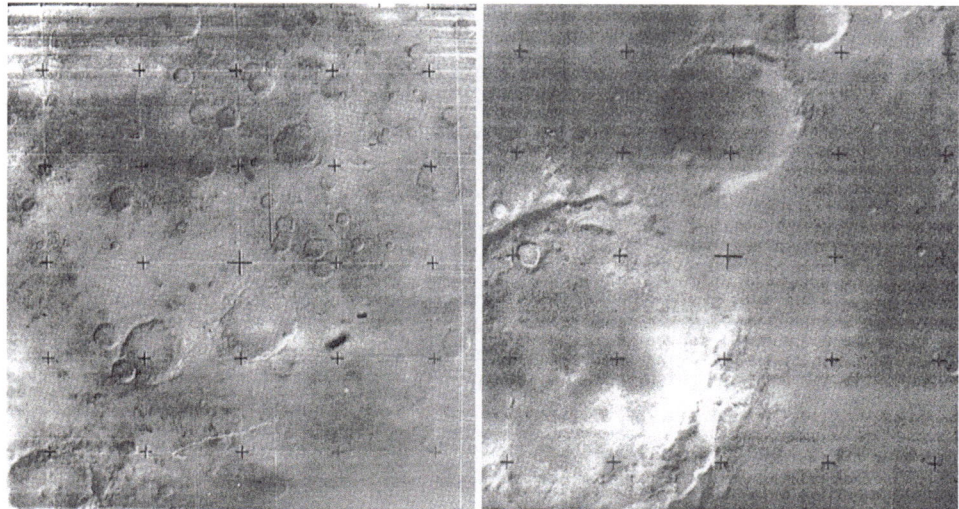

Figure 13.10 Pictures from Mars 4 during its flyby and from Mars 5 once it was in orbit. Left: cratered terrain at 35.5°S 14.5°W taken by Mars 4 from 1,800 km range through a red filter. From lower left the large craters are Lohse, Hartwig and Vogel. Right: the crater Lampland at 36°S 79°W taken by Mars 5 from 1,700 km range.

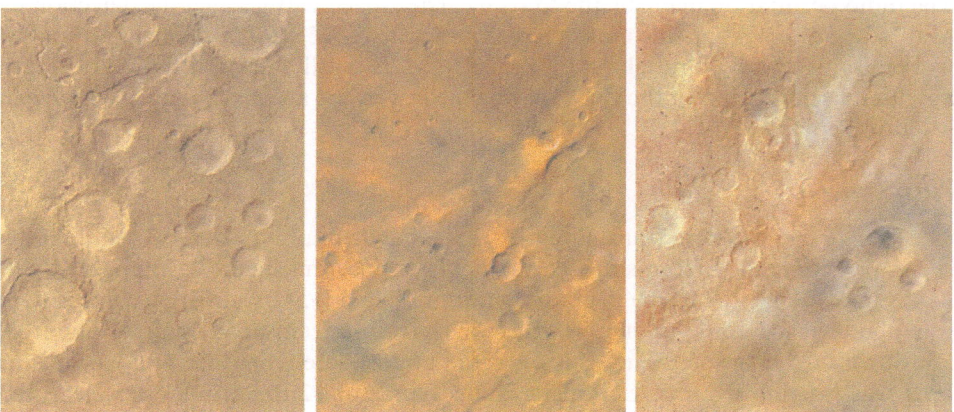

Figure 13.11 Mars 5 FTU color images.

Figure 13.12 Picture from the Mars 5 panoramic imager (processing by Ted Stryk).

Upper atmosphere and ionosphere

The Lyman-alpha instrument found the exosphere temperature to be 295 to 355K, with the temperatures in the altitude range 87 to 200 km being 10 degrees lower. No upper atmosphere emissions were noted by the visible spectrometer in the range 0.3 to 0.8 microns. Mars 5 also performed a radio occultation experiment on one orbit and its results, with those from the flyby occultation measurements for Mars 4 and 6, showed the existence of a night-side ionosphere with a maximum electron density of 4,600/cc at an altitude of 110 km, and an atmospheric pressure near the surface of 6.7 millibars.

The fields and particles instruments returned a significant data set to complement the M-71 orbiter data. Two distinct plasma zones were found inside the bow shock between the undisturbed solar wind and the planetary magnetosphere: a thermalized plasma behind the bow shock, and a small electrical current carried by protons in the magnetotail. The bow shock was held at an altitude of 350 km. The plasma results and magnetometer measurements were consistent with the planet having an intrinsic magnetic field of about 0.0003 times the strength of Earth's field, and inclined 15 or 20 degrees from the rotational axis which, as in the case of Earth, is tilted 23 degrees from the perpendicular to the plane of its orbit.

Flyby spacecraft:

The radio occultation made by the Mars 6 carrier spacecraft verified the detection by Mars 4 and 5 of a night-side ionosphere. The French STEREO instrument on Mars 7 worked satisfactorily throughout the cruise to Mars but the spacecraft did not return any useful science during the flyby. It was the last of the armada to fall silent, which it did in September 1974.

Entry systems:

The Mars 6 entry system transmitted for 224 seconds during its descent before it fell silent, providing the first in-situ measurements of this atmosphere. Although a lot of it was unreadable due to the degradation of another 2T-312 transistor, enough data was obtained to profile the temperature and pressure from an altitude of about 29 km down to the surface. The density of the atmosphere from 82 km down to this level was inferred from accelerometer data. Winds of 12 to 15 m/s were measured from an altitude of 7 km down to near the surface. A surface pressure of 6.1 millibars and a temperature of -28°C were derived from temperature, pressure, accelerometer and Doppler data. The surface winds were 8 to 12m/s. Other parameters included a lapse rate of 2.5K/km, the presence of the tropopause at an altitude of 25 to 30 km, and an almost isothermal stratosphere at a temperature of 150 to 160K. These values were consistent with measurements by the Mars 5 orbiter, and were later confirmed by the Viking missions. Instruments also indicated "several times" more atmospheric water vapor than previously reported. The mass spectrometer data were stored during the descent for transmission after landing, and so were lost. However, the current to the vacuum pump was transmitted during the descent as an engineering parameter, and a steep increase in current was interpreted as an indication of argon at an abundance of 25 to 45%, which was implausibly large; the actual value was found by the Vikings to be a more reasonable 1.6%.

Landers:

No transmissions were ever received from the Mars 6 lander and no results were obtained from the surface. The Mars 7 entry system missed the planet entirely.

THE HIATUS IN SOVIET MARS MISSIONS: 1974–1988

By early 1974 the Soviet space program was severely traumatized. Its manned lunar program had failed both to beat the US to a circumlunar flight and to introduce into service the N-1, its answer to the Saturn V launcher, meant to dispatch cosmonauts to land on the Moon. It had taken second prize with robotic lunar rovers and sample return missions. The all-out Mars effort of 1973 had been an embarrassing failure. In May, Vasily Mishin, Sergey Korolev's protégé and successor, was replaced as Chief Designer by the avowed rival to both, Valentin Glushko, who canceled the N-1 and refocused the manned program on a new Energiya launcher and the Buran reusable spaceplane to compete with the space shuttle the US had recently started to develop.

In the early 1970s a "war of the worlds" had raged in the community of Soviet scientists and engineers working on planetary exploration. The 'Venusians' argued for concentrating on Venus where they felt the USSR had a clear advantage, instead of challenging the US where it had gained the advantage. Of course, the 'Martians' argued to focus on Mars as the more interesting of the two planets. They could not compete with the sophisticated Viking landers, but studies had been underway for several years for a bold and even more prestigious mission to Mars – a sample return that would require the N-1 lunar rocket. The debate was between Roald Sagdeev, the Director of IKI, and Alexander Vinogradov, Director of the Vernadsky Institute of Geochemistry, Vice President of the Academy of Sciences and Chair of the Lunar and Planetary Section of the Academy's Inter-Department Scientific and Technical Council on Space Exploration. The ultimate arbiter was Mstislav Keldysh, who was scientifically the most acknowledged and most politically well connected member of the community. Keldysh hesitated over the very ambitious plans of the 'Martians' and eventually took the practical route by turning to Venus for the immediate launch opportunities. NPO-Lavochkin was allowed to continue designing Mars rovers and sample return missions that would use the Proton launcher, but by 1975–76 these would prove impractical. Instead, Sergey Kryukov, who had taken over from Georgi Babakin on the latter's death in August 1971, proposed to salvage the Mars program with a less ambitious mission to Phobos, the larger of the planet's two small moons. Keldysh was supportive of this concept, but it would fade after Kryukov resigned in 1977 and Keldysh died in 1978, and the 'Martians' had to stand down while Venus took center stage for the next ten years.

Nonetheless, it is interesting to describe in some detail the very ambitious plans of the 'Martians' at that time. Soviet engineers had been working on designs for Mars sample return missions in parallel with developing the Mars spacecraft for the 1971 and 1973 campaigns. Bolstered by the success of the Luna 16 sample return and the Luna 17 lunar rover missions in 1970, the "top brass" ordered NPO-Lavochkin to fly a Mars sample return mission by mid-decade. Kryukov assumed that the N-1 would be available. The first spacecraft design had a launch mass of 20 tons. The 16 ton entry system used an 11 meter aeroshell with folding petals to enable it to fit inside the payload shroud. The lander eschewed parachutes and used large retro-rockets to decelerate. A direct return to Earth was planned with a spacecraft based on the 3MV design of Venera 4 to 6, using a two-stage rocket and an entry capsule which would deliver 200 grams of Martian soil to Earth. The Soviets wrestled with the complexity of the spacecraft system and also with the issue of biological contamination of Earth. A test mission was tentatively planned for 1973 that would deliver to the surface of Mars a rover based on the successful Lunokhod.

The failure of the N-1 rocket program forced a change to a less massive design. In 1974 NPO-Lavochkin began to consider how to use the Proton to accomplish a Mars sample return mission. Two Protons would be used. The first would place a Block D upper stage and the spacecraft into Earth orbit and the second would orbit a second Block D that would rendezvous and dock. The two propulsive stages would then be fired in succession to send a flyby/lander spacecraft to Mars. Spacecraft mass would be saved by not requiring the sample return vehicle to fly directly back to

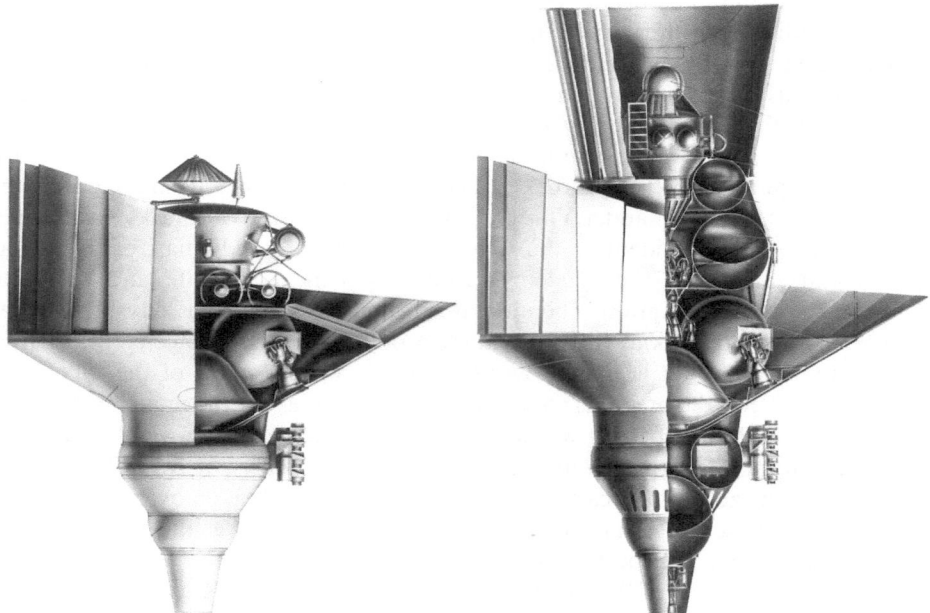

Figure 13.13 Mars rover (left) and sample return (right) concepts for launch on the N-1 rocket.

Earth but instead to enter Mars orbit, where it would rendezvous with an Earth-return vehicle that had been launched by a third Proton. And in one scenario, instead of entering the atmosphere the return vehicle would brake into a low Earth orbit for retrieval by a manned mission. Again a precursor mission for landing a rover on Mars was planned.

The project wrestled with continuing issues of complexity and mass. This led to a refinement in 1976 in which the first spacecraft would be launched into Earth orbit with its Block D upper stage dry, so as to allow for increased spacecraft mass. The second launch would deliver both a second Block D and fuel for transfer into the dry stage. The flyby/lander spacecraft launch mass was 9,135 kg. The flyby spacecraft was 1,680 kg, and the entry system 7,455 kg including 3,910 kg for the two-stage surface-to-orbit vehicle and 7.8 kg for the Earth return capsule, which in this version would pass through the atmosphere without having either a parachute or a telemetry system. The struggle to accommodate the complexity, cost and risk of this mission strained Soviet technology beyond its limits. At the same time, NPO-Lavochkin was continuing to mount complex lunar rover and sample return missions through 1976. The results from the considerable funds that were expended on designing these Mars missions were disappointing. Other programs, including a Lunokhod 3 mission, had to be sacrificed. When it became apparent that the project was impractical, it was canceled and Kryukov was transferred.

While successful at automated lunar sample return, the Soviet Union never got the chance to try a Mars sample return mission. In the mid-1970s the space ambitions of both nations were thwarted by their respective governments. In addition to losing the race to the Moon the Soviets had suffered appalling failures at Mars. Performance and cost became serious issues, and risk was less tolerable. Ironically, the result in the US was the same despite the success of the Apollo program and the Vikings at Mars. It would be a long time before either nation sent another mission to Mars but once again it was the Soviets who were the first to do so, with the Phobos missions of 1988. In the meantime, having taken the lead in planetary exploration in the 1970s by exploring from Mercury to Neptune, America fell behind again in the 1980s as their planetary launch rate dropped to zero and the Soviets reaped success after success at Venus and opened up their program to international cooperation with complex science-dense missions at Venus, Halley's comet, and finally Mars.

14

Turning from the Moon and Mars to Venus

TIMELINE: 1974–1976

After abandoning its manned lunar program, the USSR continued to send robotic missions. In May 1974 they launched Luna 22, which became their second orbiter in the new heavy series. The Luna 23 sample return mission in October 1974 damaged its drill system on landing, which ruined the mission. A year later, another sample return mission was lost to a Block D failure. But Luna 24 launched in August 1976 became the third successful sample return. This drew to a conclusion the long line of missions which began with Luna 1 to 3, small 300 kg spacecraft that used the three-stage R-7 Luna launcher, then Luna 4 to 14, with 1,600 kg spacecraft that used the four-stage R-7 Molniya, and finished with Luna 15 to 24, with 5,800 kg spacecraft that used the four-stage Proton-K. There had been many failures but this series gave the Soviets the first lunar flyby, first lunar impact, first pictures of the far side of the Moon, first landing, first orbiter, first sample return, and first surface rover.

After the objectives of the direct-entry 3MV series at Venus were achieved by the landing of Venera 8 in 1972, the Soviets stood down during the 1973 opportunity in order to develop a spacecraft capable of both orbiting the planet and dispatching a larger and more capable lander equipped with imagers. They based the new Venera on the Proton-launched Mars spacecraft that proved itself in 1971. In June 1975 two were launched as Venera 9 and 10. They both performed spectacularly by releasing their entry systems inbound and then entering orbit around the planet. Furthermore, both landers yielded atmospheric data and survived on the surface for about an hour, from where they took the very first black-and-white pictures from the surface of the planet and returned data on the composition of the rocks.

Launch date

1974

29 May	Luna 22 orbiter	Success
28 Oct	Luna 23 sample return	Landed, damaged sampler, no return

1975

8 Jun	Venera 9 orbiter/lander	Success, first images from the surface
14 Jun	Venera 10 orbiter/lander	Success
20 Aug	Viking 1 Mars orbiter/lander	Success, first successful lander on Mars
9 Sep	Viking 2 Mars orbiter/lander	Success
16 Oct	Luna sample return	Fourth stage failed

1976

9 Aug	Luna 24 sample return	Success

The US sent its two sophisticated Viking spacecraft to Mars in 1975. Both were highly successful, the orbiters as well as the landers. Reeling from the failure of their all-out assault on Mars in 1973 and the awesome performance of the Vikings, the Soviets abandoned their long and hapless campaign at Mars in favor of Venus. They would not develop the confidence to resume Mars missions until 1988. Meanwhile, their new Venera spacecraft delivered a long string of successes.

A NEW, SOPHISTICATED VENUS LANDER: 1975

Campaign objectives:

With the survival of the Venera 8 capsule on the surface of Venus in 1972, the 3MV spacecraft had reached the limit of its capability and the Soviets were ready for the next step. They now had enough data on the atmosphere of Venus and conditions at the surface to design a very capable lander that included sophisticated imaging and surface science instruments. The challenge was to enable this apparatus to operate in such harsh conditions. Also, the new heavy Proton-launched Mars spacecraft had proved itself in 1971 with the Mars 2 and 3 missions. Both orbiters were successful, and the Mars 3 lander was successfully delivered to the surface. This orbiter served as the basis for designing the Venus spacecraft. However, the entry vehicle had to be completely redesigned. For the first time since their initial launch to Venus in 1961, the Soviets skipped a Venus opportunity in October 1973 while they worked on their new spacecraft.

The main difference between the heavy Proton-launched spacecraft for Mars and for Venus was the entry system. A vehicle to enter the rarefied atmosphere of Mars needs a broad conical aerobrake for rapid deceleration in the upper atmosphere and large robust parachutes to slow to a safe speed before reaching the surface. The thick atmosphere of Venus, on the other hand, is much more forgiving and permits the use of a simpler entry system. The new system for Venus was a hollow spherical vessel that contained the heavy lander and its parachute system. The previous probes had revealed the atmosphere to be so thick that to reach the surface in a reasonable time the new system was designed to jettison its parachute high in the atmosphere and let the lander fall using an aerobrake, and since the free-fall velocity at the surface was slow enough for the impact to be survivable there was no requirement for a terminal retro-rocket.

Spacecraft launched

First spacecraft:	Venera 9 (4V-1 No.660)
Mission Type:	Venus Orbiter/Lander
Country/Builder:	USSR/NPO-Lavochkin
Launch Vehicle:	Proton-K
Launch Date/Time:	June 8, 1975 at 02:38:00 UT (Baikonur)
Encounter Date/Time:	October 22, 1975
Orbiter Terminated:	March 22, 1976
Outcome:	Successful.

Second spacecraft:	Venera 10 (4V-1 No.661)
Mission Type:	Venus Orbiter/Lander
Country/Builder:	USSR/NPO-Lavochkin
Launch Vehicle:	Proton-K
Launch Date/Time:	June 14, 1975 at 03:00:31 UT (Baikonur)
Encounter Date/Time:	October 25, 1975
Orbiter Terminated:	March 22, 1976
Outcome:	Successful.

There were minor modifications to the spacecraft, including changes in the size and position of the solar panels, the thermal systems, and increased reliability. One key change was to replace the direct-to-Earth communications system of the lander with a system to relay via the orbiter, which greatly improved the data rate from the lander. The mission plan for Venus differed from that for Mars. The orbiter/lander was first targeted at the atmospheric entry point, rather than at the orbital insertion point. Several days from Venus the passive entry system was released. Immediately afterwards, the spacecraft made a deflection maneuver for the orbital insertion point. The timing was such that by the time the lander started to transmit, the spacecraft would have just completed its orbit insertion burn and have a line of sight to relay the transmission.

The principal scientific goal of the lander was to obtain the first panoramic image on the surface of Venus. This determined the minimum time that the lander must be able to operate on the surface and also the data rate of the relay through the orbiter. These new capabilities, and the large mass available on the lander, meant a number of instruments that had never been flown before could be carried for descent science. These included instruments to measure the vertical structure of aerosols within and under clouds, the vertical and spectral distribution of solar flux penetrating through the clouds for several look angles, chemical and isotopic analysis of the atmosphere, and direct measurements of the winds at the surface. To undertake the first science from orbit, the spacecraft had experiments to report upon the plasma environment around the planet and its atmospheric structure, upper cloud layers, and outgoing thermal radiation.

Early on, consideration had been given to using a flyby spacecraft to support the lander mission but NPO-Lavochkin and IKI argued hard for the orbiter in order to obtain the additional and original science that only an orbiter could provide. And, of course, being first to place a spacecraft in orbit around the planet would be another significant first achievement in space exploration.

Spacecraft:

As the first of another generation of spacecraft, Venera 9 and 10 were five times heavier than their predecessors and were launched by the more powerful four-stage Proton-K. This launcher was introduced in 1969 for the Ye-8 lunar missions and then used for the M-69, M-71 and M-73 campaigns. These new spacecraft consisted of an orbiter with the entry system strapped on top, inside of which was the lander. The new entry system and lander were extensively tested in wind tunnels and by airdrops.

The two spacecraft for this opportunity were essentially identical, but Venera 10 was slightly heavier and required a larger fuel load for its longer orbit insertion burn.

Launch mass:	4,936 kg (Venera 9) 5,033 kg (Venera 10)
Fuel:	1,093 kg (Venera 9) 1,159 kg (Venera 10)
Orbiter dry mass:	2,283 kg (Venera 9) 2,314 kg (Venera 10)
Entry system mass:	1,560 kg
Lander mass:	660 kg

Orbiter:

The orbiter was based on the M-71 spacecraft. The UDMH and nitrogen tetroxide tanks formed a cylindrical body. At 110 cm in diameter, this was narrower than the 180 cm Mars version and it was also 1 meter shorter. Below was the KTDU-425A restartable rocket engine which could be throttled between 9,856 and 18,890 N for a total of 560 seconds. The avionics and science instruments were in a pressurized toroidal module with a diameter of 2.35 meters that was attached at the base of the cylinder, with the gimbaled engine nozzle protruding through the donut. Navigation optics attached to the exterior of the instrument module included a number of solar sensors mounted in a linear cluster, bordered either side by duplicate down-looking telescopic Canopus sensors. The Earth sensor was installed in such a way as to point in the same direction as the parabolic high gain antenna. With the entry system on top, the spacecraft was 2.8 meters tall.

Two 1.25 × 2.1 meter solar arrays extended from opposite sides of the cylinder with an overall span of 6.7 meters. These supported cold gas attitude control jets, the

Figure 14.1 Venera 9 spacecraft.

magnetometer booms, and a relay antenna for the entry system and lander. Also on the side of the cylinder were thermal control gas radiators and tanks which contained nitrogen at 350 bar for the attitude control system. During the interplanetary cruise, louvers on the entry shell provided passive thermal control of the entry system. Communications from the entry system and lander were received by the orbiter and relayed to Earth. There was a 1.6 meter diameter parabolic high gain antenna on the side of the cylinder for communicating with Earth on decimeter and centimeter bands. Six omnidirectional helical antennas were attached near the parabolic antenna, four for Earth and two for the lander. The command uplink was by the helical antennas at 769 MHz. There was a 16 megabyte magnetic tape system for data storage. The downlink to Earth was phase modulated PCM at 3 kbits/s via the

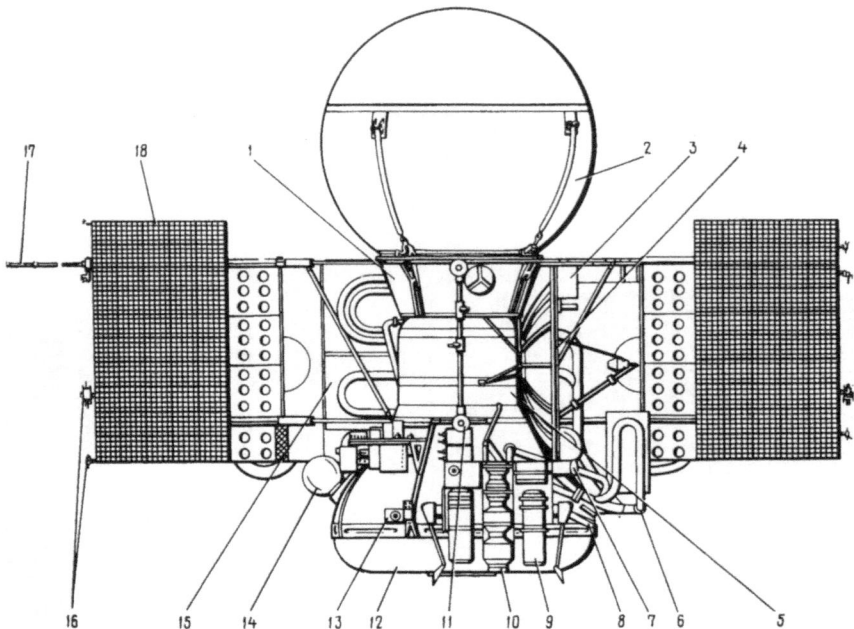

Figure 14.2 Venera 9 and Venera 10 spacecraft: 1. Orbiter bus; 2. Descent capsule; 3. Science instruments; 4. High-gain antenna; 5. Propellant tank; 6. Thermal control pipes; 7. Earth sensor; 8. Science instruments; 9. Canopus sensor; 10. Sun sensor; 11. Omnidirectional antenna; 12. Science instrument module; 13. Science instruments; 14. Attitude control gas tank; 15. Thermal control radiator; 16. Attitude control jets; 17. Magnetometer; 18. Solar panels.

Figure 14.3 Venera 9 in test at NPO-Lavochkin. The shutters on the entry vehicle are for thermal control during flight.

high gain antenna, or in an emergency at a much slower rate using the helical antennas. Data from the lander was retransmitted through the parabolic antenna to Earth in real-time and also stored on tape for later backup transmission. The spacecraft computers were similar to those carried by the M-71 missions.

As it did for Mars, this design served as the basis for all Venus spacecraft starting in 1975, and the Proton-K became the singular planetary launch vehicle in the Soviet inventory.

Entry system:

The sophisticated new lander was contained within a 2.4 meter diameter spherical entry system, and was deployed after the rate of descent through the atmosphere had been reduced to subsonic. The entry vessel was a simple sphere covered by ablative material that consisted of asbestos composite over honeycomb, and it was stabilized during entry by placing the center of mass towards leading side. The entry angle was shallower than for the 3MV capsules, reducing the peak load from around 450 G to a more modest 150 to 180 G. After entry, the sphere would split into hemispheres, releasing the lander and its parachute system.

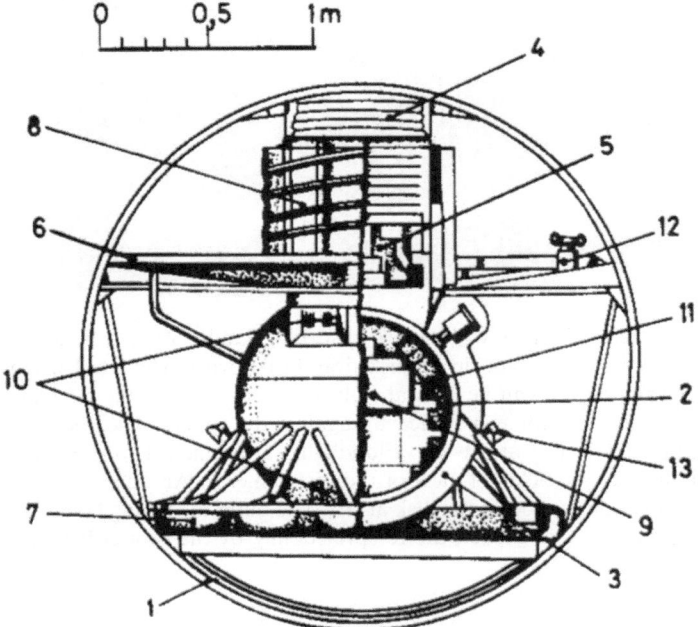

Figure 14.4 Venera 9 and Venera 10 entry capsule (from *Space Travel Encyclopedia*): 1. Heat shield; 2. Lander instrument compartment; 3. Lander insulation; 4. Parachute; 5. Descent instruments; 6. Descent brake; 7. Landing ring shock absorber; 8. Transmitter helical antenna; 9. Electronics; 10. Science instruments; 11. Panoramic camera; 12. Anemometer; 13. Illumination lamp.

Figure 14.5 Venera 9 lander with shock absorbing lander ring, spherical pressurized instrument compartment, and upper disk drag brake with cylindrical wound antenna 'top hat'. A camera pod can be seen at right under the disk brake next to the 'paint-roller' gamma-ray densitometer folded up against the sphere. The spectrophotometer housing is under the disk at the left. Floodlights are attached to the shock absorber struts to illuminate the fields of view of the two cameras. An anemometer for surface winds is mounted on the top of the disk at left. The outer insulating layers are removed. The two severed pipes on the left are for pre-cooling by the orbiter before separation.

Lander:

The lander was 2 meters in height, which was much larger than the 3MV capsules, and capable of carrying more scientific instruments. Previous Venera probes were limited to a transmission rate of 1 bit/s by their direct communications link to Earth. The lander was battery powered and transmitted to the orbiter through two VHF channels at 256 bits/s for relay to Earth using the high gain antenna.

Figure 14.6 Venera 9 lander during tests. The central band segment of the pressure vessel is removed for access. The engineer is looking at a camera.

The lander was basically a hermetically sealed titanium spherical pressure vessel 80 cm in diameter, containing most of the instrumentation and electronics. This was affixed to a ring-shaped landing cushion by a set of shock absorbers. Above was a disk-shaped aerobrake 2.1 meters diameter, to slow the rate of descent during free-fall. This disk also acted as a reflector for the cylindrically wound omnidirectional antenna above. Inside this 80 cm diameter 40 cm tall cylinder were the parachutes and some of the descent instruments. The sphere consisted of several sections bolted together with gold wire seals. It was surrounded with a 12 cm layer of honeycomb insulating material and a thin surface of titanium. The inside of the sphere was also lined with a polyurethane foam insulating material. The thermal design was similar to the earlier landers. The lander was pre-chilled to -10°C by cold air from the main spacecraft via two pipes through the entry vessel. A lithium nitrate trihydrate phase-change material which melted at 33°C absorbed heat that penetrated the insulation, and a gas circulation system distributed it evenly. These measures, and the life of the batteries, allowed operation for about an hour after landing.

Entry, descent and landing:

The midcourse maneuvers were to target the spacecraft for the entry point. The entry system would be released 2 days from Venus to make a ballistic approach and enter the atmosphere at 10.7 km/s at an angle in the range 18 to 23 degrees. Six seconds later it would experience the peak deceleration of 170 G. After 20 seconds, having slowed to 250 m/s and under a 2 G load at an altitude of about 65 km, the small pilot parachute would deploy with a ripcord to draw out the 2.8 meter drogue parachute. The spherical shell would then split into hemispheres and the drogue would pull the upper hemisphere and attached lander away from the lower hemisphere, at the same time deploying a second braking parachute. After 11 seconds, by now at an altitude of 60 to 62 km and descending at 50 m/s, the upper hemisphere would release the lander, in the process extracting the three 4.3 meter diameter main parachutes from the cylindrical section on top of the lander.

Once on its main parachutes, the lander would activate its instruments. It would spend about 20 minutes descending through the cloud layer at a rate of about 50 m/s. On reaching an altitude of 50 km, the lander would shed its parachutes and spend the next 55 minutes free falling, with the drag of the disk-shaped aerobrake slowing its rate of descent as it penetrated the thicker air close to the surface. This strategy was selected to minimize heat inflow during descent, and hence extend the lifetime of the lander on the surface. The terminal velocity on reaching the surface would be 7 m/s, and the impact would be cushioned by the compressible, metal annular landing ring.

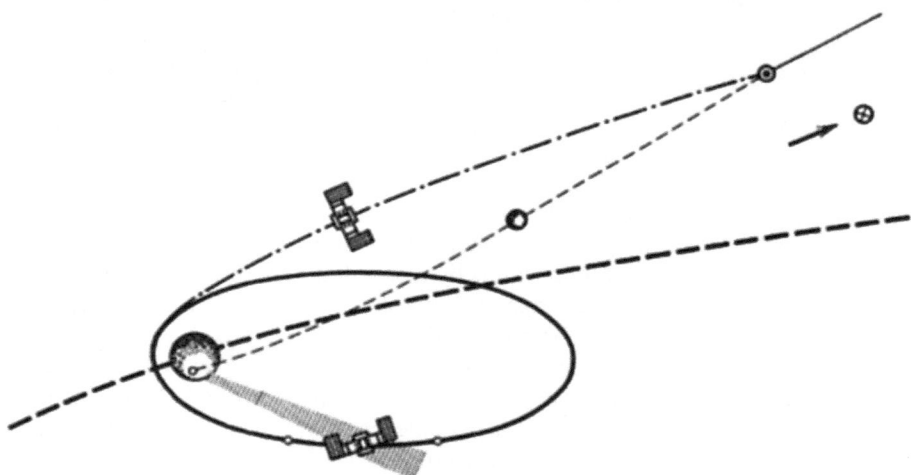

Figure 14.7 Approach geometry and relay operations for the lander. The entire vehicle is initially targeted for the impact point on the sunlit side, out of view of Earth. Two days out, the entry vehicle is released and the orbiter makes a deflection maneuver to place it in position for an orbit insertion burn before the entry vehicle arrives. Shortly after orbit insertion, the orbiter is in position above the landing site for relay operations during entry, descent and landing.

Payloads:

Orbiter:

1. Panoramic ultraviolet cloud cameras, 345 to 380 nm and 355 to 445 nm
2. Cloud infrared spectrometer (1.6 to 2.8 microns)

3. Cloud thermal infrared radiometer (8 to 28 microns)
4. Cloud ultraviolet imaging spectrometer (352 and 345 nm) [USSR-France]
5. Cloud photopolarimeters (335 to 800 nm)
6. Lyman-alpha H/D photometer
7. Airglow spectrometer (300 to 800 nm).
8. Triaxial magnetometer
9. Plasma electrostatic analyzer
10. Charged particle traps
11. Cherenkov energetic particle detectors
12. Centimeter and decimeter radio occultation experiment
13. Bistatic 32 cm radar mapping experiment

The two cloud cameras were the same as the linear scanning photometer cameras of the Mars 4 and 5 orbiters, providing cross-track scanning of 30 degrees and using the motion of the spacecraft to scan along the orbital track. The Venus cameras used violet and ultraviolet filters, scanned 500 cycles/line at 2 lines/second. Images were usually transmitted at 256 pixels/line with 6 bits/pixel. Panoramas were typically 6,000 pixels in length. For a periapsis at about 5,000 km the resolution at the cloud tops was on the order of 6 to 30 km.

Between them, the spectrometers and photometers could make measurements of the clouds throughout the ultraviolet, visible and infrared parts of the spectrum. The photopolarimeters were an improved form of those of the M-71 and M-73 missions, with design help from the French. The cloud infrared spectrometer used a circular-ramp interference filter and made high resolution spatial scans across the planet. The thermal infrared instrument used two horn radiometers for bands at 8 to 13 microns and 18 to 28 microns, both of which were relatively transparent in an atmosphere of carbon dioxide. The French-built cloud ultraviolet imaging spectrometer measured spatial profiles across the planet at two wavelengths with a resolution of 16 seconds of arc. The particle detectors included low energy electron, proton and alpha particle sensors, three semiconductor counters, two gas discharge counters, and a Cherenkov detector.

Lander:

Entry and descent:

1. Broadband photometer with three visible and two infrared channels for radiation flux
2. Narrow-band infrared photometer with three channels near 0.8 microns for radiation flux ratios in water, carbon dioxide and background bands
3. Backscatter and multi-angle nephelometers at 0.92 microns for light scattering between the altitudes of 63 and 18 km
4. Pressure and temperature measurements from 62 km to the surface
5. Accelerometers for atmospheric structure between 110 and 76 km
6. Mass spectrometer for atmospheric composition from 63 to 34 km altitude
7. Doppler experiment for wind and turbulence

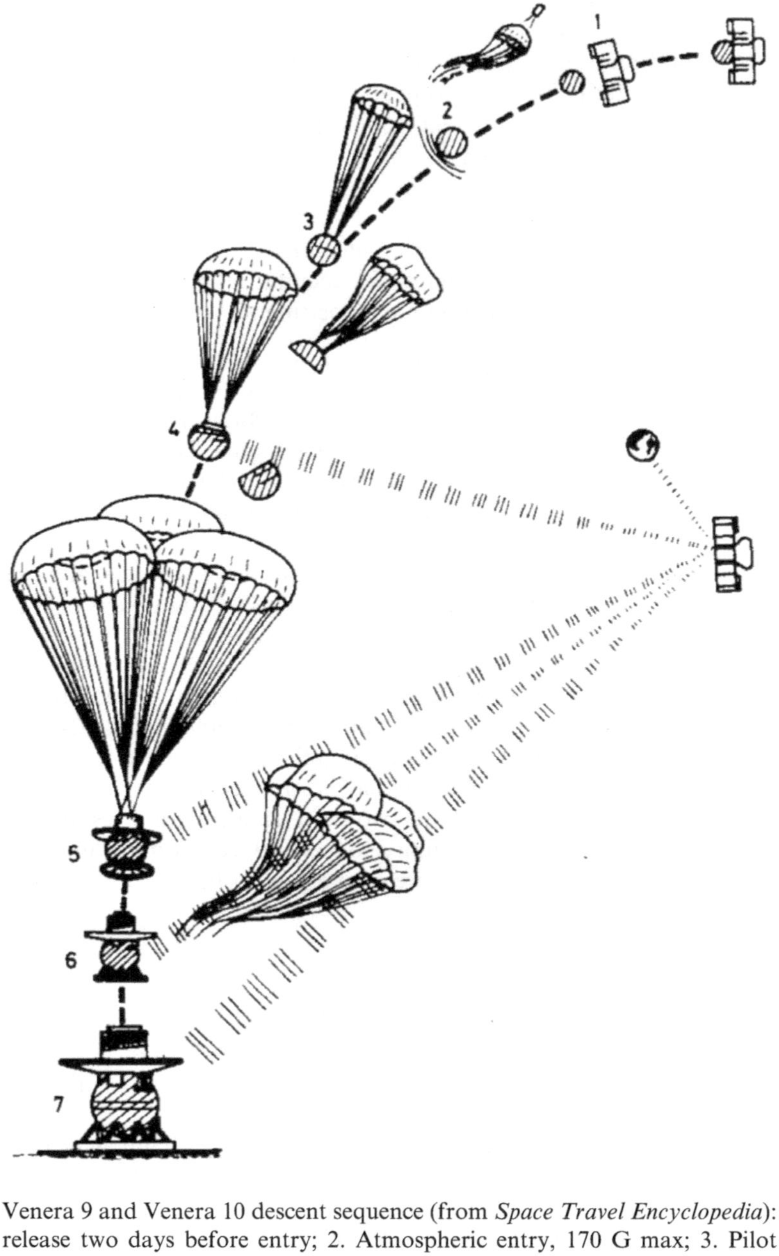

Figure 14.8 Venera 9 and Venera 10 descent sequence (from *Space Travel Encyclopedia*):
1. Capsule release two days before entry; 2. Atmospheric entry, 170 G max; 3. Pilot
chute withdraws first parachute; 4. First chute pulls top away and deploys second
braking chute. Radio and instruments activated; 5. Main chutes open at 62 km, bottom
shell jettisoned. Science investigations conducted during 20 min descent through clouds;
6. Lander released at 50 km altitude; 7. Lander on the surface 55 min later.

In comparison to Venera 8, the new photometers were greatly improved and more complex. Both upwelling and downwelling integrated radiation was measured in the range 0.440 to 1.160 microns using green, yellow, red, IR1, and IR2 glass filters for five wavelength bands with widths of 0.1 to 0.3 microns. This was complemented by an near-infrared photometer operating in three channels, one centered on the carbon dioxide band at 0.78 microns, another on the water band at 0.82 microns, and a third background channel at 0.80 microns, with each band being only 0.005 microns wide. Both back scattering and angular scattering nephelometers were carried. These were new, measuring how the atmosphere scattered light from a pulsed light source. This information could be used to infer the size distribution, refractive index, and density of cloud droplets. The sensors for the photometers and nephelometer were mounted in the external environment, had their own thermal protection, and were linked by fiber optics to the instruments inside. The mass spectrometer was a radio-frequency monopole unit with a pressure regulator designed for input pressures of 0.1 to 10 bar. The Doppler experiment was facilitated by an ultra-stable master oscillator for the transmitter.

Surface:

1. Panoramic imaging system, two cameras with floodlights
2. Surface wind rotary anemometer
3. Gamma-ray spectrometer (uranium, potassium, and thorium) for surface rocks
4. Gamma-ray densitometer

The scanning photometer imaging camera was similar to that carried by the M-71 landers and had a mass of 5.8 kg. There were two, in sealed insulated containers on either side of the lander just beneath the disk of the aerobrake to give a vantage point 90 cm off the ground. The rotational axis of the mirror system was tilted from the lander's vertical by 50 degrees in order that the center of the image was the surface directly in front of the camera at a distance of 1.5 to 2 meters, with the field of view extending 90 degrees to either side in order to include a small section of the horizon. The cameras peered through 1 cm thick cylindrical quartz windows using a lens to compensate for refraction and provide a total angular field of 40 × 180 degrees. Each 128 × 512 panorama consisted of a 115 × 512 image, with the first 13 bits of each line containing a calibration pattern. Each measurement consisted of a 6 bit picture element and 1 bit for parity checking. The quality of the imagery was limited by the projected 30 minute surface lifetime and the transmission rate of 256 bits/s. To send a panorama at 3.5 seconds per line would require 30 minutes. The panoramas were to be transmitted simultaneously on separate VHF channels. Because scientist were worried after Venera 8 that the illumination at the surface would be very weak, each camera was provided with a 10,000 lux floodlight system with two lamps in order to ensure that there would be sufficient light to acquire an image.

The deployable densitometer had a cesium-137 irradiation source and detectors to measure the gamma rays reflected back by the environment. During the descent this measured atmospheric scattering. Immediately after landing it deployed a 4 × 36 cm

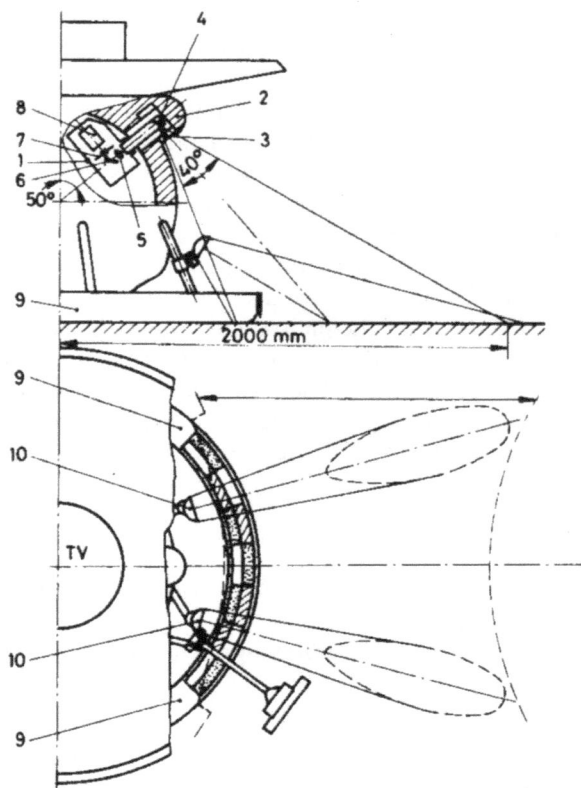

Figure 14.9 Television system mounted on Venera 9 and Venera 10 landers showing (a) camera and illuminator mounting, line-of-site FOV, and (b) imaging panorama and illuminator footprints (from *Space Travel Encyclopedia*): 1. Panoramic camera; 2. Insulation; 3. Camera port; 4. Scanning mirror; 5. Lens; 6. Mirror; 7. Pressure diaphragm; 8. Photometer; 9. Landing ring; 10. Illumination lamp.

'paint roller' on the surface in order to measure soil scattering. In addition, a sodium iodide gamma-ray spectrometer similar to that of Venera 8 was carried inside the sphere for measurement of potassium, uranium and thorium soil abundances. Two anemometers were mounted on the upper side of the aerobrake disk.

Mission description:

Venera 9 orbiter:

Venera 9 was launched on June 8, 1975. It maneuvered on June 16 and October 15 to align its trajectory with the desired entry point in the atmosphere of Venus. After releasing its entry system on October 20, it performed a 247.3 m/s deflection burn to head for the orbit insertion point, where on October 22 it made a 922.7 m/s burn and inserted itself into an orbit with a period of 48.30 hours. The relay of the transmission from the entry system and lander followed immediately thereafter. This was the first spacecraft to enter orbit around the planet. The initial orbit was 1,500 × 111,700 km inclined at 34.17 degrees. This was changed to 1,300 × 112,200 km and finally to 1,547 × 112,144 km at 34.15 degrees. The orbiter conducted 3 months of scientific observations which were terminated by the failure of its transmitter.

Venera 9 lander:

The entry system penetrated the atmosphere of Venus at a speed of 10.7 km/s and at an angle of 20.5 degrees. At 05:13 UT on October 22 the lander touched down at a speed of 7 to 8 m/s on the day-side of the planet at 31.01°N 291.64°E, where it was 13:12 Venus solar time and the solar zenith angle was 33 degrees. The site was on a slope of 15 to 20 degrees and the lander was tilted a further 10 to 15 degrees by the uneven, rocky surface. It immediately began its surface activities, relaying its data to Earth via the orbiter until that flew out of range 53 minutes later, by which time the temperatures inside the lander had risen to 60°C.

Venera 10 orbiter:

After launch on June 14, 1975, Venera 10 flew almost the same route as Venera 9, making trajectory corrections on June 21 and October 18, releasing its entry system on October 23 and then making a 242.2 m/s deflection burn. On October 25 it made a 976.5 m/s insertion burn. Its initial orbit was 1,500 × 114,000 km inclined at 29.50 degrees with a period of 49.38 hours. This was later changed to 1,651 × 113,923 km at 29.10 degrees. After relaying the transmission from the entry system and lander, the orbiter began its scientific observations. It succumbed 3 months later to the same problem as disabled its partner.

Venera 10 lander:

The entry system penetrated the atmosphere of Venus at an angle of 22.5 degrees at 01:02 UT on October 25. The lander touched down at 02:17 UT at a speed of about 8 m/s at 15.42°N 291.51°E, about 2,200 km from where Venera 9 landed. It was on the day-side, at 13:42 Venus solar time with a solar zenith angle of 27 degrees. The surface was fairly level but the lander was perched on a rocky mass that tilted it at about 8 degrees. The lander was still transmitting when the orbiter flew out of range, curtailing the relay operation after 65 minutes.

Results:

Venera 9 lander:

Entry and decent science

The Venera 9 lander inferred atmospheric density from accelerometer data between the altitudes 110 and 76 km. It directly measured atmospheric temperature, pressure, composition and light levels from 62 km to the surface. Light scattering data from the nephelometer, together with the photometer data, indicated clouds with a base at an altitude of 49 to 48 km and a lower loading of aerosols extending down to about 25 $\pm$ 5 km. The clouds were similar to a light fog, with much smaller drop size than normal for Earth and a visibility of several kilometers. Distinct layers were detected at altitudes of 60 to 57 km, 57 to 52 km and 52 to 49 km. The refractivity index was measured to be as high as 1.46; much higher than for water ice and consistent with sulfuric acid droplets. The cloud particles all scattered light but there was absorption in the blue and this, along with heavy Rayleigh scattering, led to increasing orange color with depth. The atmosphere below 25 km appeared to be free of aerosols. Red light was found to reach further towards the surface than blue, shifting the spectrum towards longer wavelengths, producing both orange colored skies and, by reflection, an orange tinted surface. Doppler data provided altitude profiles of horizontal wind speed and direction.

The detailed chemical composition measurements attempted by the first landers of this type gave poor results. The mass spectrometers did not function properly due to inappropriate cleaning procedures prior to launch and apparent clogging of the inlet system by cloud particles. The mixing ratio of molecular nitrogen to carbon dioxide was determined. Argon was detected in the atmosphere. A large ratio of argon-36 to argon-40 was measured and confirmed by later missions, but went unreported owing to mistrust of the instrument. Near-infrared photometer results for the water vapor mixing ratio proved to be spurious when spectral measurements were conducted on later missions.

Surface science

The photometers detected dust raised by the landing, but this quickly settled. The surface conditions were 455 $\pm$ 5°C and 85 $\pm$ 3 bar, and there was a light wind of 0.4 to 0.7 m/s.

Only one 180 degree panorama was taken because the cover for the other camera failed to deploy. This back-and-white image was the first picture from the surface of another planet. It showed a level landscape with a variety of flat, apparently young angular rocks without much erosion. A portion of the image extended to the horizon, and there was no indication of dust in the atmosphere. The illumination was similar to Earth mid-latitudes on a cloudy summer day, and the light scattering did not cast shadows. The floodlights for the camera were not triggered on. They were eliminated from subsequent missions. The visibility was a pleasant surprise to the scientists who, after reviewing Venera 8, had predicted a dark, murky and dusty atmosphere in which only the near field would be available for inspection.

Figure 14.10 Venera 9 lander 180 degree panorama (processing by Ted Stryk).

The indistinctness and apparent nearness of the horizon in all of the images from the Venera landers were due to the high refractivity of the dense atmosphere which made Venus appear to be a small-diameter spherical body with a horizon much less than 1 km away. The phenomenon is similar to a terrestrial mirage and is probably a function of the observer's height above the ground.

The gamma-ray composition analysis of the surface material measured potassium, uranium and thorium abundances more typical of terrestrial basalt than meteorites. The fact that the surface rocks differed from primitive meteorites in a way consistent with trends observed in terrestrial rocks indicated that Venus must have been thermal differentiated into a core, mantle and crust. The reflectivity of the surface in five wavelengths was consistent with material of a basaltic composition. The penetrometer indicated a rock density of 2.7 to 2.9 g/cc.

Venera 10 lander:

Entry and descent science

The Venera 10 lander inferred atmospheric density from accelerometer data between the altitudes 110 and 63 km. It directly measured atmospheric temperature, pressure, composition and light levels from 62 km to the surface, and structure, microphysical properties and composition of the clouds. The three distinct cloud layers observed by Venera 9 were confirmed. Doppler data profiled the horizontal wind speed and direction during the descent, and then an anemometer measured wind velocity on the surface. As the results were generally in agreement with Venera 9, some conclusions could be drawn on atmospheric convective stability and turbulence. The profile of temperature and pressure showed 33 bar and 158°C at 42 km, 37 bar and 363°C at 15 km, and 91 $\pm$ 3 bar and 464 $\pm$ 5°C at the surface.

Surface science

As in the case of Venera 9, one of the camera covers failed to deploy and this lander also only provided a single 180 degree black-and-white panoramic image. It showed a surface that was smoother with large, more eroded pancake rocks interspersed with lava or other weathered rocks. The horizon was visible and there was no evidence of dust in the atmosphere. As with Venera 9, the photometers detected some dust raised on touchdown which quickly settled. A surface albedo of 0.06 was derived from the imaging and photometers on both landers.

Figure 14.11 Venera 10 lander 180 degree panorama (processing by Ted Stryk).

The surface winds were light at 0.8 to 1.3 m/s. The gamma-ray results and surface reflectivity were suggestive of a basaltic composition. Apparently both landers came down on young volcanic shield structures with lavas close in composition to the tholeiitic basalts that emerge from oceanic spreading ridges on Earth. The penetrometer indicated a surface density of 2.7 to 2.9 g/cc, just as at the Venera 9 site. The surface of Venus appeared to be harder than the Moon or Mars.

Venera 9 and 10 orbiters:

The panoramic cameras returned 1,200 km long images taken using several different filters to distinguish cloud structures and some surface features, although the latter were poorly defined. The results included imagery of the clouds in the ultraviolet, infrared radiometry, photometry, spectrometry of both day and night sides, photo-polarimetry, radio occultation and plasma data. Orbital data suggested a cloud base at an altitude of 30 to 35 km with three distinct layers. The orbiters obtained data on the clouds above 64 km, which is the altitude at which the descent data started. The day-time temperature of the upper cloud was -35°C, warming by about 10 degrees at night. The night-time atmosphere was found to glow in the visible spectral range in bands that later investigation established to be a molecular oxygen band system that is not excited in the Earth's atmosphere owing to its lower concentration of carbon dioxide.

The airglow spectrometer on the Venera 9 orbiter found optical evidence of night-side lightning, but Venera 10 did not. Reflection spectra of clouds in the infrared at 1.7 to 2.8 microns measured the aerosol scale height near their upper boundary, and infrared wide band radiometry in the range 8 to 28 microns prompted the conclusion that outgoing radiation is systematically stronger on the night-side than the day-side.

The dual frequency radio occultations at wavelengths of 8 and 32 cm gave a set of temperature and pressure profiles for altitudes in the range 40 to 80 km that revealed details of the night-side ionosphere and the existence of a large diurnal variation of ionospheric electron density. The bistatic radar experiment mapped fifty-five strips of the surface 100 to 200 km wide by 400 to 1,200 km long. Early analysis provided one-dimensional terrain profiles at a resolution of 20 to 80 km. Later processing of the Venera 10 data produced a two-dimensional local topography for five regions at a resolution of 5 to 20 km.

Measurements of the scattering of solar Lyman-alpha radiation by the hydrogen corona that surrounds Venus, including its line width, gave an estimate of 450°C for

Figure 14.12 Mosaic of the planet from images by the Venera 9 orbiter (courtesy Ted Stryk).

the temperature of the atmosphere at the exobase. Many features of how the solar wind interacts with the ionosphere were measured. No intrinsic planetary magnetic field was detected. Nonetheless, the interaction of the solar wind with the ionosphere created a magnetic plasma tail.

As the first spacecraft to enter Venus orbit, Venera 9 and 10 were able to provide the first long-term survey of the Venusian atmosphere with a comprehensive battery of scientific instrumentation. Their landers performed marvelously, returning the first pictures of the surface of the planet. These missions began what became an unbroken string of successes running to Venera 16 in 1983 and ending with the two Vega missions that made flybys in 1985.

15

Repeating success at Venus

TIMELINE: 1977–1978

With nothing left to accomplish on the Moon, and having abandoned Mars for the immediate future, Soviet scientists and engineers focused their robotic exploration solely on Venus. In 1978 they launched a second pair of spacecraft which were near duplicates of Venera 9 and 10. Because the energetics for this opportunity were less favorable, it was not practicable to send an orbiter/lander and instead the lander was to be delivered by a spacecraft that would perform a flyby and relay to Earth the data from the entry system and lander. Although both of the Venera 11 and 12 landers touched down, they suffered a number of problems and in particular were unable to provide imagery.

The US also sent spacecraft to Venus in 1978, but these were very much smaller. The Pioneer 12 Venus orbiter was an outstanding success, reporting information on the upper atmosphere for many years. Pioneer 13 adopted a collision course and deployed one large and three small entry probes, all of which successfully returned atmospheric data during their descent.

Launch date

1977

20 Aug	Voyager 2 Outer Planets Tour	Success
5 Sep	Voyager 1 Outer Planets Tour	Success

1978

20 May	Pioneer 12 Venus orbiter	Success
8 Aug	Pioneer 13 Venus multi-probe	Success
12 Aug	International Comet Explorer	Success flyby of comet G-Z
9 Sep	Venera 11 flyby/lander	Success, lander imager failed
14 Sep	Venera 12 flyby/lander	Success, lander imager failed

DRILLING INTO VENUS: 1978

Campaign objectives:

The 1978 Venus campaign objectives were to repeat the resounding successes of the Venera 9 and 10 landers with new instruments to analyze both the atmosphere and surface. The 1976–77 opportunity was skipped in order to build the new apparatus, and a launch in 1978 dictated a much higher arrival velocity at Venus than in 1975. The larger propellant load required for the longer orbit insertion burn was unable to be accommodated together with the mass of the new instruments, so the carrier was downgraded to a flyby role. A positive outcome was that a flyby spacecraft would remain in view of the lander for longer in order to relay data from the surface. Both of the previous landers had still been operating when their orbiters flew below their horizons. The new lander investigations featured a high resolution color camera and an experiment to drill into the surface. The descent investigations included new experiments to study the chemical composition of the atmosphere, the nature of the clouds, and any electrical activity in the atmosphere. The flyby instrumentation was reduced in order to maximize the mass available for the descent and surface science.

Spacecraft launched

First spacecraft:	Venera 11 (4V-1 No.360)
Mission Type:	Venus Flyby/Lander
Country/Builder:	USSR/NPO-Lavochkin
Launch Vehicle:	Proton-K
Launch Date/Time:	September 9, 1978 at 03:25:39 UT (Baikonur)
Encounter Date/Time:	December 25, 1978
Outcome:	Successful.

Second spacecraft:	Venera 12 (4V-1 No.361)
Mission Type:	Venus Flyby/Lander
Country/Builder:	USSR/NPO-Lavochkin
Launch Vehicle:	Proton-K
Launch Date/Time:	September 14, 1978 at 02:25:13 UT (Baikonur)
Encounter Date/Time:	December 21, 1978
Outcome:	Successful.

Spacecraft:

Although assigned only to a flyby role the Venera 11 and 12 spacecraft were almost identical to their orbiter predecessors, but the lander relay was increased to 3 kbits/s per channel. After releasing the entry system 2 days prior to arriving at the planet, Venera 11 (and all later flyby spacecraft) made a deflection maneuver to establish a flyby which would enable it to relay to Earth for longer than was possible using an

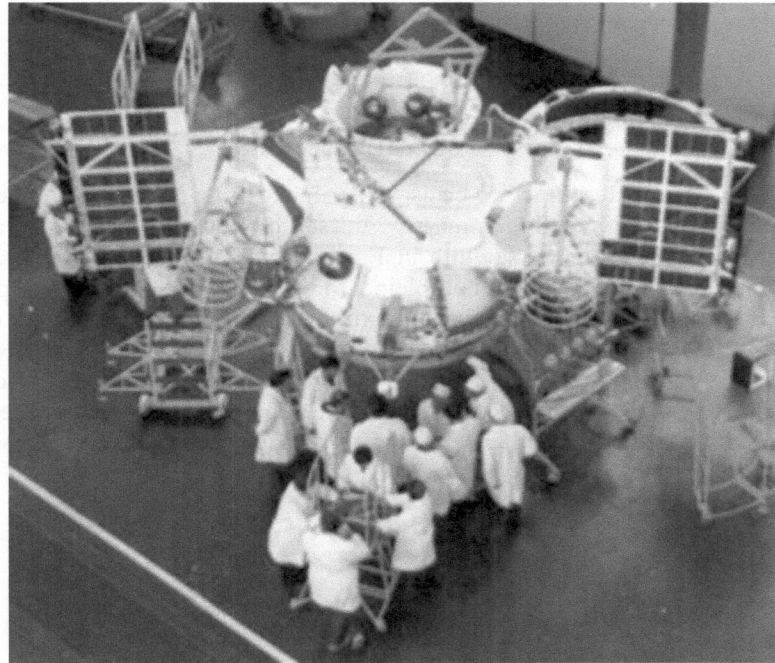

Figure 15.1 Venera 11 in test.

orbiter. The spacecraft were identical, and the landers had the same configuration as their recent predecessors but the floodlights were deleted and the camera lens cap was redesigned. The complexity and mass of the parachute system was reduced to accommodate more instruments. Only a single supersonic braking parachute was used instead of a sequence of two, and only one main parachute was used instead of a system of three. Some of the instruments were modified and new ones were added, in some cases being installed on the shock absorbing impact ring. All landers from now through to Vega 2 carried a technology experiment consisting of a set of small solar cells arranged around the lander ring.

Launch mass:	4,450 kg (Venera 11) 4,461 kg (Venera 12)
Flyby dry mass:	2,127 kg
Entry mass:	1,600 kg
Lander mass:	731 kg

Payload:

Flyby spacecraft:

1. Extreme-ultraviolet (30 to 166 nm) spectrometer (France)
2. Magnetometer

3. Plasma spectrometer
4. Solar wind detectors
5. High energy particle detectors
6. KONUS gamma-ray burst detector
7. SNEG gamma-ray burst detectors (France-USSR)

KONUS was an interplanetary cruise experiment to try and identify the source of mysterious astronomical gamma-ray bursts by having the two spacecraft coordinate with the Prognoz satellite in Earth orbit to triangulate on individual bursts. SNEG was an instrument complementary to KONUS, built in cooperation with the French. And a new French-built extreme-ultraviolet spectrometer covered the spectral lines of atomic hydrogen, helium, oxygen and other elements that it was thought might be present in the exosphere of Venus. The solar wind detector was a hemispherical proton telescope, and the high energy particle experiments used four semiconductor counters, two gas-discharge counters and four scintillation counters.

Figure 15.2 Venera 11 lander. The midriff panel is removed providing a view into the interior, and some of the instruments are labeled (from Don Mitchell).

Lander:

Entry and descent:

1. Scanning spectrophotometer (0.43 to 1.17 microns)
2. Mass spectrometer for atmospheric composition
3. Gas chromatograph for atmospheric composition
4. Nephelometer for aerosols of about 1 micron in size
5. X-ray fluorescence spectrometer for elemental composition of aerosols
6. Accelerometers for atmospheric structure from 105 to 70 km
7. Temperature and pressure sensors for 50 km to the surface
8. GROZA radio sensor at 8 to 95 kHz for electrical and acoustic activity
9. Doppler experiment for wind and turbulence

The experiments for determining the light scattering properties of the atmosphere were modified for a higher spectral resolution. The angular scattering nephelometers were deleted because they had satisfactorily measured the particle size and refractive index when carried on the Venera 9 and 10 landers. Instead, only the back scattering nephelometers were retained to examine the spatial uniformity of the cloud layers in different regions of the planet. The scanning spectrophotometer was improved for a higher spectral resolution. Every 10 seconds it measured radiation coming from the zenith using a ramp interference filter at a resolution of about 20 nm continuously over the range 430 to 1,170 nm, and the angular distribution of radiation (a full 360 degrees) in the vertical plane in the bands 0.4 to 0.6, 0.6 to 0.8, 0.8 to 1.3, and 1.1 to 1.6 microns using a rotating prism mounted on the aerobrake. The temperatures and pressures were measured by a suite of four thermometers and three barometers. The monopole radio-frequency mass spectrometer of Venera 9 and 10 was replaced by a Bennett radio-frequency design and the inlet system modified to prevent it becoming clogged by cloud particles which might then contaminate the atmospheric readings. The microscopic leak admitting the atmosphere to the instrument was replaced by a piezoelectric valve that would open a relatively large hole for a very short time in order to admit a pulse of atmosphere into a long sample tube which would trap cloud particles. In addition the instrument was not to be operated until the lander was at about 25 km, well below the aerosols. The apparatus was pumped down between atmospheric readings to purge the sample. Two other new atmospheric composition experiments were included. The gas chromatograph used neon to carry atmospheric samples through columns of porous materials and a Penning ionization detector. It had one column 2 meters long that was optimized for water, carbon dioxide and the compounds hydrogen sulfide, carbonyl sulfide, and sulfur dioxide; a second column 2.5 meters long for the volatile gases helium, molecular hydrogen, argon, molecular oxygen, molecular nitrogen, krypton, methane and carbon monoxide; and a third column just 1 meter long specifically for argon. An x-ray fluorescence spectrometer used gamma rays to excite the emission of x-rays from cloud particles collected on a cellulose acetate filter by drawing atmospheric gas through the instrument, thereby measuring the elemental composition of the aerosols.

The GROZA experiment comprised an acoustic detector and an electromagnetic wave detector using loop antennas with four narrow-band receivers at 10, 18, 36 and 80 kHz and a wide-band receiver over the range 8 to 95 kHz. This was to start at an altitude of 60 km and operate down to and on the surface. The electromagnetic wave detector was to register radio bursts from lightning and the acoustic signals could be interpreted in terms of thunder, wind speed past the lander during the descent and, while on the ground, perhaps even seismic quakes.

Surface:

1. Panoramic two-camera color imaging system
2. Soil drill with x-ray fluorescence spectrometer analysis system
3. Rotating conical soil penetrometer (PrOP-V)

The panoramic camera system had been improved by adding clear, red, green and blue filters for three-color imaging, and by increasing the image quality from 128 × 512 pixels at 6 bit encoding to 252 × 1,024 pixels at 9 bit encoding and 1 bit parity. It was capable of resolving detail as fine as 4 or 5 mm at a range of 1.5 meters. The transmission bandwidth had been increased by a factor of twelve, one reason for this being the Kvant-D upgrade to the Soviet communications facilities; in particular the introduction of 70 meter antennas at Yevpatoria and Ussuriisk. The increase in the transmission rate from the surface of Venus from 256 bits/s to 3,000 bits/s enabled a color panorama to be sent in 14 minutes, as against 30 minutes for a lower resolution black-and-white panorama previously.

The gamma-ray soil analysis instrument inside the Venera 8, 9 and 10 landers was replaced by a superior instrument. A drill mounted on the shock-absorbing impact ring was to core a sample of the surface and then pass it through a series of pressure-reduction stages to the x-ray fluorescence spectrometer carried inside the lander. The penetrometer was on a deployable arm and reported its results on a dial that was to be read by the cameras.

Mission description:

Venera 11 lander:

Venera 11 was launched on September 9, 1978, and made midcourse corrections on September 16 and December 17. After the entry capsule was released on December 23 the spacecraft made the deflection burn in order to perform a flyby of the planet at the desired relay communications altitude, and on December 25 the entry system hit the atmosphere at 11.2 km/s. After a 1 hour descent the lander touched down at a speed of 7 to 8 m/s on the day-side at 14°S 299°E. It was 03:24 UT, 11:10 Venus solar time, and the solar zenith angle was 17 degrees. The lander transmitted from the surface for 95 minutes before the relay spacecraft flew over the horizon after 110 minutes, so none of the transmission was lost.

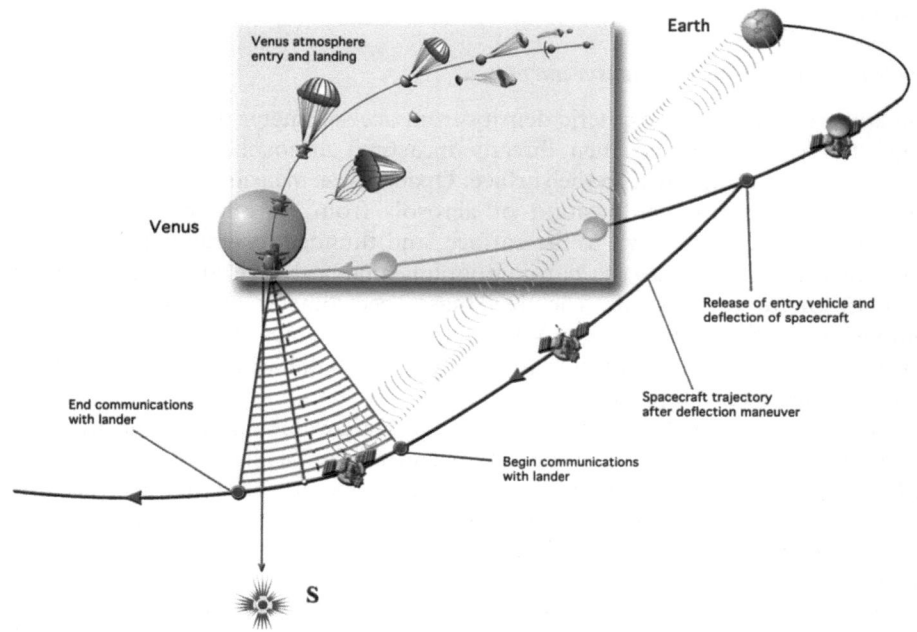

Figure 15.3 Venera 11 and Venera 12 encounter design, showing entry capsule targeting followed by flyby vehicle deflection and lander relay communications.

Venera 12 lander:

Venera 12 was launched on September 14, 1978, pursued a faster trajectory with midcourse corrections on September 21 and December 14, and arrived ahead of its partner. It released its entry system on December 19. This entered the atmosphere on December 21 at a velocity of 11.2 km/s. The parachute was jettisoned at 49 km and after a 1 hour descent the lander touched down at about 8 m/s on the day-side at 7°S 294°E. It was 03:30 UT, 11:16 Venus solar time, and the solar zenith angle was 20 degrees. Unlike Venera 11, it kicked up a cloud of dust that took about 25 seconds to settle. Both landers encountered an unexplained anomaly at an altitude of 25 km, where instrument readings went off-scale and there was an electrical discharge from the vehicle. This lander transmitted from the surface for 110 minutes until the flyby spacecraft passed below the horizon. It is therefore not known when it finally ceased to function.

Venera 11 and 12 flyby spacecraft:

After the deflection maneuver, each spacecraft flew by Venus at a range of about 35,000 km and relayed the data from its lander to Earth throughout the descent and then during the period of surface activity. The last reports from the flyby spacecraft were in January 1980 for Venera 11 and March 1980 for Venera 12.

Results:

Venera 11 and 12 decent measurements:

The landers inferred atmospheric density from accelerometer data over the altitude range 100 to 65 km and then directly measured atmospheric temperature and pressure from 61 km down to the surface. Opacity was measured from 64 km to the surface, the chemical composition of aerosols from 64 km to 49 km, aerosol scattering from 51 km down to the surface, and thunderstorm activity from 60 km down to the surface. The gas chromatograph analyzed nine atmospheric samples from 42 km to the surface. The new mass spectrometer measured atmospheric composition from 23 km to 1 km. Wind velocities were measured from about 23 km down to the surface, and altitude profiles of horizontal wind speed and direction were obtained from Doppler data.

The spectrophotometer produced the first realistic water vapor profile, identifying water vapor as the second most important greenhouse gas in the atmosphere (after carbon dioxide). The contemporary analysis indicated a profile that decreased from 200 ppm at the cloud base to 20 ppm at the surface, but a re-analysis many years later obtained a better fit for the Venera 11, 12 and 14 spectrophotometer data using a constant mixing ratio for water vapor of about 30 ppm from 50 km to the surface. The mass spectrometers on these missions reported values as high as 0.5% at 44 km and 0.1% at 24 km; these are much larger mixing ratios for water vapor than were obtained from the spectrophotometer and other remote spectral measurements from Earth, and are considered suspect.

The mass spectrometer results from Venera 11 and 12 obtained by the analysis of 176 complete spectra of 22 samples were reported as:

carbon dioxide	97%
molecular nitrogen	4.0 ± 2.0 %
argon	110 ± 20 ppm
neon	8.6 ± 4 ppm
krypton	0.6 ± 0.2 ppm

with isotopic ratios as follows:

carbon 13/12	0.0112 ± 0.0002
argon 40/36	1.19 ± 0.07
argon 38/36	0.197 ± 0.002

The gas chromatograph made eight measurements between 42 km and the surface with the following results:

molecular nitrogen	2.5 ± 0.3 %
water vapor	25 to 100 ppm
argon	40 ± 10 ppm
sulfur dioxide	130 ± 35 ppm
carbon monoxide	28 ± 7 ppm (low altitudes)
molecular oxygen	less than 20 ppm

krypton   detected
hydrogen sulfide   detected
carbonyl sulfide   detected

The x-ray fluorescence spectrometer on Venera 12 measured cloud particles from 64 km to 49 km and was then overcome by the high temperatures. It missed sulfur (< 0.1 mg/m^3), but found chlorine (0.43 ± 0.06 mg/m^3) in cloud particulates. The chlorine was suspected to be a non-volatile compound such as aluminum chloride at the time, but it was not specifically identified. The large amount of chlorine relative to sulfur was incompatible with the theory that the clouds were composed of sulfuric acid droplets, but these anomalous data were corrected by the Venera 14 mission.

Both Venera 11 and 12 detected a large number of electromagnetic pulses in the descent from 32 to 2 km similar to those produced by distant lightning flashes on Earth. The activity was more intense on Venera 11 than Venera 12, and diminished in intensity towards the surface. No such pulses were detected by Venera 11 after it touched down, but one large burst was noted by Venera 12 while on the surface. The microphones were saturated by aerodynamic noise during the descent and detected no thunder on the surface, but they did pick up sounds issued by the instruments and surface activities.

As with Venera 9 and 10, the light scattering data indicated clouds with a base at an altitude of 47 km, and a much lower loading of aerosols below that. Venera 11 and 12 found the atmosphere to be generally free of aerosols below ~ 30 km. The nephelometer on Venera 11 measured cloud particles throughout the descent and its results confirmed the uniformity of the cloud layers as reported by Venera 9 and 10. The base cloud layer was located between 51 and 48 km, with a mist below that. The nephelometer on Venera 12 did not function correctly. It was confirmed that only about 3 to 6% of the sunlight reaches the surface. Intense Rayleigh scattering in the dense atmosphere gives poor visibility. Above several kilometers altitude the surface must be invisible. At ground level the horizon will be visible, but the detail of the landscape must fade quickly into an orange haze. The Sun is not visible as a disk, merely a uniformly lit hazy sky.

Venera 11 and 12 surface measurements:

The temperature at the Venera 11 landing site was $458 \pm 5°$C and the pressure was 91 ± 2 bar. There was no surface imaging because the lens covers would not open. These had been redesigned after the problems with one camera on each of Venera 9 and 10, but with disastrous results. The transmitted pictures were uniformly black. The soil drill collected a sample, but it was not properly delivered to the instrument container and no soil analysis was accomplished.

The temperature at the Venera 12 site was $468 \pm 5°$C and the pressure was 92 ± 2 bar. The fact that this suffered exactly the same camera and soil analysis experiment failures as its partner implied a systematic design flaw. Vibrations while descending broke the sample transfer system on the drill and no soil analysis was possible. The soil penetrometers also failed on both landers.

The surface experiments were an almost total failure on both landers. It is possible that they suffered rough landings which damaged the instruments that were mounted on the impact ring. The lack of results was very disappointing, but in typical Soviet fashion this spurred the engineers on to succeed at the next flight opportunity.

Venera 11 and 12 flyby spacecraft:

The ultraviolet spectrometer detected Lyman-alpha emissions from hydrogen atoms and 584 angstrom (He-I) emissions from helium atoms. These provided exospheric temperatures and number densities. Time profiles for 143 gamma-ray bursts were obtained by Venera 11 and 12 and the results triangulated with an identical detector on Prognoz 7 in Earth orbit. On February 13 and March 17, 1980, Venera 12 used its extreme-ultraviolet spectrometer to observe Comet Bradfield.

16

Back to Venus again

TIMELINE: 1979–1981

The pace of planetary exploration slowed considerably after 1978, particularly in the US. There were no launches to the Moon or the planets in 1979–80. It was a striking contrast to the hectic 1960s and 1970s. In fact there would be no lunar or planetary launches by the US during the eleven years between 1978 (Pioneer Venus) and 1989 (Galileo). Continuing their assault on Venus using the successful Venera spacecraft, the Soviets had the field to themselves. At the next Venus opportunity in 1981 they launched another pair of Venera flyby/landers. Both were successful, and this time produced the first color images of the surface of the planet.

Launch date

1979
No missions

1980
No missions

1981

30 Oct	Venera 13 flyby/lander	Success, first color images from surface
4 Nov	Venera 14 flyby/lander	Success

COLOR PICTURES FROM THE SURFACE OF VENUS: 1981

Campaign objectives:

Venera 11 and 12 had not completed the goals set for them. Whilst the experiments carried out during the descent produced a prodigious amount of information on the atmosphere, the surface science was mostly a failure. The Soviets skipped a launch opportunity in order to develop new heat-resistant technologies in order to fix these problems. In 1981 they were ready to try again with better devices and instruments, principally aiming to obtain color images from the surface and to analyze a sample obtained by a drill. The landing sites were chosen in collaboration with US scientists using maps based on the radar imaging experiment performed by the Pioneer orbiter in 1978.

Spacecraft launched

First spacecraft:	Venera 13 (4V-1M No.760)
Mission Type:	Venus Flyby/Lander
Country/Builder:	USSR/NPO-Lavochkin
Launch Vehicle:	Proton-K
Launch Date/Time:	October 30, 1981 at 06:04:00 UT (Baikonur)
Encounter Date/Time:	March 1, 1982
Outcome:	Successful.
Second spacecraft:	Venera 14 (4V-1M No.761)
Mission Type:	Venus Flyby/Lander
Country/Builder:	USSR/NPO-Lavochkin
Launch Vehicle:	Proton-K
Launch Date/Time:	November 4, 1981 at 05:31:00 UT (Baikonur)
Encounter Date/Time:	March 5, 1982
Outcome:	Successful.

Spacecraft:

As flyby spacecraft, Venera 13 and 14 were essentially identical to their immediate predecessors. The major changes for 1981 concerned the lander. In particular, metal teeth were added to the periphery of the impact ring in an effort to reduce the spin and oscillation during the descent and prevent the rough landings experienced by the 1978 missions. The camera lens cover problem was fixed, and the soil sampler was redesigned to alleviate the problem that disabled it on Venera 11 and 12.

Launch mass:	4,363 kg
Flyby wet mass:	2,718 kg
Entry mass:	1,645 kg
Lander mass:	760 kg

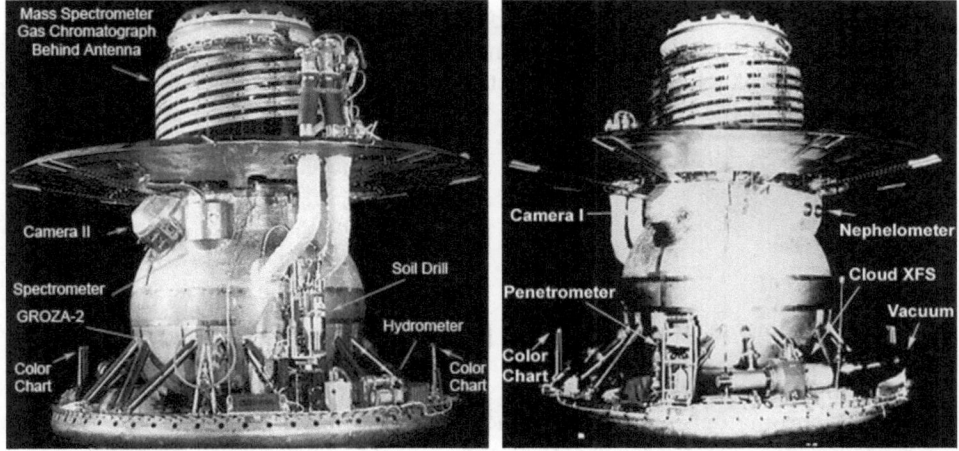

Figure 16.1 Venera 13 and Venera 14 landers and instruments (from Don Mitchell).

Figure 16.2 Venera 13 (from Don Mitchell) and lander with entry shell.

Payload

Flyby spacecraft

1. KONUS gamma-ray burst detector
2. SNEG gamma-ray burst detectors (France-USSR)
3. Magnetometer (Austria)
4. High energy particle cosmic ray detector
5. Solar wind detectors

The flyby spacecraft carried a reduced payload to facilitate an increased payload on the lander. Updated forms of the cosmic ray experiment and the two gamma-ray investigations were included, and an Austrian magnetometer on a 2 meter boom was attached to one of the solar panels.

Flyby payload mass: 92 kg

Lander

Entry and descent

1. Accelerometers for atmospheric structure (110 down to 63 km) and then lander impact analysis
2. Temperature and pressure sensors
3. Gas chromatograph for atmospheric composition
4. Mass spectrometer for chemical and isotopic composition
5. Hydrometer for water vapor content
6. Nephelometer for aerosol studies
7. X-ray fluorescence spectrometer for elemental composition of aerosols
8. Spectrophotometer for spectral and angular distribution of solar radiation
9. Ultraviolet photometer 320 to 390 nm
10. GROZA-2 radio for electrical activity and microphone for acoustic and seismic events
11. Doppler experiment for wind and turbulence

Surface

1. Panoramic color imaging system with two cameras
2. Drill and surface sampler
3. X-ray fluorescence spectrometer for surface rock elemental composition
4. Rotating conical soil penetrometer (PrOP-V)
5. Chemical oxidation state indicator

Several of the instruments were improved from their Venera 11 and 12 versions, including the camera, drill, spectrophotometer, x-ray spectrometer for aerosols, mass spectrometer and gas chromatograph. The hydrometer humidity sensor and chemical oxidation state indicator were new; the latter being a simple chemical indicator to search for traces of oxygen in the atmosphere. The spectrophotometer measured the full spectrum from 470 to 1,200 nm using a wide-angle sky view and an array of six narrow-angle directional views. The gas chromatograph was equipped with a better detector, could sense more species, and was capable of operating in the cloud layer. The mass spectrometer was improved to provide 2 to 40 times better mass resolution with 10 to 30 times greater sensitivity. The anomalous krypton reading of the 1978 missions was understood and corrected. The GROZA-2 instrument now included a modified 10 kHz detector, a new detector in the 2 kHz band, and was better able to search for seismic events after touchdown.

The panoramic imaging system was fitted with an improved lens cover. The large increase in bandwidth for the transmission to the flyby relay spacecraft allowed each camera to have clear, red, green, and blue filters. Each camera was to cycle through its 180 degree panoramic image four times, one for each filter, taking almost an hour to complete. To ensure that a color section would be transmitted if the lander lasted only 30 minutes, one camera first scanned a full 180 degrees through the clear filter and then scanned each of three 60 degree sections in turn with red, green and blue filters.

It is a testament to the technology of the Venera landers that so many instruments were mounted outside and hence exposed to the pressure, temperature and corrosive atmosphere of Venus. These included the drill, penetrometer, mass spectrometer, gas chromatograph, hydrometer, aerosol x-ray spectrometer, oxidation state indicator, and the GROZA radio sensors and microphone. The cameras and nephelometer were mounted inside with special housings attached to the pressure vessel to enable them to observe out through windows and prisms. The platinum thermometer and aneroid pressure instrument used external sensors with wires fed through the pressure vessel.

The drill sampler was on the base of the lander. The machining of this device had to take account of thermal expansion at 500°C, and the process of drilling, sampling and analysis had to complete within the 30 minute guaranteed lifetime of the lander. A telescoping drill head was to bore into the surface for about 2 minutes, reaching as much as 3 cm into solid rock and coring a 2 cc sample. Information on the speed and

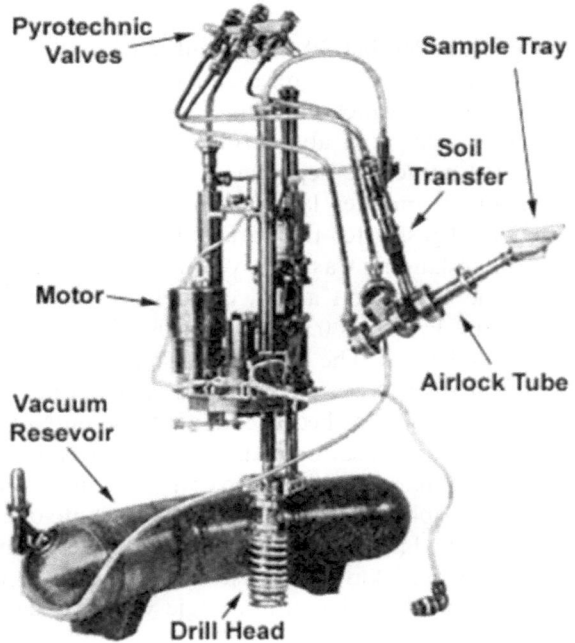

Figure 16.3 Venera 13 and Venera 14 lander drill mechanism (from Don Mitchell).

movement of the drilling rig, depth of drill penetration, and magnitude of the current drawn by the electric motor while drilling, provided information on the physical and mechanical properties of the surface. After the few grams of material gained by the drill had been deposited on a tray, this would be moved pyrotechnically through a three-stage airlock to transfer it from the ambient 90 bar pressure to just 0.06 bar in the analysis chamber of the x-ray fluorescence spectrometer, which used plutonium, uranium-235 and iron-55 sources. The gas pumping for this transfer was enabled by a vacuum reservoir on the base of the lander.

The dynamic penetrometer was designed to determine the mechanical properties of the surface material, whether rock or soil. It consisted of a cone-shaped die that could pivot on the end of a lever and fall forward onto the ground. A sighting device within the field of view of one of the cameras determined the depth of penetration. After penetration, a spring would cause the die to rotate in the soil and the angle of its turn would be displayed on the sighting device. A cable unit was also attached to one end of the die and an electronic unit within the lander could measure electrical resistance.

Lander payload mass: 100 kg

Mission description:

Landers

Venera 13 was launched on October 30, 1981, and made midcourse corrections on November 10 and February 21, 1982. It released its entry capsule on February 27 at a range of 33,000 km from the planet, and this entered the atmosphere on March 1. The accelerometer was turned on at about 100 km altitude and provided data on atmospheric density until parachute deployment. The parachute opened at 62 km and was jettisoned about 9 minutes later at an altitude of 47 km. The descent instruments were activated just after the parachute opened, and the descent time from parachute opening to landing was just over an hour. The lander hit at 7.5 m/s, bounced once, and came to rest on a flat, crumbly surface in an elevated hilly landscape at 7.55°S 303.69°E. It was 03:57:21 UT, 09:27 Venus solar time, and the solar zenith angle was 36 degrees. The surface transmission lasted 127 minutes.

Venera 14 was launched on November 4, 1981, and required three midcourse corrections to reach Venus. This was because the first one on November 14 was not executed properly. After a compensating maneuver on November 23, the final trim maneuver was made on February 25, 1982. The entry system was released on March 3, and this entered the atmosphere on March 5. The parachute opened at 62 km and was jettisoned at 47 km. The lander touched down at about 7.5 m/s on a low lying plain at 13.055°S 310.19°E, 950 km southwest of its partner. It was 07:00:10 UT, 09:54 Venus solar time, and the solar zenith angle was 35.5 degrees. The surface transmission lasted 57 minutes.

Venera 13 and 14 experienced the same electrical anomaly as had Venera 11 and 12 at 12.5 km altitude. The science sequences worked very well on descent for both

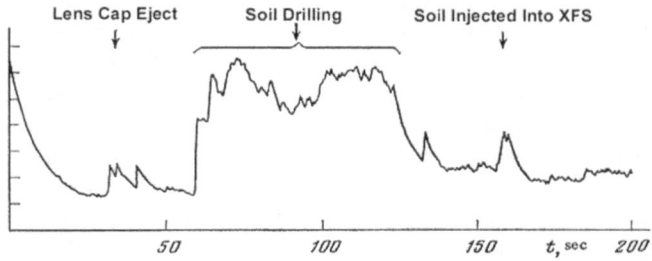

Figure 16.4 Venera 13 microphone data, t = 0 at landing (from Don Mitchell).

landers and also on the surface. Immediately upon landing, the lens covers popped off and imaging began. A black-and-white contingency image was transmitted first, followed by the color panoramas. The drill experiment was also started immediately after landing and the entire process of drilling, sampling, and analysis took a little over 32 minutes. Venera 13 obtained a 2 cc sample and Venera 14 obtained a 1 cc sample, each from a depth of 3 cm. The Venera 13 penetrometer worked well, but by bad luck on Venera 14 this experiment deployed on top of an ejected lens cover. The microphones picked up aerodynamic noise during the descent, the sounds of the landing, and then the deployment of the lens covers, the drilling noise and sampler pyrotechnics, winds at the site, and other sounds of the lander at work.

Flyby carrier spacecraft:

The flyby carriers each passed Venus at an approximate closest point of approach at 36,000 km and entered heliocentric orbit where they continued to return data on the Sun, including flares. As tests to assist in planning late midcourse maneuvers for the Halley's comet encounters by the Vega missions, Venera 13 fired its engine on June 10, 1982, and Venera 14 did so on November 14, 1982.

Results:

Venera 13 lander:

Descent measurements:

The microphysical properties reported by the nephelometers indicted three distinct layers in the main cloud system: a dense top layer from 60 km (the altitude at which the measurements began) down to 57 km, then a transparent layer from 57 to 50 km, and finally the densest layer from 50 to 48 km. Doppler measurements gave a wind profile. Interestingly, the humidity sensors on both Venera 13 and 14 indicated ten times more water in the atmosphere than was determined spectroscopically in the 46 to 50 km range. The water mixing ratio was determined from several instruments and, despite some conflicting values, it seemed to be greatest in the cloud formation region between 40 and 60 km with less water above and below the cloud layers. The amount of water vapor at 48 km was estimated at 0.2%. The mass spectrometer was not opened until below the cloud layers to preclude it being clogged by aerosols, and it provided altitude profiles of a number of atmospheric constituents between 26 km and the surface. These findings include a neon isotope ratio slightly higher than for Earth but less than for the Sun, a small krypton mixing ratio significantly less than that measured by the gas chromatograph and by Venera 11 and 12, and an argon-40 mixing ratio about four times lower than for Earth. The gas chromatograph detected some new species including molecular hydrogen, hydrogen sulfide, and carbonyl sulfide. Other species detected included molecular oxygen, water vapor, krypton, and sulfur hexafluoride.

The coronal discharge detectors of the GROZA-2 experiment established that no vehicle discharges were responsible for the very low frequency bursts interpreted as lightning by the Venera 11 and 12 experiments. No lightning or electrical discharges were detected.

Surface measurements:

Venera 13 obtained both black-and-white and the first color panoramic images from the surface of Venus. These show the lander ring base with triangular 'crown' points for lander stabilization during the descent, the ejected camera covers, the color test pattern strips, the deployed PrOP-V penetrometer, and the exhaust port of the aerosol x-ray fluorescence instrument on the lander ring just to the left of the penetrometer. The landing site appeared to be composed of bedrock outcrops surrounded by dark, fine-grained soil. The Venera 13 and 14 panoramas both showed flat, layered stones with a dark soil between them, and scattered small grains to compose a scene reminiscent of the floor of a terrestrial ocean. The photometers noted dust raised by

Figure 16.5 Venera 13 lander hemisphere 1 color panorama (processing by Ted Stryk).

Figure 16.6 Venera 13 lander hemisphere 2 color panorama (processing by Ted Stryk).

the landing, but it quickly settled and the resulting sedimentation was apparent by comparison of sequential images through different filters.

The Venera 13 site showed dark, flat rocks distributed over a darker crumbled soil surface with some low, rolling ridges in the background. The cameras returned four panoramas each, one in each filter: clear, red, green and blue; the latter three for the color image. However, precise color balancing was difficult to achieve in processing the filter data because the deployed calibration strips were affected by heat, pressure, and the orange sky, and because the radiometric response of the camera was not well known.

The drill sample analysis indicated a potassium-rich basalt of a type that is rare on Earth. The penetrometer showed the load-bearing capacity of the soil to be similar to heavy clays or compacted dust-like sand. These are both consistent with the surface characteristics derived from stress-strain profiles measured in the mechanics of the landing, which indicated a surface covered by a weak, porous material similar to the properties of weathered basalt. The electrical resistivity of the surface was surprising low, in the semi-conductor range, perhaps due to a thin film of conducting material on insulating soil particles.

The exposed-surface chemical test on both landers for the oxidation state of the Venusian atmosphere seemed to indicate a reducing carbon dioxide-rich atmosphere instead of an oxidizing one, but the experiment may have been compromised by the dust disturbed upon landing. Due to the short lifetime on the surface, the GROZA-2 acoustic experiment was designed to detect the microseismic events that commonly occur at a rate of about one every few seconds on Earth. No events were detected by Venera 13, but two such events may have been noted by Venera 14. The Venera 13 microphone recorded the aerodynamic noise of the descent, and on the surface this data gave wind speeds of 0.3 to 0.6 m/s. Successive pictures through different filters showed dust being blown off the lander ring. The temperature at the landing site was 465°C and the pressure was 89.5 bar. Only 2.4% of sunlight reached the surface. Some minor changes in illumination were observed, perhaps due to clouds, but the overall impression was of a dead calm, murky atmosphere over a flat volcanic plain.

Venera 14 lander:

Descent measurements:

The Venera 14 lander conducted the same observations as Venera 13, both during its descent and while on the surface, obtaining very similar results. The combined mass spectrometer results were:

carbon dioxide	97%
molecular nitrogen	4.0 $\pm$ 0.3 %
argon	100 ppm
neon	7.6 ppm
krypton	0.035 ppm
xenon	less than 0.020 ppm

the isotopic ratios were:

13/12 carbon	0.0108
40/36 argon	1.11 ± 0.02
38/36 argon	0.183 ± 0.003
20/22 neon	12.15 ± 0.1

and the combined gas chromatograph results were:

water vapor	700 ± 300 ppm
molecular oxygen	18 ± 4 ppm
molecular hydrogen	25 ± 10 ppm
krypton	0.7 ± 0.3 ppm
hydrogen sulfide	80 ± 40 ppm
carbonyl sulfide	40 ± 20 ppm
sulfur hexafluoride	0.2 ± 0.1 ppm

The x-ray fluorescence instrument measured the composition of the aerosols from 63 to 47 km. It detected both sulfur (1.10 ± 0.13 mg/m^3) and chlorine (0.16 ± 0.04 mg/m^3) and the chlorine abundance was considerably lower than the measurement by Venera 12. This more precisely calibrated instrument measured a sulfur/chlorine abundance ratio compatible with sulfuric acid aerosols. The aerosols in the region between 63 and 47 km were composed principally of sulfur compounds with some chlorine compounds. The abundance ratio of sulfur and chlorine varied with altitude. The highest density aerosols were in the range 56 to 47 km. The x-ray fluorescence instrument of Venera 13 did not operate properly.

Surface measurements:

Unlike Venera 13, the Venera 14 photometers detected no dust raised on landing. It came down on a smooth level plain with flat, layered rocks which had very little soil

Figure 16.7 Venera 14 lander hemisphere 1 color panorama (processing by Ted Stryk).

Figure 16.8 Venera 14 lander hemisphere 2 color panorama (processing by Ted Stryk).

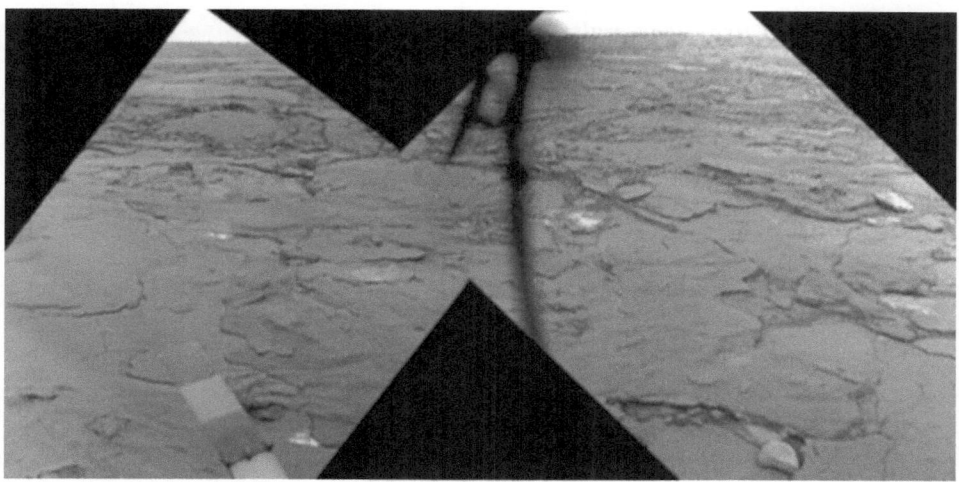

Figure 16.9 Venera 14 panoramas reprocessed to provide a horizontal human-perspective view (courtesy Ted Stryk).

between them. The composition of the drill sample indicated a low-potassium basalt similar to that of terrestrial mid-ocean ridges. A lower sulfur content than Venera 13 may well indicate that the Venera 14 site is younger. The surface properties inferred from the dynamics of touchdown indicated a surface similar to the Venera 13 site, but possibly covered with a layer of weaker, porous material. No penetrometer data was obtained from Venera 14 because the instrument swung down onto a jettisoned lens cover. The audio sensor reported two sounds that could have been distant and small seismic events. The temperature at the site was 470°C and the pressure was 93.5 bar. The amount of sunlight reaching the surface was 3.5%.

X-ray fluorescence soil analysis results were:

	Venera 13	Venera 14
silicon	45%	49%
titanium	1.6%	1.3%
aluminum	16%	18%
iron	9.3%	8.8%
manganese	0.2%	0.2%
magnesium	11%	8.1%
calcium	7.1%	10%
potassium	4.0%	0.2%
sulfur	0.65%	0.35%
chlorine	<0.3%	<0.4%

The surface composition measurements from Venera 8, 9, 10, 13 and 14 at sites around the planet were all consistent with basalts of compositions similar to those on Earth. The great variety of igneous and metamorphic rocks widespread on Earth was not evident, most likely owing to the lack of water on Venus.

Venera 13 and 14 flyby spacecraft:

Both flyby spacecraft collected data on the solar wind and solar x-ray flares. They participated in an interplanetary network to triangulate on gamma-ray bursts, in this case detecting about 150 events. Later in the decade, the Vega spacecraft were to use a flyby of Venus to set up an encounter with Halley's comet, so after leaving Venus behind Venera 13 and 14 fired their engines to rehearse the maneuvers that would be required by those later missions.

17

And back to Venus yet again

TIMELINE: 1982–1983

The final missions in the Soviet Venera series were launched in June 1983. Having achieved most of their objectives with the Venera landers, these two spacecraft were outfitted with large radar antennas replacing the entry system and sent to Venus as orbital radar mappers. Both were successful, with their radars discerning the surface through the ubiquitous clouds to map from 30°N to the north pole with a resolution of about 2 km.

Launch date		
1982		
No missions		
1983		
2 Jun	Venera 15 orbiter	Successful radar mapper
7 Jun	Venera 16 orbiter	Successful radar mapper

PIERCING THE CLOUDY VEIL OF VENUS: 1983

Campaign objectives:

After six consecutive successes of their heavy Venus landers starting with Venera 9, the Soviets decided to send radar imaging orbiters in the 1983 opportunity instead of more landers. In 1978 the US Pioneer 12 orbiter had obtained radio altimetry data of the entire planet at the very low resolution of 150 km, and operated the altimeter in a side-looking mode to obtain a narrow equatorial strip of topography at a resolution of 30 km. This data was used to target the Venera 13 and 14 landers in 1981. The 1983 Venera radar orbiters were intended to use bistatic radar techniques to improve the resolution to 2 km or better, albeit only over about 25% of the planet.

Spacecraft launched

First spacecraft:	Venera 15 (4V-2 No.860)
Mission Type:	Venus Orbiter
Country/Builder:	USSR/NPO-Lavochkin
Launch Vehicle:	Proton-K
Launch Date/Time:	June 2, 1983 at 02:38:39 UT (Baikonur)
Encounter Date/Time:	October 10, 1983
Mission End:	March 1985
Outcome:	Successful.
Second spacecraft:	Venera 16 (4V-2 No.861)
Mission Type:	Venus Orbiter
Country/Builder:	USSR/NPO-Lavochkin
Launch Vehicle:	Proton-K
Launch Date/Time:	June 7, 1983 at 02:32:00 UT (Baikonur)
Encounter Date/Time:	October 14, 1983
Mission End:	May 28, 1985
Outcome:	Successful.

Venus had become more or less a "Red" planet, left almost exclusively to Soviet exploration. After the Mariner 5 flyby in 1967 it was over a decade before the US revisited the planet, and the two small Pioneers in 1978 were primarily focused on the ionosphere and atmosphere. But at that same time the US was also developing a proposal for a Venus Orbiting Imaging Radar (VOIR) mission. NPO-Lavochkin had been working on a Venus radar mapper since 1976 and, after having pioneered local surface imaging, the Soviets wanted to conduct their radar mapping mission before the Americans. As events transpired, they did not have to compete, since VOIR was canceled in 1981 and replaced by a simpler, less costly mission named Magellan that was not launched until 1989. In essence all that NPO-Lavochkin had to do was to replace the entry system of its spacecraft with a side-looking radar to obtain imagery and electrical properties of the surface of the planet, and to add a radio altimeter to measure the topography on the ground track. But modifying the spacecraft to carry the radar was not without challenge.

Rumors of a Soviet Venus radar mapping mission began to circulate in the US in 1979, as NASA was trying to obtain funding for its VOIR mission. Familiar with the heavy nuclear-powered RORSAT orbiting radars the Soviets used to track Western navies, most observers in the US did not believe they had the technology to build a lightweight low-power synthetic aperture radar. It was indeed a struggle, particularly the data storage and computing requirements, and the launch had to be slipped from 1981 to 1983, but ultimately it performed rather well.

Spacecraft:

Venera 15 and 16 were the first in this series of carrier vehicles to be modified in a significant way since Venera 9. The bus was lengthened by 1 meter to accommodate the 1,300 kg of propellant needed to put such a heavy craft into orbit around Venus. The load of nitrogen for the attitude control system was increased from 36 to 114 kg to permit the large number of attitude changes that the orbital mission would entail. Two more solar panels were added outboard of the standard pair to provide the extra power to operate the radar system. The parabolic antenna was enlarged by 1 meter to a diameter of 2.6 meters to increase the bandwidth from 6 to 108 kbits/s and a new 5 cm band telemetry system was introduced to communicate with the 64 and 70 meter ground stations. The spacecraft were identical, and consisted of a cylinder 5 meters long and 1.1 meters in diameter. A 1.4 × 6.0 meter parabolic panel antenna for the synthetic aperture radar (SAR) was installed at the top, in place of the entry system. The entire SAR system weighed 300 kg. A 1 meter diameter parabolic dish antenna was mounted nearby for the radio altimeter. The electrical axis of the radio altimeter antenna was aligned with the long axis of the spacecraft, and the SAR was angled 10 degrees off this axis. During imaging, the radio altimeter would be lined up with the local vertical and the SAR would look off to the side by 10 degrees.

Launch mass: 5,250 kg (Venera 15) 5,300 kg (Venera 16)
Fuel mass: 2,443 kg (Venera 15) 2,520 kg (Venera 16)

Figure 17.1 Venera 15 during tests at Lavochkin.

Figure 17.2 Venera 15 museum model. SAR tilted at 10 degrees to the long axis on top of the SAR/Altimeter instrument compartment above cylindrical propellant tank.

Payload:

1. Polyus-V synthetic aperture radar (SAR) operating at a wavelength of 8 cm
2. Omega radiometric altimeter
3. Thermal infrared (6 to 35 microns) Fourier emission spectrometer (IFSE, DDR-USSR)
4. Cosmic ray detectors (6)
5. Solar plasma detectors
6. Magnetometer (Austria)
7. Radio occultation experiment

All of the components of the SAR and radio altimeter were shared except for the antennas. The electronics cycled the 80 W traveling wave tube oscillator between the antennas every 0.3 seconds. An onboard computer controlled their sequencing and operation. The SAR antenna would illuminate the surface over 3.9 milliseconds with 20 cycles of 127 phase shifts for cross-track encoding. Spacecraft motion over that same interval swept out a 70 meter virtual antenna. After each transmission, the antenna was switched to the receiver, which digitized the magnitude and phase of the reflected radar pulses and stored the data as 2,540 complex numbers in a solid-state memory buffer. To keep up with the radar illumination cycle of 0.3 seconds, the data were read out alternately onto two tape recorders to complete a period of 16 minutes

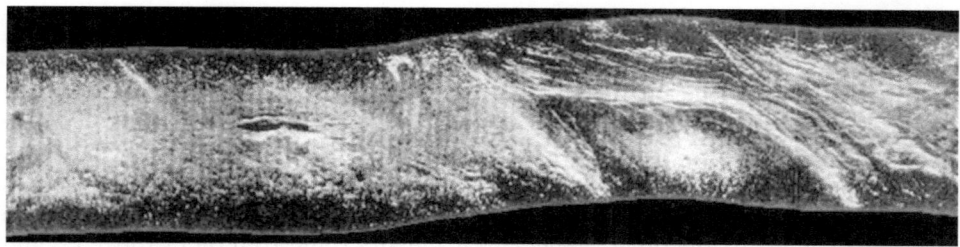

Figure 17.3 Venera 15 SAR strip taken during a single periapsis pass (from Don Mitchell).

of mapping during a periapsis pass. Each such pass produced about 3,200 return images to compose a data strip approximately 120 × 7,500 km. Once the data were received on Earth, each individual 3.9 millisecond return was divided by time delay into 127 ranges across-track and 31 ranges along-track and then processed to correct for atmospheric, geometric, and orbital effects. The individual return images for a pass were then assembled to yield an image strip representing the slope, roughness, and emissivity of the surface of Venus.

During altimetry, the antenna would transmit a code sequence of 31 pulses, each of 1.54 microseconds duration. After transmission, the antenna was switched to the receiver, which recorded the reflection of the pulses from the surface over a period of 0.67 millisecond. The oval footprint of the altimeter radio beam was 40 km cross-track and 70 km along-track. After onboard processing of the return waveform, the data were stored on the tape recorder for later transmission to Earth, which further processed the data to correct for atmospheric, geometric, and orbital effects to yield altitudes. A low resolution mode was used until the orbital elements were precisely determined, and then it was switched to a high resolution mode. In combination with Doppler analysis, the high resolution mode reduced the footprint to 10 × 40 km with an error of about 1 km. The vertical accuracy was about 50 meters.

It was also decided to include an infrared Fourier-transform spectrometer supplied by East Germany. This weighed 35 kg and was intended to provide a higher spectral resolution than the infrared radiometer operated by the Pioneer 12 orbiter. It divided the spectrum into a continuous set of 256 channels over the range 6 to 35 microns. It had a field of view of 100 × 100 km, and provided 60 complete spectra along each periapsis pass. The objectives were to obtain atmospheric temperature profiles from the 15 micron carbon dioxide band in the 90 to 65 km altitude range, the temperature of the upper cloud deck, the abundances of aerosols, sulfur dioxide and water vapor in the atmosphere, and data on the thermal structure and dynamics of the clouds and atmosphere.

The cosmic ray and solar wind experiments were similar to those flown on every Venus mission since Venera 1.

Mission description:

Venera 15 was launched on June 2, 1983, and conducted midcourse corrections on June 10 and October 1 before entering orbit around Venus on October 10. Venera 16 was launched on June 7, conducted midcourse corrections on June 15 and October 5, and entered orbit on October 14. Their orbital planes were inclined about 4 degrees relative to one another, so that any area that was missed by one spacecraft should be able to be imaged by the other. Venera 15 made an orbital trim on October 17, and Venera 16 did so on October 22. Each operating orbit was inclined at 87.5 degrees to the equator, with the periapsis at 1,000 km and the apoapsis at ~65,000 km and a period of 24 hours. The periapsis was positioned at about at 62°N and each periapsis passage would image the surface on a 70-degree arc. Both spacecraft began science operations on November 11. Small burns were made from time to time to preserve the periapsis, accommodate high gain antenna position changes as the Sun-vehicle-Earth angle decreased, and maintain the 3 hour interval between the periapses of the two spacecraft.

Mapping and altimetry would typically begin at 80°N on the inbound side of the pole and continue over the pole down to 30°N on the retreating side. Radar imaging was conducted continuously with a best resolution of about 1 km. The data collected on each 16 minute periapsis pass was stored on the tape recorders, then replayed to Earth during a daily 100 minute communications window prior to the next periapsis. During each 24 hour interval Venus would rotate on its axis by 1.48 degrees, and so successive mapping passes partially overlapped one another. At that rate, 8 months was required to cover all longitudes. The 24 hour orbit was necessary to enable the spacecraft downloads to be synchronized with the receiving stations in the USSR. Several orbital corrections were made during the mission to maintain the period and shape of each orbit. In June 1984, Venus went through superior conjunction and no transmissions were possible while it passed behind the Sun as seen from Earth. This provided an opportunity to conduct radio occultation experiments to study the solar and interplanetary plasma. After conjunction, Venera 16 rotated its orbit backwards 20 degrees relative to its partner to map areas missed prior to superior conjunction, and mapping was concluded shortly thereafter, on July 10.

Between them, the two spacecraft were able to image all of the planet from 30°N to the north pole, or about 25% overall. The resolution of 1 to 2 km was similar to what could be achieved by the 300 meter Arecibo radio telescope dish operating as a radar, but it was limited to equatorial latitudes and could not get the accompanying altimetry.

Venera 15 reportedly exhausted its supply of attitude control gas in March 1985, but Venera 16 continued to transmit data from its other instruments until May 28 of that year. No attempts were made to change orbits for higher resolution or increased coverage.

Results:

Together, the two spacecraft imaged from 30°N to the north pole at a resolution of 1 to 2 km. The primary product consisted of 27 radar mosaics at a scale of 1:5,000,000 of the northern 25% of the planet. The results confirmed that the highest elevations, meaning those which stand more than 4 km above the plains, have greatly enhanced radar reflectivity.

The radar experiments produced major discoveries about the surface of the planet, imaging new types of terrain that included:

Coronae – large circular or oval features with deep concentric rings
Domes – flat, nearly circular raised features some with central calderas
Arachnoids – collapsed domes with radial cracks
Tessera – large regions of linear ridges and valleys

Prior to Venera 15 and 16, the coronae glimpsed by Arecibo had been thought to be impact features filled with lava. About 30 coronae and 80 arachnoids were in the area mapped. As no evidence of plate tectonics was evident, the coronae, domes and arachnoids were all postulated to be surface expressions of mantle plumes heating an immobile crust. There were no direct terrestrial analogs. The tessera appeared to be the oldest crustal regions on the planet, and were often overlapped by lava flows.

Even if large objects that penetrate the thick atmosphere are destroyed before they can reach the ground, they can create a shock wave that leaves an impression on the surface. There were about 150 craters in the area surveyed. Analysis of the cratering data led to a very young age of 750 ± 250 million years, consistent with the idea of catastrophic resurfacing making the tessera, and large scale 'blistering' over mantle plumes between resurfacing events.

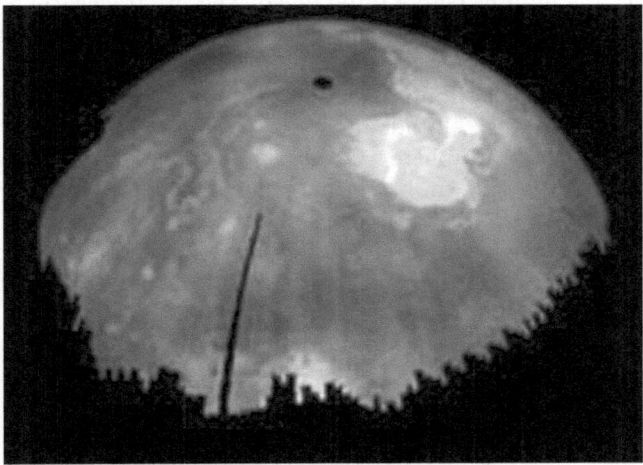

Figure 17.4 Venera 15 and 16 global imaging at about 1 km resolution. The elevated Lakshmi planum is at upper right with Maxwell Montes (from Don Mitchell).

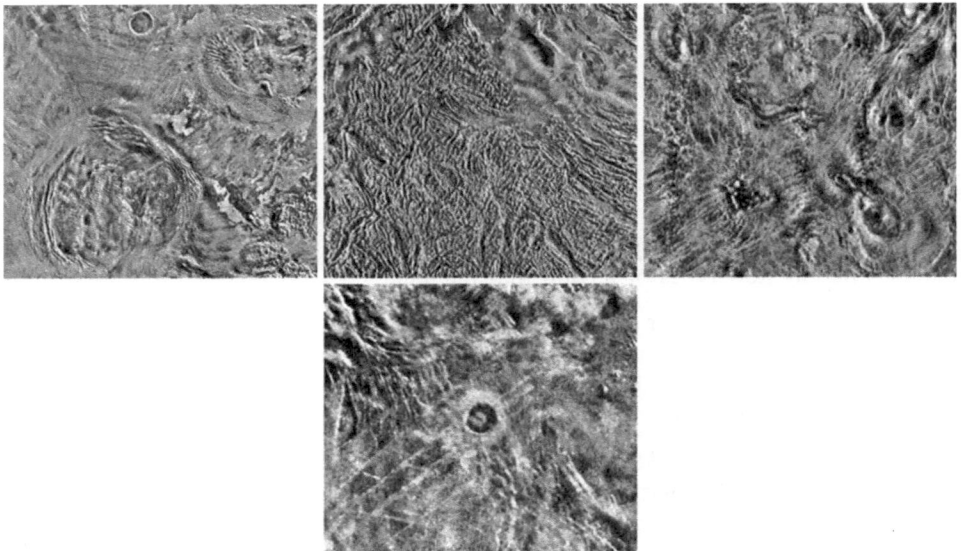

Figure 17.5 Landforms found by Venera 15 and Venera 16. From upper left clockwise: Anahit and Pomona Coronas, Fortuna Tessera, Arachnoids in Bereghinya, and Duncan crater.

The altimeter produced extensive data on topography in the northern hemisphere. In combination with the radar data, scientists were able to produce detailed maps of the surface.

The infrared spectrometer on Venera 16 malfunctioned, but the one on Venera 15 worked in orbit for 2 months before it too failed. The spectra clearly resolved carbon dioxide, water vapor, sulfur dioxide, and sulfuric acid aerosol. This data was strong confirmation that the particles in the upper cloud layer were a 75 to 85% solution of sulfuric acid. The aerosol distribution and mixing ratios for sulfur dioxide and water vapor were determined in the altitude range 105 to 60 km. The thermal structure and optical properties of the atmosphere were also determined in this altitude range. The clouds ranged from 70 to 47 km, but in the polar region the clouds were 5 to 8 km lower and the air above 60 km was warmer than in equatorial regions. The average surface temperature was measured at 500°C, but some warmer spots were detected along with some cooler regions. There were no features in the spectrum to suggest the presence of organic compounds.

The two orbiters produced 176 radio occultation profiles between October 1983 and September 1984.

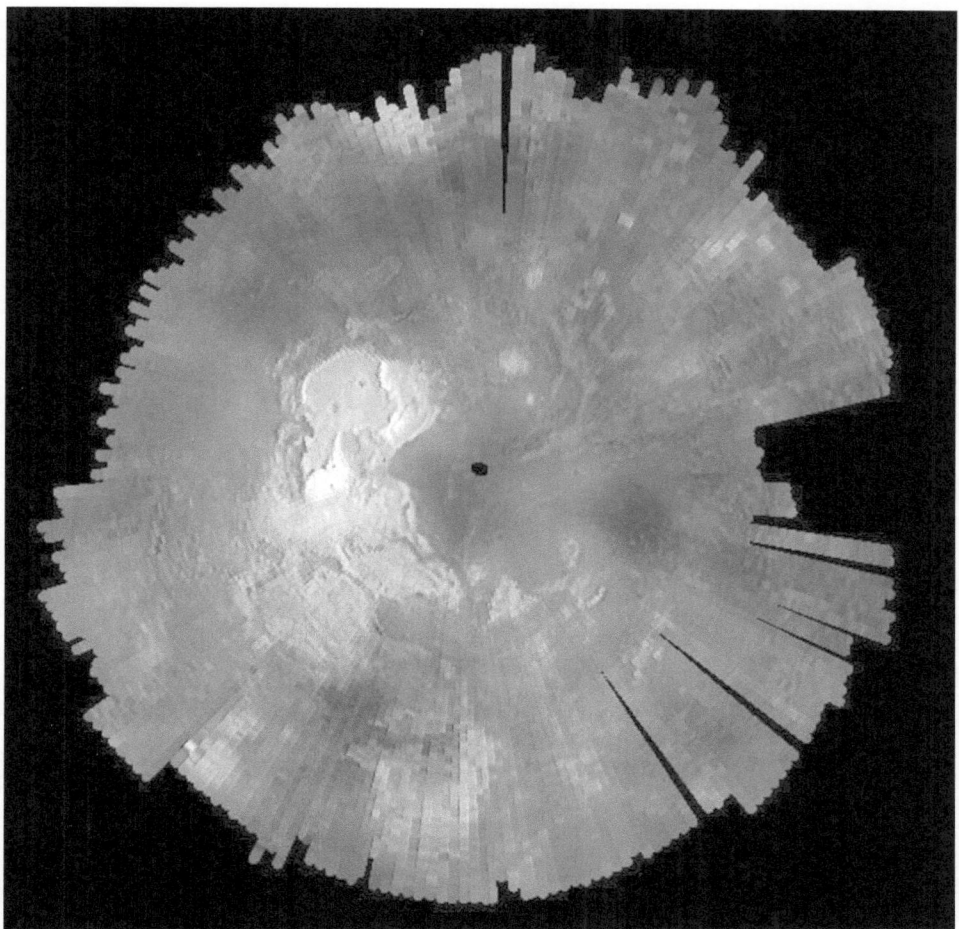

Figure 17.6 Venera 15 and Venera 16 altimeter data. Lakshmi planum at left (from Don Mitchell).

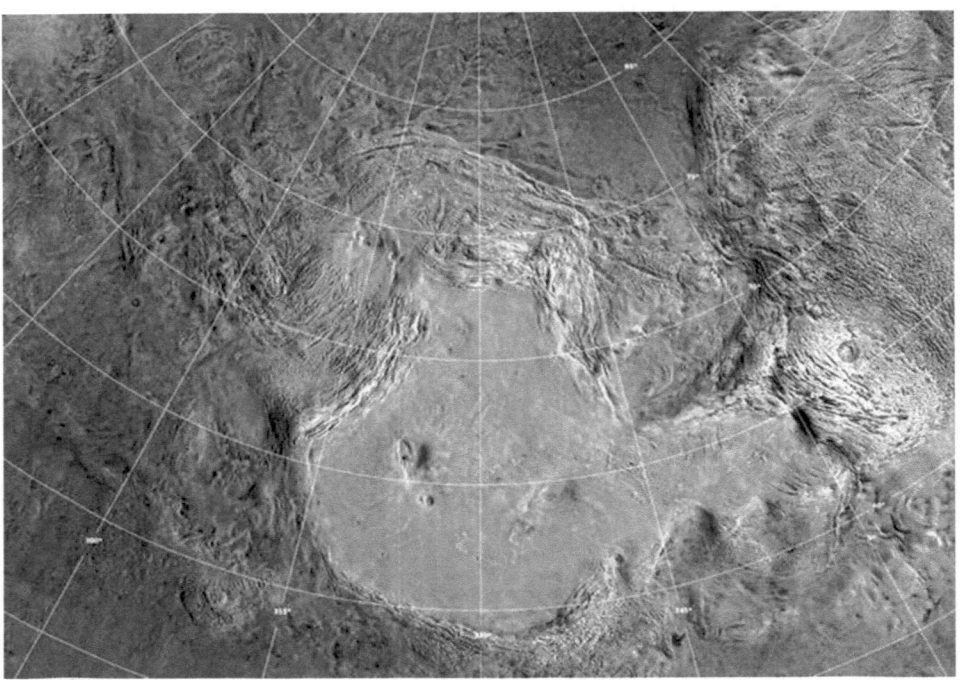

Figure 17.7 Venera 15 and Venera 16 cartography of Lakshmi planum with Maxwell Montes and caldera at right (from Don Mitchell).

18

The International Comet Halley campaign

TIMELINE: 1984–1985

During the 1970s there was mounting interest in the coming apparition in 1986 of the famous Comet Halley. The US was developing various plans for intercepting the comet at close range. The fledgling European Space Agency was considering doing the same. And the very small and academically oriented Japanese Institute for Space and Astronautical Science had decided to send two small spacecraft equipped with plasma instrumentation.

At the start of the 1980s the USSR began to develop a large balloon mission for Venus with the French. But with the development of the mission advancing well the Soviets became interested in Comet Halley when they realized it would be possible for a spacecraft to use a Venus flyby to redirect itself towards the comet. The two missions were partially combined, and in an unprecedented move the Soviets issued an international call for instruments to fly on their spacecraft to Halley. As a result, their planetary exploration program suddenly became international outside the Iron Curtain. Not even the US had opened its more public space exploration program to such extensive international cooperation.

The spacecraft for this Venus-Halley (Vega) mission was very similar to a flyby Venera, but with an instrument scan platform added to track Halley. But instead of a large balloon consuming essentially the entire mass limit for the entry system, it was decided to carry a lander and augment this with a smaller balloon package. The balloon package would be released from the top portion of the entry sphere and the lander from the bottom portion. The balloon was to be inflated after release and then float at an altitude of about 50 km with a battery life of about 50 hours. The lander would be of the standard configuration.

Launched in December 1984, both of the Vega spacecraft performed well. Their landers and balloons were successful, the balloons drifting thousands of kilometers from the night-side to the day-side after more than 40 hours of flight. The spacecraft proceeded to encounter Halley, imaging the nucleus, taking data on the surrounding

Launch date

1984

15 Dec	Vega 1 Venus/Halley	Success at Venus and Halley
21 Dec	Vega 2 Venus/Halley	Success at Venus and Halley

1985

7 Jan	Sakigake Halley flyby	Japanese mission success
2 Jul	Giotto Halley flyby	ESA mission success
18 Aug	Suisei Halley flyby	Japanese mission success

environment, and supporting the rest of the Halley 'armada' that comprised the two Soviet spacecraft, two Japanese spacecraft at far encounter, and the European Giotto spacecraft which, with navigational assistance from the Vega missions, was able to refine its trajectory to achieve a very close approach to the nucleus of the comet. The US was notably absent from the armada having failed, in a major embarrassment, to fund a mission to the comet.

THE VENUS-HALLEY CAMPAIGN: 1984

Campaign objectives:

For the Soviets this campaign combined a Venus flyby/entry mission with a flyby of Comet Halley, and it was their first (and thus far only) multiple-target mission. After releasing their entry systems at Venus in June 1985 the two flyby spacecraft were to be re-targeted by the gravity-assist of their encounter with the planet onto a course to intercept Comet Halley in March 1986.

In addition to a lander, the entry system carried an atmospheric balloon. The idea to float a balloon in the atmosphere of Venus grew from French-Soviet cooperation initiated after the successful Venera 4 mission in 1967. France and the Soviet Union had come to a rapprochement of sorts in the Cold War, opening a breach in the Iron Curtain by establishing cooperation in space science. In 1974 Dr. Jacques Blamont of CNES and Boris Petrov, Chairman of the Intercosmos Council, began to discuss a joint mission consisting of an entry probe to deliver a large French balloon into the atmosphere of Venus, and a Soviet orbiter to provide the communications relay. By 1977 a date had been tentatively set for a 1984 launch of the 'Venera-84' mission to mark the bicentennial of the Montgolfier brothers' invention of the hot-air balloon, and the division of work had been established. Jacques Blamont and Mikhail Marov were named as science co-chairs for the mission. The French would supply the two 10 meter diameter balloons with their 50 kg gondolas, including transponders for very long baseline interferometry (VLBI) tracking, and the Soviets would supply the spacecraft, entry systems, and the remaining mission support. But events changed these plans.

In the late 1970s the world's space science community was beginning to plan for the eagerly awaited apparition of Comet Halley in 1986. The US offered to carry a French ultraviolet instrument on one of its spacecraft. When the US withdrew from this effort in 1979 the Soviets offered to fly the French instrument on Venera-84 to enable it to observe the comet from Venus orbit – which would be a more favorable vantage point because although the comet would approach that planet no closer than 40 million km, that was much closer than it would approach Earth. In the process of investigating how to improve observations of Halley from Venus the Soviets found that it would be possible to utilize a gravity-assist during a flyby of Venus to set up an encounter with Halley. The science value of a mission to both Venus and Halley as argued by Jacques Blamont intrigued Roald Sagdeev, Director of IKI, who set out to have it supersede the Venera-84 mission. The new project was called 'Vega' as a Russian contraction of 'Venera' and 'Galley', with the name of the comet using a 'G' because there is no 'H' in the Cyrillic alphabet. Valery Barsukov, Director of the Vernadsky Institute, was far more interested in Venus than he was in the comet, but Sagdeev sold the mission to him by including a lander, albeit at the cost of reducing the size of the balloon package to enable both to fit inside the standard entry system. Three years of intensive development of the Venera-84 mission, including partially manufactured hardware, was lost. When the furious French declined to participate further, the small balloon became a Russian project. Nevertheless, Sagdeev managed to coax the French into providing several instruments for the lander and balloon, as well as two key remote sensing instruments for the Halley encounter. And by taking advantage of their bridging position between the East and the West, the French were able to gain the participation of the Deep Space Network in the VLBI network that would measure the dynamics of the balloons as they drifted in the atmosphere of Venus. For the first time, therefore, the archrival Americans became a participant in a Soviet planetary mission, albeit by providing tracking resources. The University of Chicago supplied an instrument to investigate dust particles during the Halley flyby, but this was arranged through the science community as a private venture rather than at government level and the principal investigator had to assure the US military that he was using only commercial parts from his local Radio Shack store! He dismissed the military's concerns with, "Let them [the Soviets] copy this, it will set them back years."

Sagdeev, by enthusiasm, energy, and personal effort, instituted the new project as a broadly international venture by offering 120 kg on the spacecraft for instruments originating from countries outside the USSR. This extensive internationalization was unprecedented for the historically closed Soviet space program. And internally the *perestroika* initiative enabled him to overcome resistance by the Soviet bureaucracy.

But the final credit must go to Chief Designer Vyacheslav Kovtunenko and the NPO-Lavochkin scientists and engineers who, by building the most comprehensive and successful deep space mission in their history, created a legacy for Soviet lunar and planetary exploration.

Spacecraft launched

First spacecraft:	Vega 1 (5VK No.901)
Mission Type:	Venus Flyby/Lander/Balloon and Halley Flyby
Country/Builder:	USSR/NPO-Lavochkin
Launch Vehicle:	Proton-K
Launch Date/Time:	December 15, 1984 at 09:16:24 UT (Baikonur)
Venus Encounter:	June 11, 1985
Halley Encounter:	March 6, 1986
Outcome:	Successful.
Second spacecraft:	Vega 2 (5VK No.902)
Mission Type:	Venus Flyby/Lander/Balloon and Halley Flyby
Country/Builder:	USSR/NPO-Lavochkin
Launch Vehicle:	Proton-K
Launch Date/Time:	December 21, 1984 at 09:13:52 UT (Baikonur)
Venus Encounter:	June 15, 1985
Halley Encounter:	March 9, 1986
Outcome:	Successful.

The Vega missions became an integral part of the International Halley Mission (IHM) organized initially by the European Space Agency to coordinate operations and data analysis for the various Halley missions being planned by Europe, Japan, the US and the Soviet Union. An Interagency Consultative Group consisting of high level representatives of the space agencies overseeing the IHM provided cover for US participation in the midst of the Cold War, effectively circumventing the absence of a formal agreement between the US and USSR. Ironically spacecraft were sent to Halley by all these nations except the US, whose formal involvement was ultimately limited to providing tracking and science support.

With lander, balloon, and flyby components the Vega missions were both very ambitious, and by involving a host of international interfaces including a large array of international instruments were extraordinarily complex. The nations participating included Austria, Bulgaria, Czechoslovakia, East Germany, France, Hungary, West Germany, Poland, and the United States. The Hungarians built part of the navigation system and the Czechs supplied the optical system for the automated scan platform. Foreign investigators were allowed into the country to participate fully and actively in the project from beginning to end; not passively as previously by delivering their completed instruments in advance and waiting at home to find out what happened to them. Team meetings were held in the USSR and foreign contributors were allowed into Soviet facilities for development, testing and integration activities. This style of cooperation with the USSR was unprecedented. An organization called Intercosmos had existed since the 1960s for coordination of cooperation in space research mainly among Eastern Bloc nations and with France, but this was the first time the activity assumed such a large scale and included Western nations to such a degree.

The 1984 launches gave the Soviets enormous influence in the international space community. With such a bold move to internationalization, leadership in planetary exploration passed to the USSR. After the busy era of Mariner, Pioneer, Viking and Voyager launches in the 1970s, the US launch rate had fallen precipitously to zero in the 1980s. The USSR continued to reap a harvest from its Venera series, and began its transition from a closed program to an open program far more international than any flight project in the US. The Soviets now issued open calls for participation in its science missions. US science missions would not become more international than "participation by invitation only".

The Vega missions were highly successful in meeting all their science objectives, and a major achievement for the Soviet robotic lunar and planetary program. They concluded the run of ten consecutive highly successful heavy-class Venera missions that started with Venera 9 in 1975, and they were the final Soviet missions to Venus after twenty-nine launch attempts since 1961. During this 24 year period only three of sixteen windows for Venus were not used. Nineteen of the twenty-nine launches sent spacecraft on trajectories to Venus, of which fifteen successfully delivered three entry probes, ten landers, two balloons, and four orbiters. The Soviet scientists and engineers participating in the Vega missions would have dismissed as ridiculous the prospect of there being only two more campaigns in the Soviet planetary exploration program, both of which would be embarrassing failures.

Spacecraft:

Flyby spacecraft:

The flyby spacecraft was nearly identical to Venera 9 to 14 but used the larger solar panels of Venera 15 and 16 to handle the power demand and was loaded with 590 kg of propellant instead of the usual 245 kg. It was protected from hypervelocity comet dust impacts by an aluminum shield consisting of an outer multi-layer sheet of 100 micrometers thickness mounted at a standoff distance of 20 to 30 cm.

A data rate of 65 kbits/s was provided for the comet encounter, but a slower mode would be used in the cruise phase. Approximately half of the spacecraft was devoted to the Halley science instruments and half to the Venus entry system. In making the flyby of Venus in the manner required to set up the Halley encounter, the spacecraft would relay to Earth the transmission from the lander during its descent and surface operations as previously. However, the balloon would transmit its telemetry directly to Earth.

The spacecraft was fitted with an 82 kg articulated scan platform that could rotate from -147 to +126 degrees in azimuth and from -60 to +20 degrees in declination for a pointing accuracy of 5 minutes of arc and a stability of 1 minute of arc per second. Its automated tracking would enable instruments to be continuously pointed at the nucleus of the comet during the rapid flyby while the spacecraft held an orientation that permitted its high gain antenna to point at Earth for real-time transmission. The pointing was controlled either by an eight-element photometer or by using the wide angle camera, and gyroscopic attitude control was provided as a

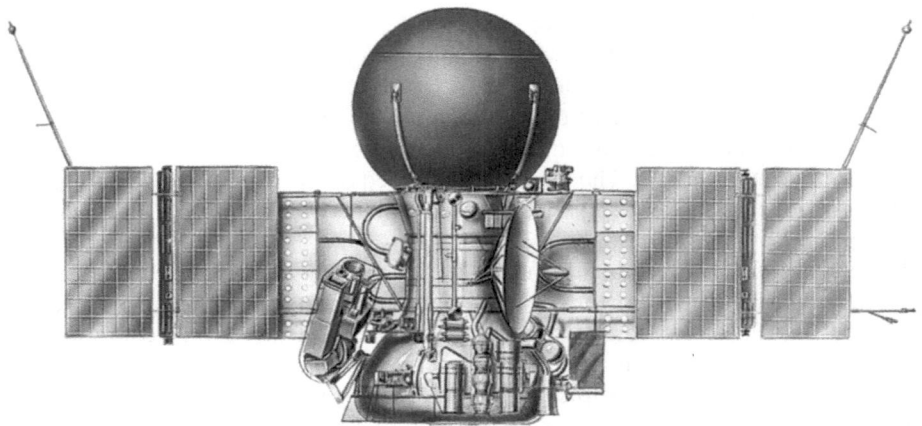

Figure 18.1 Vega spacecraft (courtesy NPO-Lavochkin). Scan platform folded on left, parabolic antenna on the right, toroidal instrument compartment on the bottom with external instruments.

Figure 18.2 Museum model Vega spacecraft without insulation and dust shields. Front side at right shows solar panels, parabolic antenna, and navigation instruments. Back side at left shows camera platform hanging down below toroidal instrument section, radiator panels and black disks where helical lander relay antennas were mounted.

precaution against comet dust upsetting the optical sensors. The scan platform carried the narrow and wide angle cameras, an infrared sounder, and a three-channel spectrometer. All other experiments were body-mounted except for two magnetometer sensors and various plasma probes and plasma wave analyzers which were mounted on a 5 meter boom. The total science payload for Halley weighed 130 kg.

Figure 18.3 Vega 1 folded and ready to launch. Note scan platform, insulation and metal shielding.

Entry system:

The entry system was virtually identical to the recent Venera missions, consisting of an insulated sphere 2.4 meters in diameter whose upper and a lower hemispheres were joined non-hermetically. In this case, however, the lander was installed in the lower half and the balloon in the upper half.

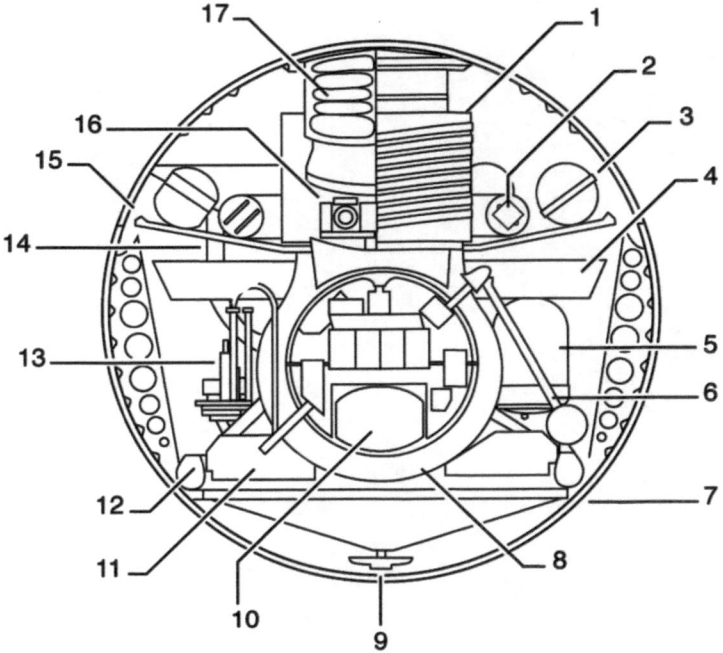

Figure 18.4 Entry capsule cross-section (by James Garry): 1. Antenna; 2. Balloon compartment; 3. Helium inflation tank; 4. Lander aerodynamic stabilizer; 5. Gas chromatograph; 6. Spectrophotometer; 7. Entry heat shield; 8. Thermal insulation; 9. Oscillation damper; 10. Battery; 11. Stabilizing vanes; 12. Crushable impact torus; 13. Drill and sample collector; 14. Coolant delivery piping; 15. Balloon aerobrake; 16. Science instrument bay; 17 Parachute.

Lander:

The Vega landers were almost identical to the Venera 13 and 14 landers with some aerodynamic modifications for increased stability while free falling. These included spoke-like blades interior to the landing ring to reduce spinning and a thin collar-like sleeve installed beneath the disk of the aerobrake to minimize the turbulence which would be induced by the externally mounted instruments.

Figure 18.5 (left) shows the sleeve and the blades. In view on the landing ring are the two white hygrometer compartments, the temperature and pressure unit offset to its right, and also the drill. Figure 18.5 (right) shows the large shiny cylindrical gas chromatograph on the ring to the left, the horizontal drill vacuum reservoir, and the penetrometer and the hydrometers on the far right. The impact velocity of 8 m/s was to be cushioned by the shock absorbers that support the main spherical pressurized compartment.

Figure 18.5 Venera 13 and Venera 14 landers during tests at Lavochkin.

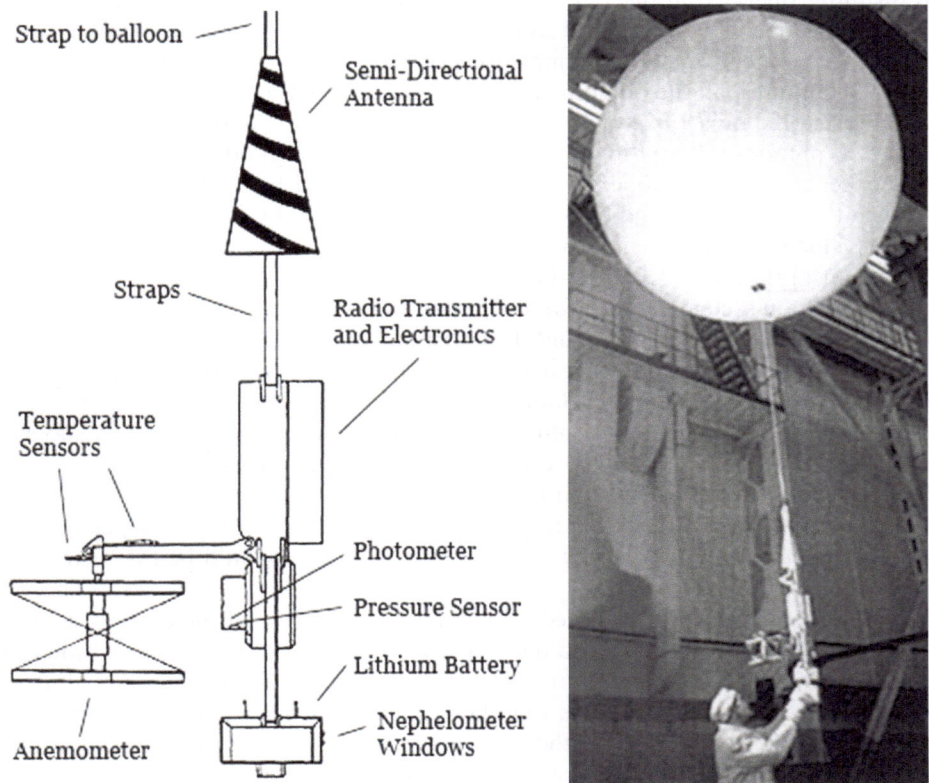

Figure 18.6 Gondola diagram (from Don Mitchell) and testing on a short tether.

Balloon:

The balloons were a new component, and were to be carried in and deployed by the upper hemisphere of the entry system. The super-pressure helium aerostat with its attached gondola was designed to float in the middle layer of cloud at an altitude of 54 km, where the temperature was a mild 32°C and the pressure was 535 millibars.

Each radio-transparent balloon had a mass of 11.7 kg and when inflated it was 3.4 meters in diameter and held 19.4 cubic meters of helium that had a mass of 2 kg. A 13 meter long tether suspended the 7.0 kg gondola (including 1.6 kg for the tether). The entire system weighed a little over 20.7 kg. The rate of helium diffusion was sufficiently low to sustain pressure for about 5 days.

The 1.2 meter long 14 cm wide gondola contained a transmitter with a stabilized oscillator for Doppler tracking, a conical antenna, a vertical anemometer, sensors for ambient temperature and pressure, a light photometer, a nephelometer, a control and ballast system, and sixteen lithium batteries for 300 watt-hours of power. The 1 kg battery package was designed for 46 to 52 hours of life. To simplify the task for the network of radio telescopes which would track the balloons, the 4.5 W transmitter operated in the 18 cm astronomical band at 1.6679 GHz. It transmitted direct to Earth via the conical antenna at either 1 or 4 kbits/s. Except for the lightning counter which was sampled every 10 minutes and the photometer twice every 30 minutes, all the other instruments were sampled once every 75 seconds. The data was stored on a 1,024 bit memory. A 5.5 minute burst of data was sent to Earth every 30 minutes, alternating between two transmission modes in a predetermined sequence. In the first mode, 852 bits of data collected from the instruments were transmitted in a 270 second burst preceded and followed by 30 seconds of carrier for VLBI velocity measurements. In the alternative 330 second mode, only two tones were transmitted for VLBI position and velocity.

The balloon system had to be folded up during cruise and entry, survive the forces of deployment, and then withstand the corrosive atmosphere of sulfuric acid aerosol. The envelopes were made using a woven teflon and cloth matrix, the gondola was covered with a white paint resistant to sulfuric acid, and the tethers were made of a type of nylon. Timing, as determined by pressure sensors, was critical to successful deployment: if the envelope were inflated at too high an altitude it would burst in the low pressure; if it were inflated at too low an altitude it would not gain the necessary buoyancy, would penetrate too deep and be destroyed by the high temperature. The inflation system had 2 kg of helium, and altitude control would be by the release of ballast.

The balloon system was carried in the upper hemisphere of the entry system, in a toroidal canister that surrounded the helical antenna of the lander. In addition to the folded balloon and gondola, this canister contained a 35 square meter parachute and the spheres of pressurized helium to inflate the balloon. The deployment began at an altitude of 64 km by separating the hemispheres while on the drogue parachute. This released the lower hemisphere containing the lander. Separation deployed a braking parachute for the lander, which then performed its own deployment sequence as on previous missions. The upper hemisphere then released the toroidal balloon package

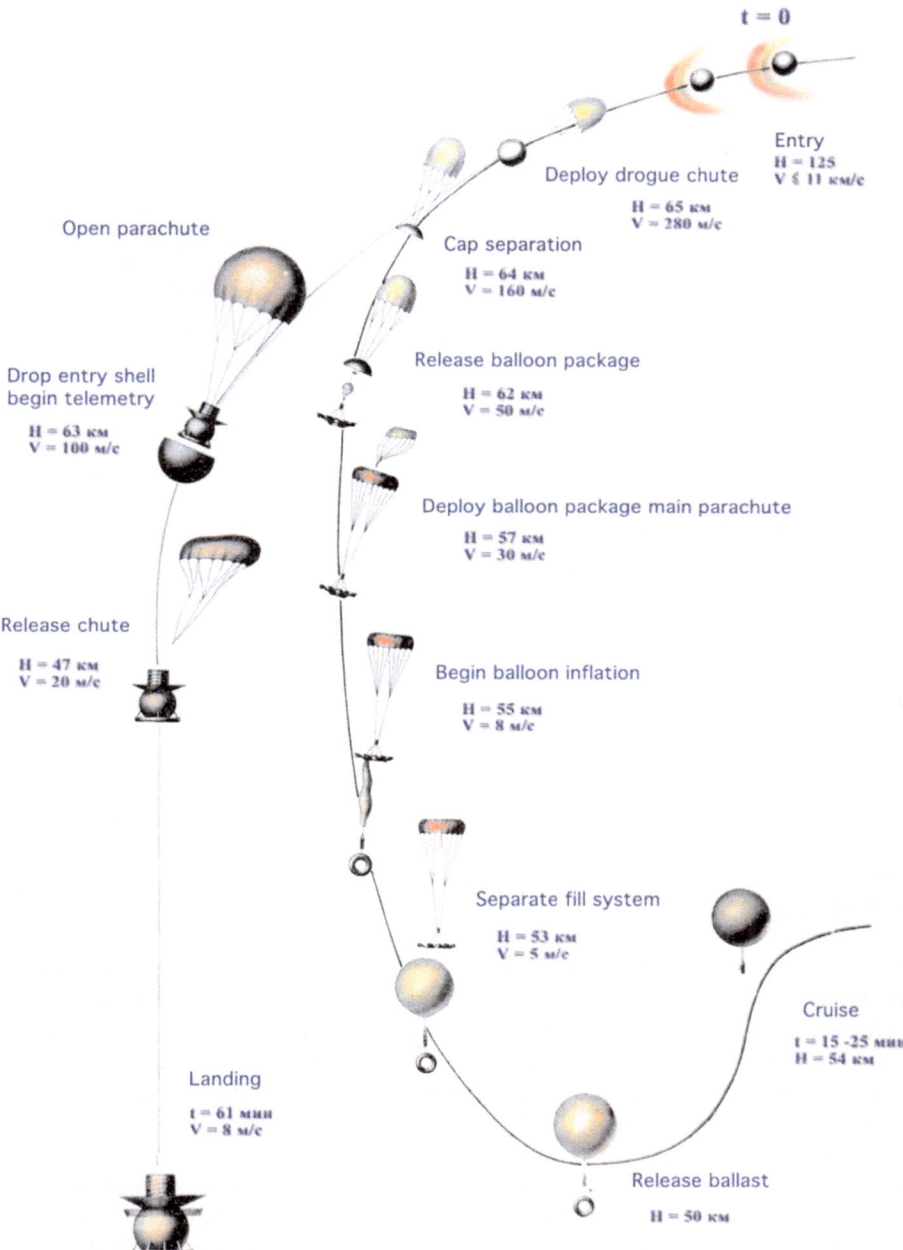

t = 0

Entry
H = 125
V ≤ 11 км/с

Deploy drogue chute
H = 65 км
V = 280 м/с

Cap separation
H = 64 км
V = 160 м/с

Open parachute

Release balloon package
H = 62 км
V = 50 м/с

Drop entry shell
begin telemetry

H = 63 км
V = 100 м/с

Deploy balloon package main parachute
H = 57 км
V = 30 м/с

Release chute

H = 47 км
V = 20 м/с

Begin balloon inflation

H = 55 км
V = 8 м/с

Separate fill system

H = 53 км
V = 5 м/с

Cruise

t = 15 -25 мин
H = 54 км

Landing

t = 61 мин
V = 8 м/с

Release ballast

H = 50 км

Figure 18.7 Vega entry system deployment sequence (courtesy NPO-Lavochkin).

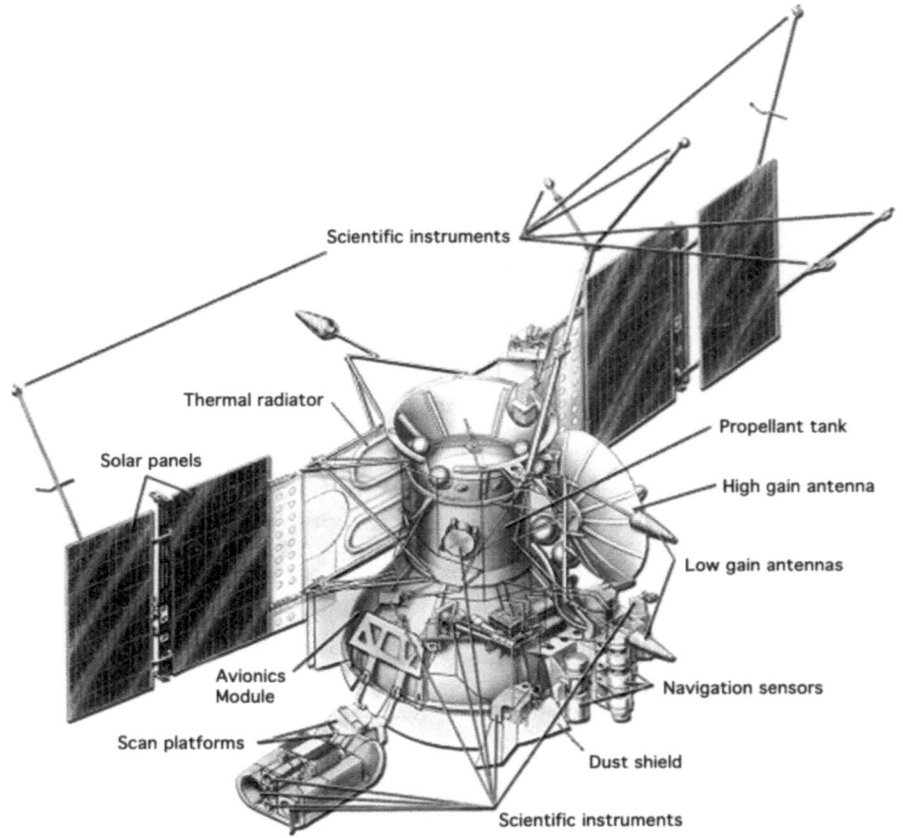

Figure 18.8 The Vega spacecraft configured for Halley encounter (courtesy NPO-Lavochkin).

at 62 km, deploying its parachute in the process. At 57 km the package deployed the balloon system. At 55 km the inflation system was activated. By the 53 km level the envelope had inflated and the package on the parachute released the balloon system. At 50 km the balloon system released its ballast, deployed the boom that carried the temperature sensors and anemometer, and then rose to 54 km to travel wherever the prevailing wind took it. Since the temperatures in this altitude range were benign there was no requirement for thermal control.

On Earth, a global distribution of twenty international antennas consisting of two networks was ready to perform Doppler tracking and receive the scientific data from the balloons – one at a time, as they were to arrive at the planet several days apart. One network was led by IKI and used six Soviet antennas including a new 70 meter dish that was built for the Vega missions. The second network was led by CNES and used the three 64 meter antennas of the Deep Space Network in the US, Australia and Spain, and astronomical antennas in Brazil, Canada, England, Germany, Puerto

Rico, South Africa, and Sweden. Doppler tracking by each antenna gave the range and velocity along the Earth-Venus line, but lateral motion of the balloons required interferometry that combined phase information from antennas located far apart and linked electronically. In addition, the network simultaneously tracked carrier wave signals provided by one or other of the two flyby spacecraft to provide a third leg to greatly increase the precision of distance and velocity measurements in a differential interferometry technique developed by the US for the Apollo lunar missions.

Vega spacecraft system mass

Launch mass:	4,924 kg (Vega 1, *fuel mass* 755 kg)
	4,926 kg (Vega 2, *fuel mass* 766 kg)
Carrier mass:	3,222 kg (Vega 1, *dry mass* 2,466 kg)
	3,228 kg (Vega 2, *dry mass* 2,462 kg)
Entry mass:	1,702 kg (Vega 1)
	1,698 kg (Vega 2)
Lander mass:	716 kg (both)
Balloon mass:	122.75 kg at entry with parachutes, fill system, ballast etc.
	21.74 kg at float

Payload:

Flyby spacecraft:

Mounted on the scan platform:

1. TV imaging system (TVS, USSR-France-Hungary)
2. Three-channel (ultraviolet, visible and near-infrared) spectrometer (TKS, France-USSR-Bulgaria)
3. Infrared spectrometer (IKS, France)

Body mounted:

1. Dust mass spectrometer (PUMA, FRG-USSR-France)
2. Dust particle counter (SP-1)
3. Dust particle counter (SP-2)
4. Dust particle detector (DUCMA, USA)
5. Dust particle detector (FOTON)
6. Neutral gas mass spectrometer (ING, FRG)
7. Plasma spectrometer (PLASMAG)
8. Energetic particle analyzer (TUNDE-M, Hungary-USSR-FRG-ESA)
9. Energetic particles (MSU-TASPD)
10. Magnetometer (MISCHA, Austria)

11. Low frequency wave and plasma analyzer (APV-N, USSR-Poland-Czecho-slovakia)
12. High frequency wave and plasma analyzer (APV-V, USSR-France-ESA)

Three instruments for remote sensing of Halley were mounted on the ASP-G scan platform; the 32 kg TVS camera, the 14 kg TKS three-channel spectrometer, and the 18 kg IKS far-infrared spectrometer. The camera was Russian and the spectrometers were provided by France. The far-infrared spectrometer was cryogenically cooled by a Joule-Thompson cryostat and operated in the range 2.5 to 12.0 microns. The three-channel instrument operated in the ultraviolet 120 to 290 nm, visible 275 to 715 nm, and near-infrared 950 to 1,200 nm. The flyby range at Halley was deliberately large to avoid damaging to the spacecraft, so to obtain the desired view of the nucleus the camera required a narrow angle optical system capable of a resolution of 150 meters at a range of 10,000 km. The computers for the science instruments were enhanced by using Western electronics. However, because CCD technology was restricted the Soviets had to develop their own 512 × 512 device for the camera. The optics were built by the French, and comprised a 150 mm f/3 wide angle lens that was limited to the red, and a 1,200 mm f/6.5 narrow angle lens with six filters from the visible to infrared. The Hungarians were responsible for the camera electronics with assistance from the Soviets.

The five instruments to study the dust issued by the nucleus of the comet were the 19 kg PUMA dust particle impact mass spectrometer to measure the composition of

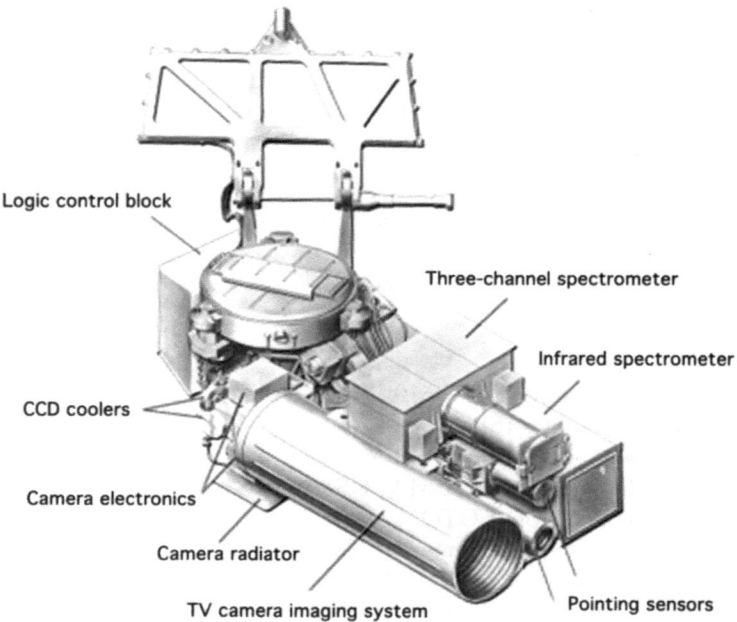

Figure 18.9 The Vega spacecraft scan platform (courtesy NPO-Lavochkin).

individual dust particles, the 2 kg SP-1, 4 kg SP-2, and 3 kg DUCMA dust particle counters to determine the flux and mass distribution of dust particle in different size ranges, and the FOTON dust particle detector that was installed to measure the large particles that punched through the standoff shield. In-situ measurements by the 7 kg ING neutral gas mass spectrometer would analyse gas in the space through which the spacecraft was traveling. The composition and energy spectrum of ions would be determined by the 9 kg PLASMAG plasma spectrometer, and the flux and energy of ions would be measured by the 5 kg TUNDE-M energetic particle analyzer. It also had the 4 kg MISCHA magnetometer, two plasma wave analyzers, the 5 kg APV-N for ion flux and frequencies below 1 kHz, and the 3 kg APV-V for plasma density, temperature and frequencies in the range 0 to 300 kHz.

Including the scan platform and its supporting structure, the instrument payload of the flyby spacecraft was 253 kg.

Lander:

Entry and descent:

1. Temperature, pressure and wind sensors (METEO, USSR-France)
2. Hydrometer for water vapor concentrations (VM-4)
3. Ultraviolet spectrometer for atmospheric SO2 and sulfur measurements (ISAV-S)
4. Optical nephelometer-scatterometer for aerosol size and properties (ISAV-A)
5. Particle size spectrometer for aerosols (LSA)
6. X-ray fluorescence spectrometer for aerosol elemental analysis (IFP)
7. Gas chromatograph for aerosol chemical analysis (SIGMA-3)
8. Mass spectrometer for aerosol chemical analysis (MALAKHIT-V, USSR-France)
9. Doppler experiment for wind and turbulence

The descent instruments focused on aerosols in particular. There were two particle size instruments for measuring the physical properties of aerosols, two instruments for aerosol chemistry, and one instrument for an elemental analysis of the aerosols. These five instruments had externally mounted components with limited insulation from the ambient temperature and pressure, but since the aerosols were confined to the upper atmosphere they were required to function only above 35 km. The aerosols were carried into the instruments by inlet tubes. Some instruments analyzed the light scattered by the aerosol particles in these tubes to determine their size. The ISAV-A instrument also included a nephelometer to determine the cloud density by shining a beam of light through a window in the pressure vessel and measuring the light returned through this window. It shared electronics with the ultraviolet spectrometer.

The gas chromatograph instrument was specifically designed for Vega to measure sulfuric acid aerosol by trapping the droplets in a carbon saturated filter that reacted with sulfuric acid to produce sulfur dioxide and carbon dioxide.

The x-ray spectrometer was a significant improvement on the ones carried by the

Venera 13 and 14 landers. It distinguished grain size using laser imaging. The mass spectrometer sampling system used an aerodynamic inertial separator to segregate grains into small and large sizes on two separate filters. These were then vaporized and analyzed in the mass spectrometer.

The ultraviolet spectrometer was an active experiment, particularly effective for a descent in darkness. It had an ultraviolet lamp and a 1.7 meter path length absorption cell into which the atmosphere was admitted in order to measure the absorption at 512 points between 230 and 400 nm. The objective was to determine the nature of the mysterious 'ultraviolet absorber' deduced from remote sensing measurements. The spectrometer was inside the lander, but there was a pipe through the hull to allow the atmosphere into the instrument. It was operated from 62.5 km down to the surface.

The temperature and pressure instruments were similar to those of the Venera 13 and 14 landers but revised for greater accuracy. They comprised two platinum wire thermometers and three pressure sensors covering the ranges 0 to 2, 0 to 20 and 2 to 110 bar. The hydrometer was also improved.

Surface:

1. Drill and surface sampler (SSCA)
2. X-ray fluorescence spectrometer (BDRP)
3. Gamma-ray spectrometer (GS-15STsV)
4. Dynamic penetrometer (PrOP-V)

As both the gravity-assist to deflect the flyby trajectory for Halley and the mission of the balloons required the Vega entries to occur on the night-side, the landers were not given cameras or optical instruments, and those instruments they did carry were similar to those utilized previously. The Vega landers were focused mainly on solving mysteries about the atmosphere and rectifying problems with instruments on previous missions that were caused by the hostile atmosphere. The gamma-ray soil spectrometer had been deleted after Venera 9 and 10 in favor of the combined drill and x-ray fluorescence spectrometer; this time they were all carried. And since there were no imagers the penetrometer was upgraded to provide an electrical readout.

Lander instrument mass   117 kg.

Balloon:

1. Temperature and pressure sensors (USSR-France)
2. Vertical wind velocity anemometer
3. Nephelometer for density and particle size of local aerosols (USA)
4. Light level photometer and lightning detector
5. Stable oscillator for VLBI measurements

A boom was deployed from the side of the gondola to expose sensors. One was a propeller anemometer. It measured vertical winds as fast as 2.0 m/s. The horizontal winds were measured by VLBI analysis of radio tracking. The ambient temperature was measured by two thin-film resistance thermometers with a range of 0 to 70°C and an accuracy of 0.5°C mounted at separate positions on the boom. Pressure was measured by a vibrating quartz beam sensor with a range of 0.2 to 1.5 bar and an accuracy of 0.25 millibar. The photometer consisted of a silicon PIN diode sensitive in the 400 to 1,100 nm range with a 60 degree field of view at the nadir. It was also designed to detect lightning by counting short bursts of abnormally bright intensity. The nephelometer was a simple backscatter instrument similar to those of previous missions.

Mission description:

Flyby spacecraft:

In keeping with the international nature of the project, Westerners were allowed to visit Baikonur and view the launches of Vega 1 on December 15, 1984, and Vega 2 on December 21. This was also the first time that Soviet television showed a Proton launch. And although the US routinely tracked Soviet spacecraft, this was the first time that this was done officially. The announcement that an American instrument was onboard prompted a small furor in the US. One of the booms for the plasma wave experiment initially failed to deploy on each spacecraft, but these both sprung out after the first midcourse maneuver.

Vega 1 arrived at Venus in early June 1985, only weeks after Venera 16 had been switched off. The spacecraft released their entry systems 2 days out from the planet,

Figure 18.10 Vega 1 launch.

on June 9 for Vega 1 and on June 14 for Vega 2. The points at which they were to enter the atmosphere were on the night-side in order to enable the spacecraft to head for Halley and to maximize the cruising lifetime of the balloons before they suffered solar heating. After releasing its entry system Vega 1 maneuvered to pass the planet at a range of 39,000 km for the gravity-assist to Halley and to relay the data from its lander. Vega 2 did likewise at a range of 24,500 km. Each spacecraft turned off-Sun to receive the transmission from its lander at a rate of 3,072 bits/s in the meter band and to relay it to Earth in the centimeter and decimeter bands. They did not conduct any science observations at Venus. On finishing the relay, each spacecraft resumed cruise operations. The gravity-assist of the flyby did most of the work in deflecting the path of each spacecraft toward Halley, but maneuvers were needed to refine the final approach.

Vega 1 flew past the nucleus of the comet at a range of 8,890 km on March 6, 1986, and Vega 2 did likewise at a range of 8,030 km on March 9. Both made highly successful scientific measurements. Two Japanese spacecraft had been observing the comet at extreme distance and Europe's Giotto was scheduled to arrive on March 13 for a daring close flyby at a range of only 500 km. By combining tracking data with imaging, the Vega spacecraft gave a more precise position for Halley in space than was possible using terrestrial telescopes. This was used to improve the accuracy of Giotto's terminal maneuvers, both to reduce the targeting error in order to obtain the intended observations and to reduce the potential risk to that spacecraft. Both Vega spacecraft flew through the tail of the comet and were pummeled by small grains impacting at 80 km/s. The shields installed on one side of each vehicle protected it from damage. The solar panels suffered both dust impacts and electrical discharges induced by the comet plasma. Vega 1 lost 40% of its power supply and Vega 2 lost 80%. After a circuit around the Sun, both spacecraft passed through the tail again in 1987, providing further data. Vega 1 ran out of attitude control gas on January 30 of that year and then on March 24 contact with Vega 2 was discontinued.

Entry system:

The Vega 1 capsule entered the night-side atmosphere at 01:59:49 UT on June 11, 1985, at a speed of 10.75 km/s and at an angle of 17.5 degrees. The Vega 2 capsule entered at 01:59:30 UT on June 15, at 10.80 km/s and at an angle of 18.13 degrees. The pilot parachutes were deployed at an altitude of 65 km. Eleven seconds later, at 64.5 km, the capsules split into hemispheres and the pilot parachutes drew the upper hemispheres containing the balloon systems away, in the process deploying the main parachutes of the landers in the lower hemispheres. Four seconds later, at 64.2 km, the landers shed their hemispheres. Having slowly descended to 47 km, each lander released its parachute in order to free-fall to the surface. The new aerodynamic drag devices successfully reduced both vibration and spin, thereby increasing the stability of the descending landers.

Meanwhile, the balloon packages were released from their hemispheres at 62 km, in the process deploying the pilot parachute of each balloon package. At 57 km the main parachute was deployed. At 55 km the inflation of the envelope was initiated. With the balloon fully inflated, the main parachute was released at 53 km. At 50 km the inflation system and ballast was released and the balloon system rose to 54 km in the middle of the cloud layers with its gondola deployed to make measurements.

Landers:

The Vega 1 lander settled at 7.11°N 177.48°E, just north of eastern Aphrodite Terra and 0.6 ± 0.1 km below the planetary mean radius. It was 03:02:54 UT on 11 June, 0:24 local time, and the solar zenith angle was 169.3 degrees. The measured surface temperature was 467°C and the pressure was 97 bar. The transmission was curtailed 20 minutes after landing in order to conserve energy on the flyby spacecraft, which was not facing its solar panels at the Sun, and to ensure readiness for the subsequent Halley trajectory maneuver.

At an altitude of 17 km the Vega 1 lander experienced electrical spikes and the Doppler tracking data showed violent upward excursions. This shock triggered the accelerometer that was to indicate contact with the ground, causing a premature start to the surface activity sequence, including deployment and operation of the drill and x-ray spectrometer. As the x-ray soil analysis instrument had failed its pre-launch tests and been flown regardless, this may not have mattered. Venera 11 to 14 and the four US Pioneer probes also experienced electrical anomalies in the altitude region 12 to 18 km, but Venera 9 and 10 and Vega 2 did not. The cause of these anomalies remains unknown.

The Vega 2 lander touched down at 7.52°S 179.4°E, 1,300 to 1,500 km southeast of Vega 1 and 0.1 ± 0.1 km above the planetary mean radius. It was 03:00:50 UT on 15 June, 1:01 local time, and the solar zenith angle was 164.5 degrees. The surface temperature was 462°C and the pressure was 90 bar. The transmission was truncated 22 minutes after landing to preserve energy on the flyby spacecraft. There were no anomalies during the descent and the surface operations were performed nominally.

Balloons:

The Vega balloons were both successfully deployed at the anti-solar point (i.e. local midnight) and drifted with the wind at an altitude of about 53 km where the pressure was about 0.5 bar, right in the middle of the three cloud layers. They were carried longitudinally by zonal winds through the night-side atmosphere for 30 hours before crossing the dawn terminator. No latitude measurements could be made, and it was assumed that the balloons remained at a constant latitude, 8°N in the case of Vega 1 and 7°S for Vega 2. Each balloon transmitted for 46.5 hours until its batteries were exhausted. Loss of signal occurred in the early morning hours on Venus after having traveled some 10,000 km, about one-third the way around the planet. The balloons continued silently into the day-side where they would eventually have succumbed to solar heating and burst their envelopes.

Results at Venus:

Landers on descent:

A telemetry problem prevented Vega 1 temperatures from being transmitted during the descent, but the Vega 2 data indicated the presence of a sharp thermal inversion that reached a minimum temperature of -20°C at an altitude of 62 km. The optical spectrometers operated between 63 and 30 km and reported an atmospheric structure similar to that seen by earlier landers and confirming a three layer cloud deck. But on this mission, as for Venera 8, no sharp lower cloud boundary was observed. Aerosol particle size measurements were taken down to 47 km, and were in general agreement with earlier Soviet results and the data from the Pioneer entry probes and confirmed that there were at least two layers of differing particle sizes. The measurements from Vega 1 and 2 were highly consistent, indicating the cloud layers to be very similar at their entry points except in the uppermost layer where Vega 2 found less dense aerosols than Vega 1. The smallest 'mode 1' particles were speculated to be aluminum and/or ferric chloride. About 80% of the larger 'mode 2' particles were shown to be spherical with a refractive index of 1.4, a characteristic consistent with sulfuric acid, while the remaining 20% had a refractive index of 1.7, suggestive of solid sulfur. The highest particle counts were in the altitude range 58 to 50 km. The Vega instruments were insensitive to the largest 'mode 3' particles reported by Pioneer probes.

The Vega 1 and 2 gas chromatographs and the Vega 1 mass spectrometer were the first to make an in-situ detection of sulfuric acid, confirming remote sensing results and yielding a density for the altitude range 63 to 48 km of about 1 milligram of sulfuric acid per cubic meter. The Vega 1 mass spectrometer heavy particle sample contained sulfur trioxide (sulfuric acid anhydride) and chlorine. Unfortunately, the Vega 2 mass spectrometer failed. The x-ray fluorescence spectrometer on Vega 2 detected sulfur (~ 1.5 mg/m^3), chlorine (~ 1.5 mg/m^3), and iron (0.2 ± 0.1 mg/m^3). It also made the first detection of phosphorus (~ 6 mg/m^3), this possibly in the form of phosphoric acid, and explaining the persistence of a small amount of aerosol in the sub-cloud region with a base at 33 km. Iron was also reported by the x-ray

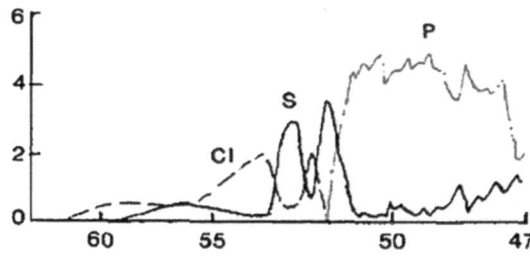

Figure 18.11 Chlorine, sulfur and phosphorus profiles from the descent x-ray aerosol analyzer (from Don Mitchell).

analysis, perhaps as ferric chloride in the aerosols. The Vega 1 x-ray fluorescence instrument failed. The ultraviolet spectrometers gave vertical profiles for sulfur dioxide mixing ratios with upper region abundances in general agreement with remote sensing and other sources, and generally decreasing towards zero at the surface. The possibility of elemental sulfur vapor was also noted. Solar ultraviolet was completely absorbed below an altitude of 10 km, although this was probably due to aerosols coating the instrument. The hydrometer reported a water vapor abundance of 0.15% at high altitudes (60 to 55 km) decreasing by a factor of ten at lower altitudes (30 to 25 km). The fact that this large abundance is inconsistent with other measures may indicate that the instrument was confused by other atmospheric constituents. The water vapor profile on Venus remains poorly determined.

Landers on the surface:

The Vega 1 lander conducted a gamma-ray soil analysis but the drill had failed and so no x-ray soil analysis could be performed. The Vega 2 gamma-ray spectrometer, drill, and x-ray fluorescence experiments all worked well.

X-ray fluorescence results from Vega 2 (as oxides):

silicon	47%
titanium	0.2%
aluminum	16%
iron	8.5%
manganese	0.14%
magnesium	11%
calcium	7.3%
potassium	0.1%
sulfur	4.7%
chlorine	<0.3%

These analyses showed rocks poor in iron and magnesium but rich in silicon and aluminum, indicating a composition similar to lunar highland rocks. The fairly high sulfur abundance may be an indicator of older rocks.

Gamma-ray results:

	Vega 1	Vega 2
potassium	0.45 ± 0.22 wt%	0.40 ± 0.20 wt%
uranium	0.64 ± 0.47 wt%	0.68 ± 0.38 wt%
thorium	1.5 ± 1.2 wt%	2.0 ± 1.0 wt%

The potassium, uranium and thorium values were very similar to Venera 9 and 10, in contrast to the Venera 8 results that showed significantly higher concentrations of all three elements.

Balloons:

Even although this was the first attempt at deploying a planetary aerostat, both of the balloons succeeded. They made the first measurements of the horizontal structure of the atmosphere to complement the many vertical profiles from descent probes. The temperature in the Vega 1 air mass was a constant 40°C. It was about 6°C cooler for the Vega 2 balloon. The atmosphere was more turbulent than expected. At times the balloons precipitously plunged in downdrafts of 1 to 3 m/s by hundreds of meters, sometimes several kilometers. The Vega 1 balloon encountered heavy turbulence at the start of its run and then again towards its end. Shortly after sunrise, passing over the Aphrodite Terra highlands, the Vega 2 balloon plunged more than 3 km to a pressure level of 0.9 bar, very close to the lower limit of its buoyant zone, before it rebounded.

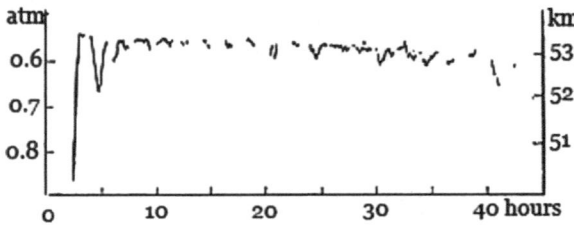

Figure 18.12 Flight profile of the Vega 1 balloon (from Don Mitchell).

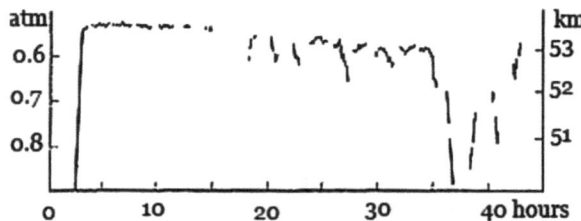

Figure 18.13 Flight profile of the Vega 2 balloon (from Don Mitchell).

The nephelometer on the Vega 1 balloon was hard to interpret due to calibration problems but generally seemed to agree with particle data from the nephelometers on the descent probes, showing the middle cloud in which the balloon drifted to be horizontally homogeneous with no clear regions. Unfortunately, the Vega 2 balloon nephelometer failed. In their cruise to the dawn terminator, the photometers noted some variation in light levels that may have been due to variations in the underlying clouds, and although there were some light flashes there was no strong evidence for lightning. Vega 1 crossed the terminator into daylight 34 hours into the flight, and its photometer registered dawn 2 hours prior to sunrise. The Vega 2 photometer did not function correctly, but indicated dawn 3 hours before the terminator crossing. The anemometers reported downdrafts of 1 m/s. The VLBI Doppler measurements found horizontal winds of up to 240 km/hr, made the first in-situ observations of the 'super-rotation' of the atmosphere at this altitude, and made measurements of atmospheric turbulence.

Results at Halley:

The results of the Vega Halley encounter were more than just scientific, they were also cultural and political. The project would be the first to image the nucleus of the world-famous Comet Halley. For the first time, a Soviet mission and its purpose was made known well in advance. The portion of the Vega mission at Venus went barely noticed outside scientific circles, but the whole world was waiting in expectation for the spectacle of the Halley encounter and the Soviets were well aware that this was unlike any other space mission they had ever conducted.

The Vega 1 spacecraft closed in on the comet at the blazing speed of 79.2 km/s in early March 1986. It performed a final trajectory correction on February 10. Its scan platform locked onto the comet on February 14 and began tracking. Far encounter images on March 4 and 5 demonstrated the camera's performance. On March 6, the day of close encounter, the world's press was present in the IKI control room for the first time, disturbing the usual professional calm with a bustling jumble of people eager to experience a Soviet mission event as it happened, including US television and media with both Roald Sagdeev and Carl Sagan providing commentary. Sagan as commentator for a Soviet spacecraft encounter in real-time was clear evidence that *perestroika* had become reality. Vega 1 switched to high rate telemetry 2 hours before closest approach and took over 500 images during the 3 hour encounter. The raw images looked overexposed and fuzzy. It was hard to pick out the nucleus from the obvious dust jets. But the IKI press room was filled with awe and applause. The images and other data streamed in for another 2 days.

Vega 2 closed in 3 days later at 76.8 km/s. It did not require a final correction but 30 minutes before the encounter on March 9 it gave its controllers a scare when the computer guidance system failed. However, the spacecraft quickly switched over to the backup system and the observations began as planned. By the time the encounter was over on March 11 the spacecraft had provided over 700 images.

The images of Halley revealed a potato-shaped nucleus 14 × 7 km with a very dark albedo of 4%, a rotation rate of 53 hours, and at least five dust jets that could be counted on its sunward side. The environmental sensors on board the two spacecraft made pioneering measurements of the plasma fields in the vicinity of the comet, and defined the interaction of the solar wind with the out-flowing cometary gases. Some of the constituents of the gas were identified and measured. The size and flux of dust particles varied enormously as the spacecraft flew through and in between the jets of dust and gas. A number of instruments were lost during the encounter, and the solar panels were extensively damaged by impacts and the electrical discharges that were induced by the cometary plasma.

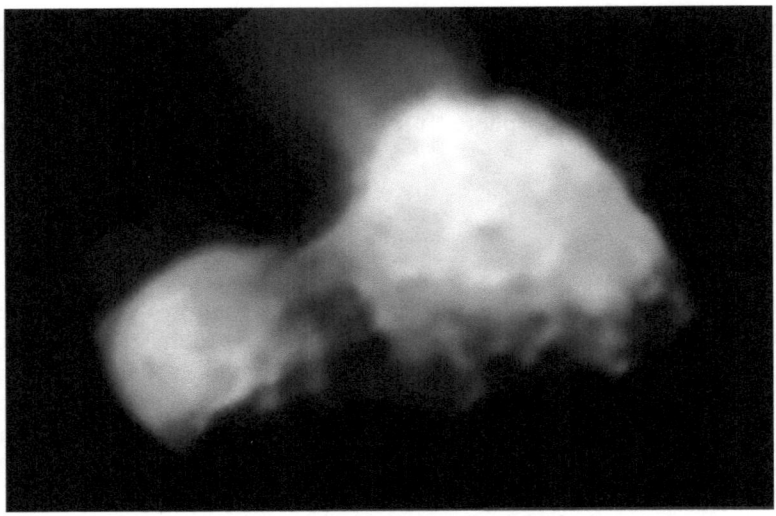

Figure 18.14 Vega 2 image of Halley (processing by Ted Stryk).

The infrared spectrometer on Vega 2 failed due to a leak in the cryogenic system. The Vega 1 infrared spectrometer was sent an erroneous command which put it into calibration mode during the 30 minutes at closest approach, which was unfortunate, but it did report data taken at greater distances. The C-H band of hydrocarbons was detected. The fact that the temperature of the nucleus was 300 to 400K meant that it had an insulating layer at its surface. The dust and gas were jetting through fissures in this crust opened by the heating of volatiles contained within. The three-channel spectrometer on Vega 1 was crippled by an electrical fault, and despite its partner on Vega 2 losing the ultraviolet channel this was able to detect water, carbon dioxide, the hydroxyl radical and the cyano radical, various other products of hydrocarbon photolysis, ammonia and other organic materials in the coma. It was concluded that the principal components of the gas were water containing carbon monoxide and carbon dioxide molecules, as well as photo-produced radicals and atomic hydrogen, oxygen and carbon.

Analysis of the dust in the jets revealed grains in the submicron-to-micron size range of compositions varying from metallic to siliceous to carbonaceous. The dust mass spectrometers returned results showing three families of materials: one very similar to the carbonaceous chondrite meteorites which are thought to be the most primitive of Solar System material, another enriched in carbon and nitrogen, and the third enriched with water and carbon dioxide ice.

Thus ended one of the most daring, innovative, complex and successful missions in the history of robotic space exploration to that time. It established the USSR as the leader in the field; a distinction that was sadly short-lived and later forgotten.

19

Another try at Mars and its moon Phobos

TIMELINE: 1986–1988

Bolstered with confidence as a result of the extremely successful Vega missions and leading the internationalization of robotic planetary exploration after the Americans had sidelined themselves, the Soviets decided to make another attempt at the Red Planet in 1988. As approved in 1976 by Mstislav Keldysh after the demise of the very ambitious rover and sample return proposals, this time the focus would be on the moon Phobos. The spacecraft would enter Martian orbit and after several weeks of orbital phasing during which it would study the planet, it would make a very slow pass just 50 meters over the surface of Phobos to deposit two landers and undertake not only passive remote sensing by imagers and spectrometers but also active remote sensing with radar, ion beams and laser beams. In addition to the new power hungry active remote sensing instruments, the massive spacecraft would be equipped with a variety of other scientific instruments. Once again the Soviets invited the world's scientific community to provide investigations for the mission, and this time even American instruments were accommodated.

The Phobos project was a model for international cooperation, but in the end also turned out to be a lesson in the international dissonance caused when such a mission fails. Phobos 1 and Phobos 2 were successfully launched in July 1988, but Phobos 1 was lost early in its interplanetary cruise owing to an elementary operational error. Phobos 2 reached Martian orbit and in just a few weeks conducted enough first-class observations of the planet to make up for all the flawed Soviet missions in the past, but then, just days prior to the close encounter with Phobos, the spacecraft failed to respond to a scheduled communications session and was lost.

Launch date		
1986		
No missions		
1987		
No missions		
1988		
7 Jul	Phobos 1 Mars orbiter	Lost enroute
12 Jul	Phobos 2 Mars orbiter	Failed in orbit before Phobos encounter

BACK TO MARS: 1988

Campaign objectives:

The contrast between the Soviet space program in the mid-1970s and the mid-1980s was stark. In the mid-1970s the Soviets were in the deep shadow of the Americans. They had convincingly lost the race to land a man on the Moon. The Mars fleets of 1971 and 1973 were failures, whereas the Mariner 9 and Viking missions were total successes. In the 1970s the Americans launched inward to Mercury and outward to Jupiter, Saturn, Uranus, and Neptune. The Soviets were confined to Venus. But in just 10 years all this was turned on its head. By the mid-1980s the US launch rate for planetary missions had fallen to zero and their ambitious plans for Venus, Mars and Jupiter orbiters were troubled with delays and funding problems. By failing to fund a spacecraft for the Halley armada, the US had yielded the leadership it had worked so hard to achieve to Europe and Russia.

While the American program was struggling in the 1980s, the Soviet program was flourishing. Since 1975 they had achieved a string of unbroken successes at Venus with ever more complex spacecraft. They had taken the international lead from the US by opening the Vega missions to international participation on an unprecedented scale. Their Proton-launched spacecraft could carry a large, comprehensive payload of sophisticated and complex instruments. With landers and balloons at Venus and a gravity-assist to Halley the Vega mission had demonstrated boldness, ambition and success on a grand scale that silenced Western criticism of the quality of the Soviet program. The USSR was launching a space station, Mir, launching the largest rocket ever built, Energiya, developing a competitor to the US Space Shuttle, Buran, and launching almost 100 spacecraft per year. The *National Geographic* magazine gave a profile on the massive Soviet space industry and its successes, and *Time* magazine ran a cover story 'Surging ahead – Soviets overtake the US as the No.1 spacefaring nation'. By the mid-1980s, therefore, Soviet confidence and optimism were running high.

After the Vega missions there was a consensus that the Venera series had run its course, having accomplished about all that it could. Of course there were ideas for long-lived landers and aerostats, but these were for the future. As the Vega missions were underway, the 'Martians' at IKI were insistent that it was now time to revive interest in Mars. The early proposals for a Mars sample return mission had fallen by the wayside, with even the plan designed to use the Proton launcher being discarded in 1977. Since then, however, the 'Martians' had been investigating more practical missions and had revived interest in a proposal to investigate Phobos, the larger of the planet's two small moons. Strategic planning at IKI was still to seek to outdo the Americans, and since the US had not considered Phobos as a target there was scope for another Soviet showplace. They would not be repeating or competing with the US. After the Viking landers did not find any strong evidence for life on the surface of Mars the Americans had all but abandoned the planet. In addition, the continuing development of the Mars 2 to 7, Venera 9 to 16, and Vega 1 and 2 lines of heavy Proton-launched spacecraft provided the technology to undertake very capable Mars missions on a scale that the US was no longer able to fund. The lack of immediately compelling Venus missions, the lack of US interest in Mars, and the availability of proven technology, were compelling reasons to mount another campaign at Mars. So Soviet engineers decided to develop a new-generation planetary spacecraft based on the Venera-Vega heritage and to switch their attention from Venus to Mars. It was expected that the new spacecraft would serve as the baseline for the next 20 years of Soviet planetary exploration. The three Vega spacecraft spares were refurbished for astronomy missions in Earth orbit, two of which flew as Astron in 1983 and Granat in 1989.

Spacecraft launched

First spacecraft:	Phobos 1 (1F No.101)
Mission Type:	Mars Orbiter, Phobos Flyby/Landers
Country/Builder:	USSR/NPO-Lavochkin
Launch Vehicle:	Proton-K
Launch Date/Time:	July 7, 1988 at 17:38:04 UT (Baikonur)
Mission End:	September 2, 1988
Outcome:	Failed in transit due to command error.
Second spacecraft:	Phobos 2 (1F No.102)
Mission Type:	Mars Orbiter, Phobos Flyby/Landers
Country/Builder:	USSR/NPO-Lavochkin
Launch Vehicle:	Proton-K
Launch Date/Time:	July 12, 1988 at 17:01:43 UT (Baikonur)
Encounter Date/Time:	January 29, 1989
Mission End:	March 27, 1989
Outcome:	Lost in Mars orbit prior to Phobos encounter.

The concept of a mission to Phobos actually preceded the Vega missions, but the latter had priority because their launches were dictated by the apparition of Comet Halley. The intention to send spacecraft to Phobos was first announced in November 1984, a month before the launch of the Vega missions. At that time the schedule was to launch in 1986, but this was slipped to 1988. Several mission scenarios had been considered including a Phobos landing, an outpost on Phobos for remote sensing of Mars, and a Phobos landing with sample return. The mission design selected was no less audacious, but less risky to arrange. The plan was to hover a large spacecraft about 50 meters above Phobos and conduct both passive and active remote sensing using lasers and particle guns. Later the mission expanded to include deploying two small landers for Phobos, one built by IKI and the other by the Vernadsky Institute. This campaign was to be the first in a Mars-focused exploration program which the Soviets expected to be every bit as successful as their exploration of Venus.

The objectives of the Phobos missions were to:

1. Conduct studies of the interplanetary environment
2. Perform observations of the Sun
3. Characterize the plasma environment in the Martian vicinity
4. Conduct surface and atmospheric studies of Mars
5. Study the surface composition and environment of the moon Phobos.

Having reaped major scientific and political results by the internationalization of the Vega missions, this policy was repeated for the Phobos campaign. This time the involvement of the US and Western countries was even more extensive. Many of the investigations and instruments were supplied by European countries; not only Soviet Bloc but also Western Europe. Given its history of cooperation in Soviet missions starting with the 1971 Mars missions, France was the primary Western contributor. The fact that the US supplied one instrument, a number of science co-investigators, and tracking support, was a major breakthrough initiated by the Soviet Union as the Cold War thawed. As a result, each spacecraft had a record number of instruments with which to perform a comprehensive scientific investigation of Mars and Phobos. In fact, each spacecraft carried twenty-four experiments provided by the USSR, the European Space Agency, and thirteen other countries including the US. The Phobos missions constituted an aggressive, innovative, and very impressive scientific attack on both Mars and Phobos, especially when compared to the modest orbiter that the US was then planning to send to Mars in 1990.

Spacecraft:

Orbiter:

The spacecraft was similar to the Vega design but represented a new generation in technology, and was the first major design change in Soviet planetary missions since Mars 2 and 3. It was the heaviest planetary spacecraft yet devised, with a total mass over 6,200 kg including 3,600 kg for the separable ADU propulsion system. The scientific payload capacity was an incredible 500 kg. It was constructed around a pressurized toroidal compartment for electronics, with four spherical outrigger tanks containing hydrazine monopropellant for the onboard propulsion system. The solar panels were mounted with their flat planes parallel to the toroid, perpendicular to the earlier Mars and Venus spacecraft. Attached to the outrigger tanks were twenty-four

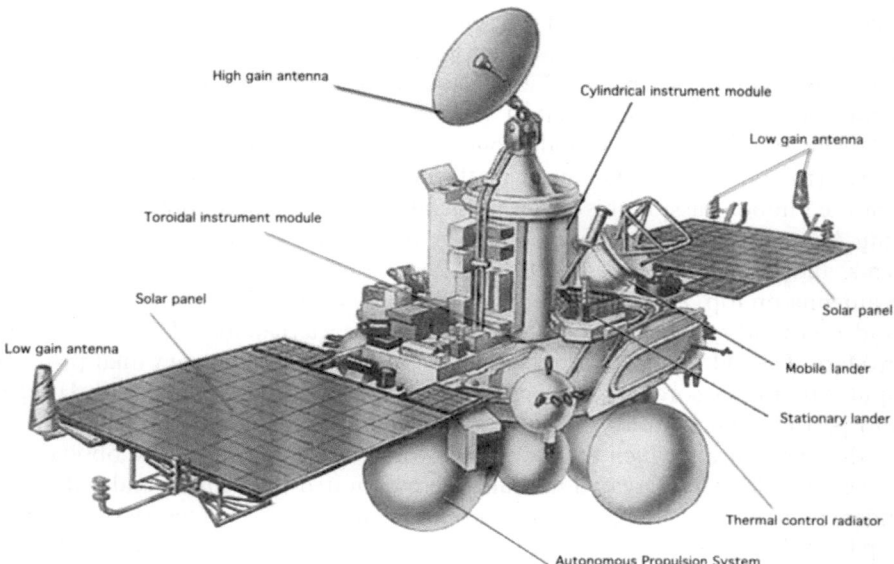

High gain antenna

Cylindrical instrument module

Low gain antenna

Toroidal instrument module

Solar panel

Solar panel

Low gain antenna

Mobile lander

Stationary lander

Thermal control radiator

Autonomous Propulsion System

Figure 19.1 Phobos spacecraft (courtesy NPO-Lavochkin).

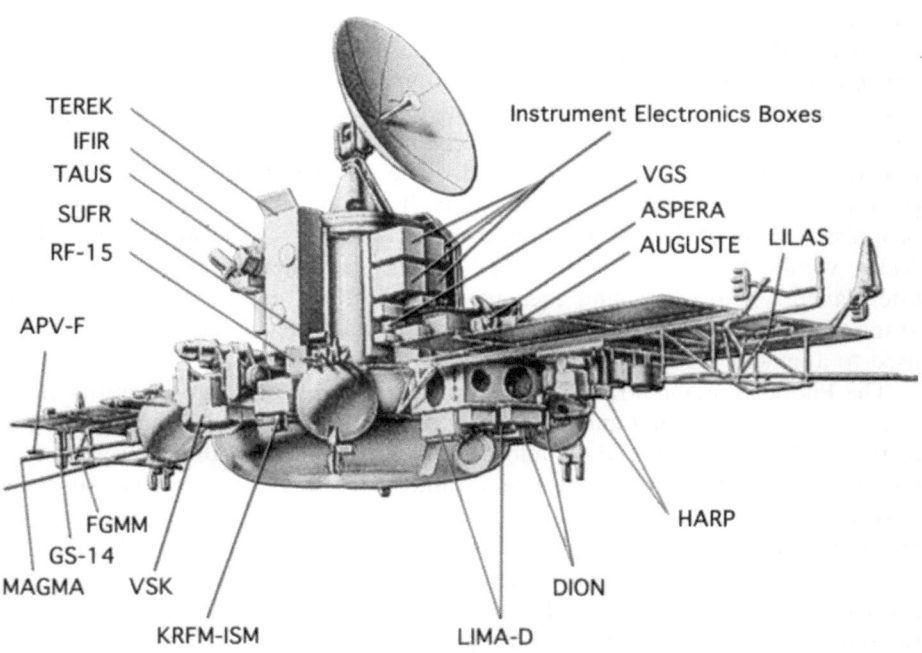

TEREK

IFIR

TAUS

SUFR

RF-15

APV-F

Instrument Electronics Boxes

VGS

ASPERA

AUGUSTE   LILAS

FGMM

GS-14

MAGMA   VSK

KRFM-ISM

DION

LIMA-D

HARP

Figure 19.2 Phobos spacecraft and instruments, after the ADU propulsion system has been jettisoned (courtesy NPO-Lavochkin).

engines rated at 50 N of thrust and four engines rated at 10 N for maneuvering, and a dozen thrusters rated at 0.5 N for attitude control; and there were additional thrusters on the body and the solar panels. Two basic attitude control modes were provided: 3-axis control, and 'drift mode' control with the solar panels facing the Sun and the vehicle spin stabilized. The attitude control avionics consisted of Sun sensors, star sensors, gyroscope and accelerometer platforms, and a triply-redundant computer.

For the Phobos mission this basic spacecraft had a cylindrical pressurized module mounted above the toroid. This module contained spacecraft control avionics, radio systems, and science experiments. It had a 2-axis 1.65 meter diameter parabolic high gain antenna on top. A 30 megabit data recorder was provided as a buffer. With the upgraded ground systems, the new 50 W transmitter was capable of 65 to 131 kbits/s from Mars. The old Venera control system was abandoned and a new dual-processor computer that incorporated a 4.8 gigabyte memory developed jointly with Hungary was to perform the complex maneuvers to match orbits with the target moon and make the flyby during which the spacecraft would undertake its closest observations and release its landers. Payload modules were mounted both inside and outside the cylinder. Unlike Vega, there was no scan platform; to make planetary observations the spacecraft had to be reoriented away from Earth-pointing. While cruising it would employ the drift mode in which it spun slowly with its solar panels pointed at the Sun.

Another innovation was the Autonomous Propulsion System (ADU) for the final escape burn, midcourse corrections, and orbit insertion. It was to be discarded after completing the major orbital maneuvers. This large assembly was carried below the toroidal section. It comprised eight tanks, four 1.02 meters in diameter and four 0.73 meters in diameter, nested around a single KTDU-425A restartable engine that could be throttled between 9.8 and 18.6 kN, burning UDMH and nitrogen tetroxide. It had a total burn time of 560 seconds. The ADU weighed 600 kg dry and was capable of carrying 3,000 kg of propellants. It was an extensive redesign of the main propulsion stage of the old Ye-8 lunar spacecraft. Part of the strategy was that the ADU would serve as a fifth stage of the launcher to achieve the desired interplanetary trajectory, and then perform the most energy intensive maneuvers at the target planet. It would eventually become the 'Fregat' stage and be extensively used to augment the Soyuz launcher.

The Phobos mission strategy was to place the spacecraft into an initial elliptical near-equatorial orbit that maneuvers over several weeks would circularize very close to that of Phobos. The ADU would be jettisoned, and from this intermediate orbit the moon would be approached several times at low velocity over a period of weeks at distances varying from several hundred kilometers to 35 km. Observations of the moon would be used to refine knowledge of its orbital parameters sufficiently for a maneuver to be calculated which the onboard propulsion system would perform to produce the desired close flyby. The radar would be activated at a range of 2 km and the spacecraft would be controlled using the radar to perform a 20 minute pass at an altitude of about 50 meters at a relative velocity of 2 to 5 m/s. Considering the large topographic variation, this low altitude flyby would need a very robust and complex automated control system. Two types of lander were to be deployed, one of which would be stationary and the other capable of 'hopping' around in the weak gravity. Two active remote sensing experiments would be attempted using laser and ion guns in a manner that would enable the spacecraft to fly through the material evaporated off the surface for analysis with mass spectrometers. An active radar system would map the regolith material down to 2 meters depth. Passive remote sensing included both imaging and spectrometry. After the low pass over Phobos, the spacecraft was to adjust its orbit and spend its remaining time observing Mars. If the Phobos flyby were a success, there was the possibility of maneuvering the spacecraft to make a similar flyby of Deimos.

Launch mass:	6,220 kg
ADU propulsion system:	3,600 kg
Orbiter wet mass:	2,620 kg
Instrument payload:	540 kg including landers

Mobile lander:

The PrOP-F hopper ('Frog') resembled a small, flattened ball consisting of a 50 cm diameter hemisphere on top of a semi-cylindrical base. It would be ejected from the side of the spacecraft at the time of closest approach, and fall to the surface in the moon's weak gravity – some 2,000 times weaker than that of Earth. It was designed

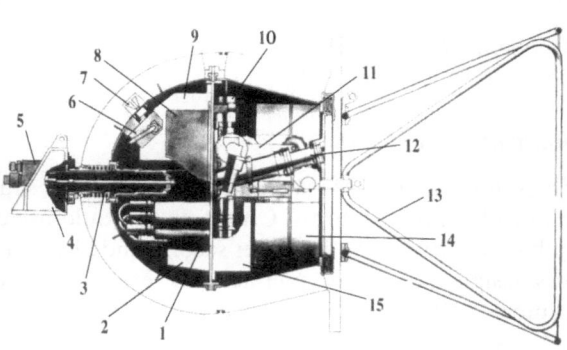

Figure 19.3 Phobos PrOP-F 'Hopper': 1. Sequencer; 2. Data unit; 3. Separator; 4. Spacecraft mount; 5. Pyro device; 6. Transmitter; 7. Antenna; 8. Battery; 9. Accelerometer; 10. Attitude control; 11. Penetrometer; 12. Spring device; 13. Damper; 14. X ray spectrometer; 15 Controller.

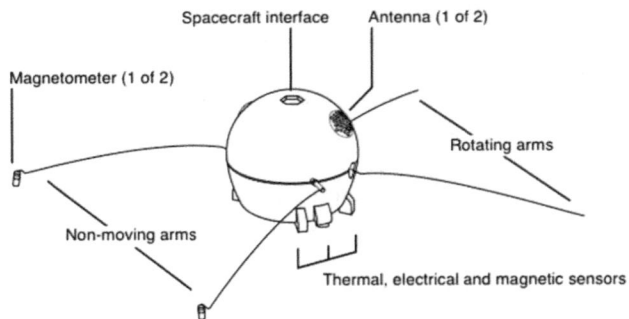

Figure 19.4 Phobos PrOP-F diagram as deployed on the surface (from Ball et al.).

to survive contact at a horizontal speed of 3 m/s and a vertical speed of 0.45 m/s. A damper truss would reduce the time taken to settle on the surface after initial impact. This would be ejected after settling, and a set of four long levers ('whiskers') on the base, two of which could be rotated, would deploy in order to orient the ball flat-side down. After about 20 minutes of sensing and the transmission of results, the hopper would flex its levers to propel itself to another location about 20 meters away. Each such hop would peak at a height of about 20 meters. The levers would right the ball after each landing. Ten hops were planned. The hopper was powered by a 20 amp-hour battery, and a 0.3 W transmitter would send to the spacecraft at 224 bits/s. Operations would begin immediately upon landing, and run for the 4 hour battery lifetime, by which time the spacecraft would have traveled about 300 km from the moon. Due to mass limitations the hopper was assigned only to Phobos 2.

Mobile lander mass: 50 kg
Payload: 7 kg

Stationary lander (DAS):

The stationary lander was carried on top of the spacecraft and was to be deployed at 2.2 m/s by a pair of arms. Once free, it would use cold gas thrusters to spin itself for stability at about 2 radians per second and also to propel itself towards the surface. It was designed to survive contact at a vertical speed of 4 m/s and a horizontal speed of at most 2 m/s. Contact probes underneath the lander were to ignite hold-down solid rockets, and simultaneously fire a harpoon on the underside down into the surface. The harpoon was tethered and nesting motors would wind up the tension to hold the lander firmly in place on the surface. After allowing the dust to settle for 10 minutes, the lander would extend three legs to raise its instrument platform 80 cm above the surface while maintaining the tension in the harpoon tether, and deploy and point its solar panels and antenna. Because Phobos rotates relative to the Sun, the orientation of the solar panels would be controlled by Sun sensors.

The lander was to undertake science investigations for 3 months, communicating directly with Earth by a transmitter and receiver at 1,672 MHz using antennas on the

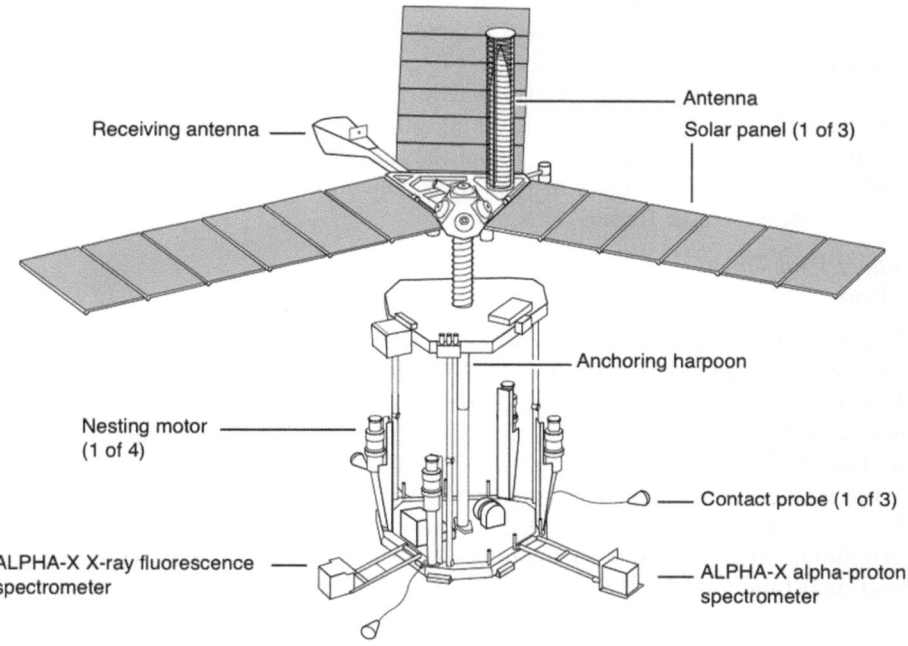

Receiving antenna

Antenna

Solar panel (1 of 3)

Anchoring harpoon

Nesting motor
(1 of 4)

Contact probe (1 of 3)

ALPHA-X X-ray fluorescence
spectrometer

ALPHA-X alpha-proton
spectrometer

Figure 19.5 Phobos DAS stationary lander diagram as deployed on the surface (from Ball et al.).

Figure 19.6 DAS stationary lander folded for attachment to the Phobos spacecraft (courtesy NPO-Lavochkin).

instrument platform. It was realized that there might be shadowing problems on the solar panels, but there was insufficient time to develop mitigation options even using a secondary battery. Also, the Soviets and French developed different algorithms for data compression and because of insufficient capacity in the computer to implement both, the landers on the two spacecraft had different algorithms. The dual-processor computer was supplied by Hungary under the supervision of IKI. At a data rate that varied from 4 to 16 bits/s, three or four communication sessions would be needed to transmit a single image frame.

Stationary lander mass: 67 kg
Payload: 20.6 kg

The number of science instruments and the complexity of the spacecraft and its operations was unprecedented. A special MORION central interface was provided to handle the complexities. In the end, the mass budget was exceeded and some of the instruments had to be deleted from each spacecraft. The PrOP-F hopper was deleted from Phobos 1, as were the TERMOSKAN and ISM infrared instruments. Phobos 2 lost the RLK radar, the TEREK solar telescope, and the IPNM neutron detector. The complexity of the mission was daunting, both to develop with all of its international interfaces and to operate with all of its instruments competing for operational time including spacecraft targeting and data transmission.

Payload:

Orbiter:

Phobos active remote sensing:

1. Laser mass spectrometer for elemental surface composition (LIMA-D, USSR-Bulgaria-Finland-FRG-DDR-Czechoslovakia)
2. Ion gun mass spectrometer for elemental surface composition (DION, USSR-Austria-Finland-France)
3. Radar system for subsurface structure and mapping (RLK, USSR), Phobos 1 only

The 80 kg laser mass spectrometer was to conduct active remote sensing. It would fire 150 laser pulses, each of 10 nanoseconds, to evaporate material in the uppermost 1 mm of the surface of Phobos. The mass spectrometer would analyze the ions in the resulting plasma cloud to provide an elemental analysis for ranges up to 100 meters. The 24 kg ion gun mass spectrometer was to fire krypton ions at the moon and then measure the ions scattered back from its surface. Between them the two instruments were to measure about 100 sites on Phobos. The 41 kg radar system was to operate after the landers had been deployed and after the remote sensing at closest approach was done. When the spacecraft had opened the range to 2 km, the radar would map the surface of the moon and sound its subsurface to a depth of 2 meters.

Phobos and Mars passive remote sensing:

1. CCD camera and spectrometer for surface mapping at three wavelengths (VSK, USSR-Bulgaria-GDR)
2. Thermal infrared radiometer and ultraviolet-visible spectrometer for surface temperature, thermal inertia, stratospheric temperature and aerosol characteristics (KRFM, USSR-France)
3. Near-infrared mapping spectrometer for mineralogy and atmospheric structure (ISM, USSR-France); Phobos 2 only
4. Thermal infrared mapping radiometer for surface temperature mapping (TERMOSKAN, USSR); Phobos 2 only
5. Gamma-ray spectrometer for surface radioactive element content (GS-14, USSR)
6. Neutron spectrometer to search for water in surface layers (IPNM-3, USSR); Phobos 1 only
7. Solar occultation ultraviolet and near-infrared spectrometer for distribution of minor constituents and aerosols (AUGUSTE, USSR-France)

The 52 kg VSK imaging system comprised a spectrometer and three cameras with 288×505 pixel CCD arrays. There was a narrow angle camera with a clear filter across the range 400 to 1,100 nm, a wide angle camera with a blue-green filter (400 to 600 nm), and a wide angle camera with a near-infrared filter (800 to 1,100 nm). The solid-state memory supplied by East Germany could store over 1,000 images. Bulgaria provided the electronics and full assembly and test with the help of France, Finland and the US. The KRFM multi-wavelength instrument would measure the reflectivity and thermal properties of the regolith, optical properties of atmospheric aerosols, and temperature of the Martian stratosphere using ultraviolet, visible and thermal infrared wavelengths. The ISM near-infrared mapping spectrometer would obtain spectra in a single pixel of the surface, providing data on mineralogy and, for Mars, the column depth of carbon dioxide which was related to the elevation of the surface. The pixel was scanned across-track, and along-track scan was by spacecraft motion. It would provide a 1,600 km strip map of the planet's surface at a resolution of 5 km during early Mars operations and a resolution of 30 km later in the orbit for the Phobos encounter.

The 28 kg TERMOSKAN infrared multispectral imager was a new line-scanning photometer camera with better detectors than on the Venera 9 and Mars 5 missions, one of them cryogenically cooled with liquid nitrogen from a Stirling refrigerator for thermal wavelengths between 8.0 and 12.5 microns. The other detector was for the red and near-infrared between 600 and 950 nm. TERMOSKAN recorded thermal emission (essentially the temperature) of the surface in $512 \times 3,100$ pixel panoramas at a resolution of about 2 km. Only 384 of the 512 pixels were image data, the rest provided calibration information. The image width was about 650 km and the length was on the order of 1,600 km at a resolution of about 1.8 km/pixel. Images could be presented as surface temperature, thermal inertia, and texture. The 18 kg AUGUSTE experiment used a combination of two spectrometers and an interferometer to obtain vertical profiles for ozone, carbon dioxide, water, and oxygen by observing the limb of the planet at orbital sunrise and sunset and measuring atmospheric absorption in the solar spectrum. The IPNM gas scintillation neutron detector would identify areas on Mars (or indeed Phobos) that had hydrogen atoms in the regolith that were almost certainly due to water. This could provide evidence of habitable areas on Mars. The GS-14 gamma-ray spectrometer was mounted 3 meters away from the spacecraft on the edge of one of the solar panels and was to measure the elemental composition of the surfaces of both Mars and Phobos.

Solar wind and the plasma environment of Mars:

1. Plasma scanning analyzer for ion composition and direction, electron distribution, magnetosphere structure and dynamics (ASPERA, Sweden-USSR-Finland)
2. Plasma wave analyzer for plasma density and frequency spectrum of plasma waves (APV-F, ESA-Poland-Czechoslovakia-USSR)
3. Flux gate magnetometer for Martian magnetic field (FGMM, USSR-GDR)
4. Flux gate magnetometer for Martian magnetic field (MAGMA, USSR-Austria)
5. Electrostatic analyzer for energy and angular distribution of ions and electrons (HARP, Austria-Hungary)
6. Electrostatic and magnetic analyzer for direction and velocity of protons, alpha-particles and heavy ions (TAUS, Austria-Hungary)
7. Energy, mass, and charge spectrometer for ion composition, energy distribution and plasma structure (SOVIKOMS, USSR-Austria-Hungary-FRG)
8. Low energy telescope for solar wind and cosmic rays (LET, USSR-Hungary-ESA-FRG)
9. Energetic charged particle spectrometer for low-energy cosmic rays (SLED, USSR-Hungary-Ireland-FRG)

Of the two flux gate magnetometers, MAGMA was derived from the instruments used by the Venera and Vega missions and FGMM was a new one from a USSR-Germany collaboration. Both were mounted on a 3.5 meter boom, MAGMA at the tip and FGMM a meter from the tip. The APV-F plasma wave instrument included a dipole antenna and Langmuir probe for electron fluxes, and two 10 cm spheres at a separation of 1.45 meters to measure electromagnetic waves and plasma instabilities. The ASPERA instrument had two spectrometers on a scanning platform to measure plasma properties around the entire spacecraft. LET could measure the flux, energy spectrum, and composition of the solar wind and cosmic rays from atomic hydrogen up to iron. It was to complement measurements from a similar instrument on ESA's Ulysses mission, but that launch was delayed when the Space Shuttle was grounded after the Challenger accident. TAUS was to measure the energy and distribution of ions in the environment of Mars. HARP would measure electrons and ions from eight different directions.

Solar physics and astrophysics:

1. Solar telescope to observe the solar corona in x-ray and visible light (TEREK, USSR-Czechoslovakia); Phobos 1 only
2. Solar high precision photometer for solar oscillations (IFIR, Switzerland-France-ESA-USSR)
3. Ultraviolet photometer for solar extreme-ultraviolet monitoring (SUFR, USSR)
4. Solar x-ray and gamma-ray analyzer (RF-15, USSR-Czechoslovakia)
5. Gamma-ray burst monitor for high energy solar and galactic bursts, 100 keV to 10 MeV (VGS/APEX, USSR-France)
6. Gamma-ray burst monitor for low energy solar and galactic bursts, 3 keV to 1 MeV (LILAS, USSR-France)

The 36 kg TEREK solar telescope had three sets of optics, one a coronagraph to view the corona in the visible and the others the whole Sun in different x-ray bands using CCD detectors. The plan was to make observations in conjunction with Earth-based telescopes in order to compose a 360 degree view of the Sun. A three-channel photometer was to precisely measure solar irradiance in order to detect oscillations. The SUFR photometer was to monitor the ultraviolet flux of solar irradiation. The RF-15 instrument was similar to those of the geostationary meteorological satellites operated by America, and would be able to view different hemispheres of the Sun simultaneously. The two gamma-ray burst monitors were on the tip of one of the solar panels. The high energy instrument would have some utility for measuring the composition of the surface of Mars.

DAS small stationary Phobos lander:

1. CCD camera for surface imaging and microstructure (France)
2. Alpha, proton and x-ray spectrometer for surface elemental composition (FRG)
3. Harpoon anchor penetrometer with accelerometer and temperature sensor
4. Seismometer for internal activity and structure
5. Sun angle position sensor for determination of libration (France)
6. VLBI celestial mechanics experiment for orbital motion (USA-USSR-France)

The French were major partners in providing instruments and operational support for the DAS. They supplied the CCD camera and the optical sensor for tracking the Sun in order to determine the libration motions of the moon, and were participants in the VLBI experiment. The harpoon was instrumented with a temperature sensor and accelerometer to serve as a penetrometer to determine surface properties during the anchoring operation. The seismometer on the first DAS to land on the moon would have the opportunity to record the arrival of its partner. Its sensitivity was sufficient to detect the hopping activities of the PrOP-F.

PrOP-F small mobile Phobos lander, Phobos 2 only:

1. X-ray fluorescence spectrometer for surface elemental composition
2. Magnetic susceptibility and electric resistance probes for surface properties
3. Dynamic penetrometer for surface mechanical properties
4. Temperature sensors for measurement of surface layers
5. Radiometer for surface thermal flux
6. Magnetometer for surface magnetic field and permeability
7. Gravimeter (pendulum) to determining gravity field during descent
8. Accelerometer for surface properties

Mission description:

Phobos 1

Phobos 1 was launched on July 7, 1988. Another precedent was set when the launch was attended by the press, the international group of scientists participating in the mission, and even a delegation of US military. The rocket was also adorned with advertising for Italian and Austrian steel companies! The spacecraft used its ADU to achieve interplanetary injection for Mars, and again on July 16 for its first midcourse maneuver. However, on September 2 the spacecraft failed to respond to a scheduled communications session. Attempts to re-establish contact in September and October were unsuccessful and Phobos 1 was abandoned on November 3. Prior to its loss, it had returned data from its solar physics, plasma, and cosmic radiation instruments.

An investigation followed immediately. The communications failure was traced to a software upload made on August 29. An error was discovered in a command that was intended to turn on the gamma-ray spectrometer. An omitted hyphen created an unintended command to deactivate the attitude thrusters. The loss of Sun-pointing left the vehicle free to tumble and it depleted its batteries. The erroneous command was part of a test program that was encoded in software PROMs which had not been removed and replaced prior to launch owing to time pressure. This was extremely humiliating. For a spacecraft this complex, expensive, and international in scope it was unimaginable that there would not be sufficient operational checks in place to prevent something as simple as a human coding error. Adding to the embarrassment was a battle that summer between the Moscow and Yevpatoria control centers over responsibility. Moscow had been given responsibility, and Yevpatoria was to check everything for transmission. When Moscow provided this command on August 29 the Yevpatoria checking equipment was out of order and so the command was sent to the spacecraft unchecked. Compounding the problem was that the spacecraft had not been programmed to undertake its own checks and reject fatal commands. In the midst of the operations team's fear of reprisal and anxiety over the loss of Phobos 1 and additional problems on Phobos 2, no one was shot as might have been the case in earlier times but the Yevpatoria commander lost his job.

Phobos 2

Phobos 2 was dispatched towards Mars on July 12, 1988, and performed midcourse maneuvers on July 21 and January 23, 1989; the latter occurring six days from Mars and on the same day as its silent partner passed the planet. During the interplanetary cruise Phobos 2 experienced serious problems. Its primary transmitter failed, and it continued on a less powerful backup transmitter that reduced the data rate. Also, one of three independent attitude control processors in the onboard computer failed and a second was occasionally giving spurious results. The three-fold redundancy of the computer system required two of the three processors to function properly. If two failed, the remaining functional processor could be outvoted by the failed ones! This was a serious design flaw, and it would ultimately decide the fate of the mission. In spite of its mishaps, controllers had been able to operate the spacecraft nominally. The solar telescope instrument had pointing problems but nevertheless returned a fair amount of good data. The gamma-ray instruments detected hundreds of bursts and measured their fine structure. The various other solar plasma, solar physics, and astrophysics instruments all operated well.

The spacecraft fired its ADU near Mars at 12:55 UT on January 29, 1989, and successfully entered into orbit. This initial orbit was 876 × 80,170 km inclined at an angle of 0.87 degrees to the equator with a period of 77.91 hours. Observations of the plasma environment were made during this time. A burn on February 12 raised the periapsis to 6,400 km and increased the period to 86.5 hours. There was some anxiety when the spacecraft temporarily fell silent on February 14. The apoapsis was gradually reduced until the final ADU maneuver on February 18 almost circularized the orbit near 6,270 km, several hundred kilometers above the orbit of Phobos. This maneuver also reduced the inclination to 0.5 degree and trimmed the period to 7.66 hours, a few minutes longer than that of its target. The ADU was then jettisoned. All future maneuvers were to be made by the onboard propulsion system. Observations of both Mars and Phobos were made from this orbit while the final maneuvers were being planned to produce the Phobos encounter in early April.

High resolution images were taken during two relatively close passes of Phobos on February 23 at 860 km range and on February 28 at 320 km range. These enabled the orbit of the moon to be refined to an accuracy of 5 km. On March 7, the plane of the orbit was aligned to precisely 0 degrees. The orbit was trimmed twice more on March 15 and March 21 into a 5,692 × 6,276 km path that was nearly synchronous with the moon, with the separation varying periodically between 200 and 600 km. A third close pass at 191 km was made on March 25 with all passive remote sensing instruments operating to identify landing points for the two DAS landers. Sufficient data having been gained to design the maneuver that would produce a low pass over the surface of the moon, the encounter was scheduled for April 9.

In the meantime Phobos 2 was suffering additional degradation. Both the backup transmitter and another attitude control processor were experiencing malfunctions. Pictures and thermal images were taken of Phobos on March 26. The following day additional navigation images were returned during communications sessions at 8:25 and 12:59 UT. Each such session required the spacecraft to turn towards Phobos for imaging and then to turn the high gain antenna back to Earth for transmission. At the next scheduled communication session (15:58) the spacecraft failed to respond. A weak signal, perhaps from the omnidirectional antenna, was detected between 17:51 and 18:03 but no telemetry was received. Analysis of the signals indicated that the vehicle had lost attitude control and started to tumble. Deprived of solar power, the spacecraft would die after 5 hours when its batteries drained. Efforts to re-establish contact were futile and the mission was officially declared lost on April 15.

A board of inquiry was established on March 31. The cause of the failure was attributed to the omission of fail-safe software capable of automatically dealing with onboard emergencies, most notably to orient the spacecraft to the Sun in the event of a critical power shortage. The most likely immediate cause was failure of the second attitude control processor. Other design failures cited included the failure to enable the surviving good processor to outvote two failed ones, and the failure to undertake command uplink checks. It was evident that the systems and software for this new UMVL spacecraft had not been developed to maturity prior to launch. In the view of some, a contributing factor was that IKI scientists had not been able to participate at the top level of project management with the engineers at NPO-Lavochkin who were charged with building the spacecraft, as they had in past missions. The Ministry of General Machine Building had deleted the "supervising science team" and assigned management of the mission exclusively to the manufacturer. It is interesting to note that in the US, scientists were generally excluded from critical project management decisions and often still are today for major missions.

These lost spacecraft were the first after an unbroken string of successes for the Soviet program starting with Venera 9 in 1975 and culminating with the fabulously successful Vega investigation of Comet Halley in 1986. It was a shock, and set off an acrimonious debate of blame between Soviet scientists and engineers in the full light of international attention. The team of international scientists was summoned to Moscow in May 1989 for a post-mortem in which the Deputy Director of Lavochkin delivered a fog of excuses unrelated to the spacecraft that fooled no one and angered everyone. He could not resort to the old Soviet habit of concealment, not admitting a fault with the system. However his colleagues and the scientists at IKI, including its Director, Sagdeev, were more forthcoming about the faults with the systems on the spacecraft and the reasons for their failures. The questions from the audience were angry and pointed, something to which the Soviets were wholly unaccustomed. The dejected community of scientists that departed Moscow left behind a demoralized Soviet project team.

The loss of these complex spacecraft, especially Phobos 2 shortly prior to the culmination of its mission, was a staggering disappointment for the Soviet side and a great loss to the international community of planetary scientists eager for the results. After two decades of near secrecy in pursuing their planetary exploration program, the Soviets had opened it up for the Vega campaign, complete with international participation and coordination with missions by other nations. The Vega experience was a watershed for the Soviet program and brought positive results and widespread recognition. Built on the same implementation model, the Phobos campaign had the potential to propel the Soviets to an insurmountable lead in international planetary exploration. The abrupt and humiliating end of the missions brought a major crisis in confidence both internally and externally and presaged a swift decline in the Soviet planetary exploration program as the Soviet Empire came to an end in 1991.

Results:

Despite its premature loss, Phobos 2 provided a significant scientific return both on Mars and on Phobos. The mission failed to meet its primary objectives at Phobos, but during its 2 month lifetime as an orbiter the spacecraft returned more data than all previous Soviet Mars missions combined. Furthermore, this data was of a quality never before obtained for the planet.

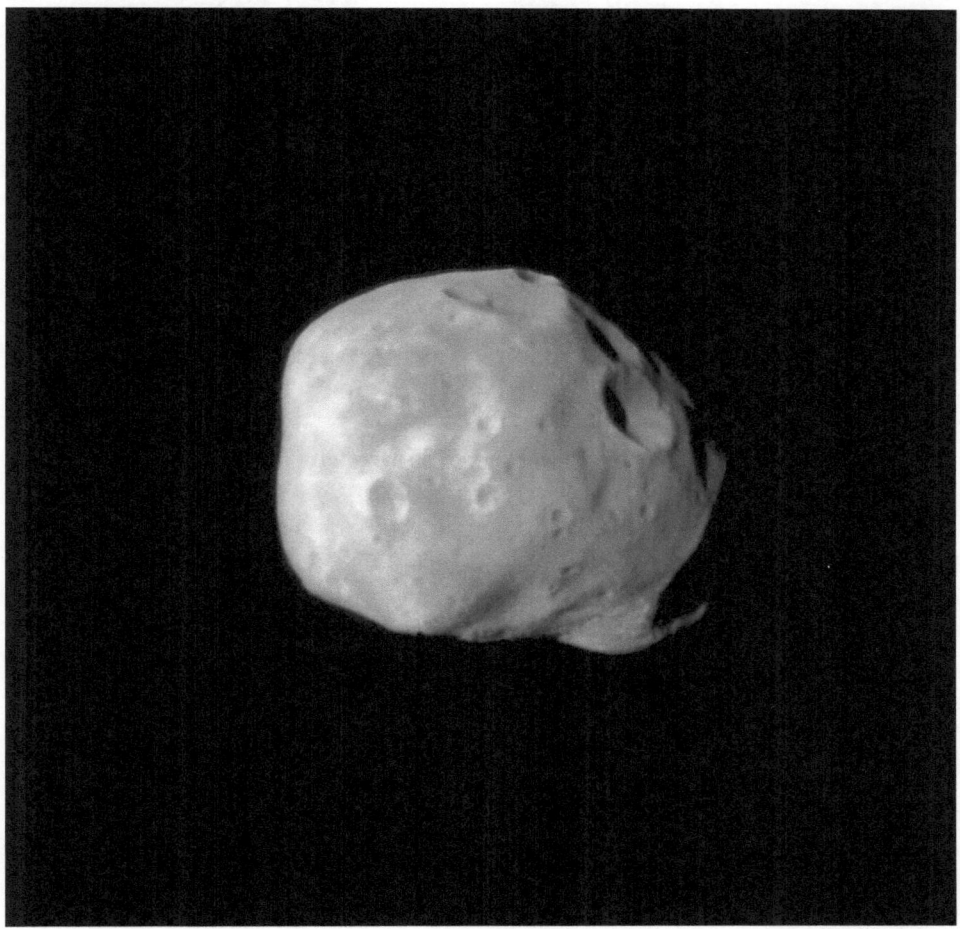

Figure 19.7 Phobos image from Phobos 2 (processing by Ted Stryk).

Figure 19.8 Phobos over Mars from the Phobos 2 spacecraft (processing by Ted Stryk).

Phobos:

Thirty-seven images of Phobos were obtained with a coverage and resolution that complemented the Mariner 9 and Viking results. About 80% of the surface of the moon was documented. Thermal inertia values for the surface were determined from the thermal emission spectrum, highlighting some inhomogeneities. The reflectance from the ultraviolet through the near-infrared (i.e. 0.3 to 0.6 and 0.8 to 3.2 microns) at a spatial resolution of 1 km appeared to indicate a more carbonaceous chondrite composition than a water-rich chondrite, and also indicated strong inhomogeneities on the surface. Perturbations of the spacecraft's orbit by Phobos provided a value for the mass of the moon which, together with a volume derived from imagery, gave a density in the range 1.85 to 2.05 g/cc. This was a low value even for a primitive volatile-rich meteorite, and suggested that Phobos might be more porous, or contain more ice, than expected. The temperature of the sunlit surface was 27°C. There were enticing indications from the magnetometer during close passes that the moon might possess a weak magnetic field.

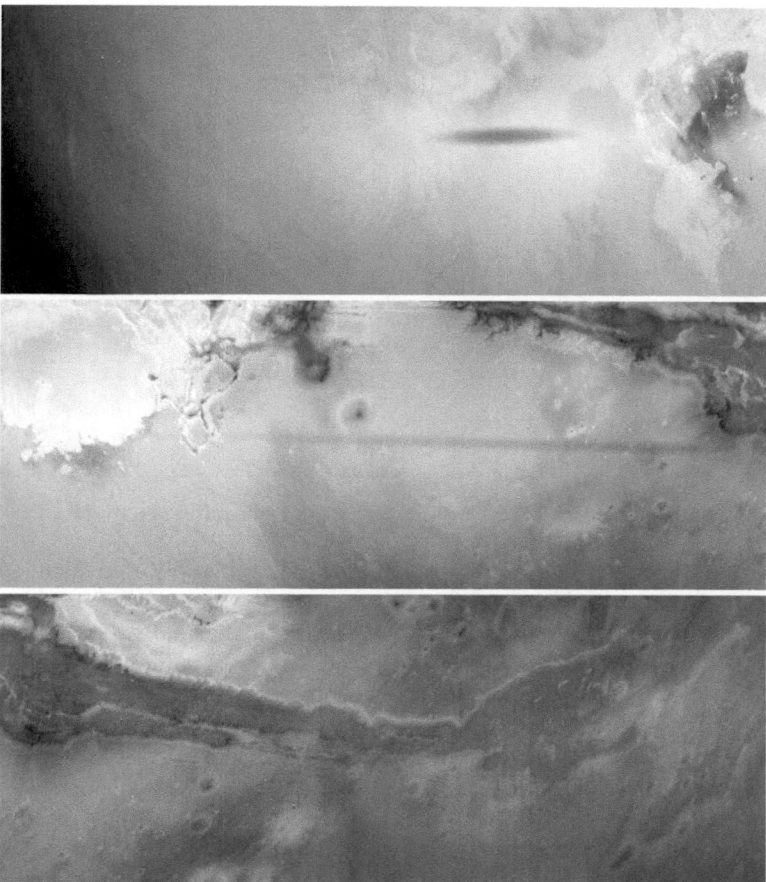

Figure 19.9 TERMOSKAN panorama of the Martian equatorial region from Olympus Mons to Valles Marineris through the red filter. Note the 'trailed' shadow of Phobos as it moves across the surface during the exposure.

Mars:

Phobos 2 had an array of instruments for multispectral investigations. Limb-to-limb photometric profiles were obtained using a prism spectrometer in the visible at about 25 km resolution, with matching profiles in the thermal infrared (5 to 60 microns) at about 50 km resolution. The data from these two instruments provided information on clouds and aerosols in the atmosphere and temperature profiles at altitudes in the 10 to 30 km range. Information was gained on optical depths, the sizes of particles, and the vertical distribution of aerosols. The occultation spectrometer saw a day-to-day variability in the atmospheric ozone profile, and a large altitude variation in the water vapor mixing ratio. Water vapor vertical profiles between 20 and 60 km were measured for the first time. The vapor content of the atmosphere was only 0.005%. Mars appeared to be losing its atmosphere at a rate of 2 to 5 kg per second, which

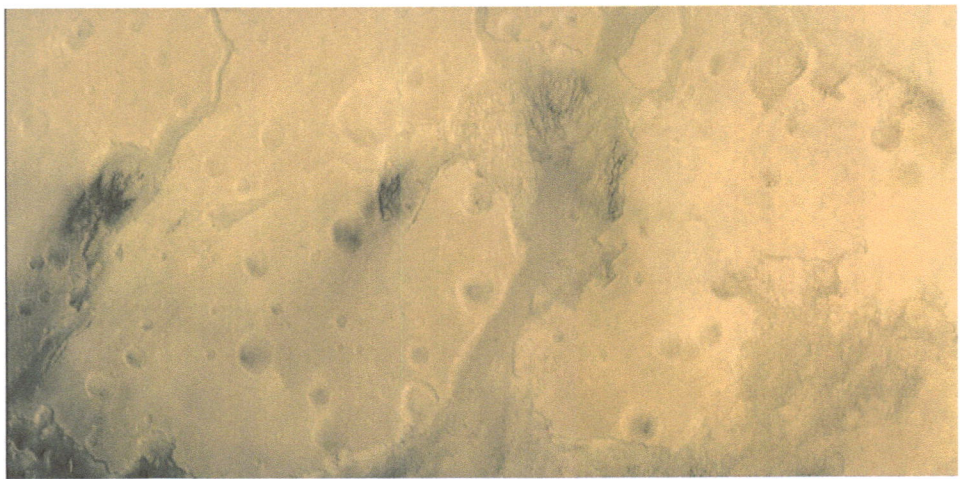

Figure 19.10 Detail from a TERMOSKAN image in the far-infrared (courtesy Ted Stryk).

was fairly significant given the low atmospheric mass; this was equivalent to losing a global ocean to a depth of 1 to 2 meters.

Imaging of the surface of Mars was conducted at thermal wavelengths (8 to 12 microns) with about 2 km resolution using pixel-by-pixel coincident visible-infrared (0.5 to 0.95 micron) images, and the near-infrared mapping spectrometer provided multispectral images at spatial resolutions between 5 km at the low periapsis of the initial orbit and 30 km at the higher altitude of the circular orbit. The thermal data covered most of the equatorial region. The thermal imagery was less sensitive to the atmospheric haze, and showed the same surface features as in the visible range but with sharper contrast, emphasizing the pervasiveness of atmospheric haze on Mars. The measured surface temperatures ranged from -93°C to +30°C. Surface thermal inertia from the passage of the shadow of Phobos indicated that there must be a good insulating material down to a depth of 50 microns and poorer insulation below. The near-infrared mapping spectrometer provided data for most of the major geological formations on Mars except the polar regions. Data on the mineralogy of the surface and its local variations revealed a more pronounced variability than was apparent in Earth-based studies. Two bands were of particular interest, the 3.1 micron band of hydrated minerals and the 2 micron band of carbon dioxide, and some contour maps were derived from carbon dioxide measurements.

Gamma-ray spectroscopy conducted during the first four low periapsis passages in the initial orbit provided data on bulk elemental abundances that were consistent with results from Mars 5 and also the x-ray fluorescence measurements made by the Viking landers. Due to its extensive instrumentation, Phobos 2 was able to make a detailed survey of the plasma environment around Mars and how this interacted with the solar wind. No permanent intrinsic planetary magnetic field was measured, even at the low periapsis of the initial orbit. The SLED detector showed levels of radiation in orbit around Mars to be less than limits that would be dangerous to humans.

20

The last gasp: Mars-96

TIMELINE: 1989–1996

The Soviet Union had planned to follow up their Phobos mission with a surface investigation of Mars. This was originally scheduled for launch in 1992 but funding was delayed and the launch date had to be postponed until 1994. The plan called for two orbiters to be launched in 1994, each of which would deploy a balloon into the atmosphere and small landers onto the surface, the launch of two orbiters in 1996 to deploy rovers onto the surface, and the launch of a sample return mission in 1998. In a revision, the plan was descoped to a single orbiter in 1994 carrying small landers and penetrators and a second orbiter in 1996 carrying the balloon and a rover. After the fall of the Soviet Union, further funding difficulties in the new Russia resulted in the launch of the 1994 mission being postponed to 1996 and the launch of the 1996 mission being postponed to 1998. But the technical and funding problems involved in building Mars-96 made it obvious that the 1998 mission would never materialize. The continuing problems with Russian suppliers and with government funding for development and test were frustrating to the Russians, and a source of consternation to the international community supplying science investigations for the mission. All of which led to massive disappointment when the launch failed on November 16, 1996. The propulsion sequence involving the second burn of the Block D stage and subsequent boost by the spacecraft's own Fregat propulsion module went awry.

The loss of Mars-96 was tragic for the Russian planetary exploration program that had been losing support in the fiscally strapped government. The US had suffered its first major tragedy at Mars in 1993 with the loss of Mars Observer shortly prior to arrival at the planet. But the US had recovered by establishing a new series of Mars missions and had launched the first one, Mars Global Surveyor, 9 days prior to the Mars-96 debacle. The Mars Pathfinder mission was launched on December 4, and went on to successfully land on the planet and deploy a small rover, thereby erasing a goal that the Russians had been working towards for more than a decade.

Launch date		
1989		
4 May	Magellan Venus orbiter	Successful radar mapper
18 Oct	Galileo Jupiter orbiter/probe	Success
1990		
24 Jan	Hiten lunar flyby/orbiter	Multiple flybys, Hagoromo orbiter lost
1991		
No missions		
1992		
25 Sep	Mars Observer orbiter	Lost shortly before arriving at Mars
1993		
No missions		
1994		
25 Jan	Clementine lunar orbiter	Success at Moon, failed on departure
1995		
No missions		
1996		
17 Feb	Near Earth Asteroid rdv	Orbited Eros, touch-down finale
7 Nov	Mars Global Surveyor	Successful orbiter
16 Nov	Mars-96 orbiter/landers	Fourth stage failure
4 Dec	Mars Pathfinder lander/rover	Successful lander, first Mars rover

Mars-96 was the last gasp in the storied history of Soviet lunar and planetary exploration in the 20th Century.

A DEBILITATING ATTEMPT AT MARS IN 1996

The tortuous path from Phobos to Mars-96:

Turmoil reigned after the failure of the Phobos missions. The long-held tolerance in the Soviet space program for failure in the attempt of bold initiatives collapsed. One result of international exposure was to open the Soviet space program to scrutiny at high political levels, where it was punished by a severe budget cut in 1990, a year of general economic gloom in the USSR.

Inside IKI a debate ensued about whether to repeat the Phobos mission in 1992 with the backup spacecraft, or to devise a new mission to the Martian surface. There were competing priorities at IKI, at the Vernadsky Institute, and with the French who were still pursuing their balloon ideas with the Soviets albeit this time at Mars instead of Venus. While the Phobos missions were being developed, a follow-on mission plan had been studied. Named 'Columbus', this called for dual orbiters in 1992 and 1994 with entry vehicles to drop French balloons into the atmosphere and Soviet rovers onto the surface. By 1989 the government had not provided the money required to launch the project in 1992 so the proposal was delayed with two orbiters in 1994 carrying small landers and the French balloons, two orbiters in 1996 with the rovers, and a sample return mission in 1998. A meeting was held in Moscow in November 1989 to solicit international participation. The first funding for Mars-94 was in April 1990, and both Germany and France agreed to contribute investigations equivalent to over $120 million. Ultimately twenty countries, including the US, were to provide science investigations.

The French balloon was a bold and exciting aspect of the 1989 plan. Its envelope was a 6 micron thick film in the shape of a cylinder 13.2 meters in diameter and 42 meters tall. It was to be inflated with 5,000 cubic meters of helium at an altitude of about 10 km during the parachute descent. After being released, it would float at an altitude of 2 to 4 km during the warm day and descend during the cold night to drag a 7 meter long instrumented tail along the surface in order to ensure that the balloon remained airborne. The tail, also known as the snake, carried 3.4 kg of instruments including a gamma-ray spectrometer, thermometer, and a subsurface radar that used the titanium segmented snake as its antenna. A 15 kg gondola suspended below the balloon and above the snake carried a camera, infrared spectrometer, magnetometer, reflectometer and altimeter, and a meteorological package to measure temperature, pressure and humidity. It was expected that the balloon would last 10 to 15 diurnal cycles and travel several thousand kilometers. In another first, tests of the balloon were carried out in the American Mojave desert in 1990 by a joint team of Russian, French and American scientists and engineers.

The other exciting feature was the Mars rover. The USSR had sent two successful rovers to the Moon in the early 1970s, and now modified this technology for Mars. The Mars rover was smaller than its lunar predecessor, but at 200 kg was still large. It had a clever new chassis and wheel design, and with a top speed of 500 meters per hour it was expected to drive 500 km during the lifetime of 2 to 3 years facilitated by its RTG power supply. The rover was equipped with four cameras for panoramic coverage, a quadrupole mass spectrometer for atmospheric analysis, a laser aerosol spectrometer, a visible-infrared spectrometer for surface analysis, magnets to reveal the magnetic properties of the soil, a radio sounder to probe subsurface structure to a depth of 150 meters, a meteorology package, and a manipulator arm that would dig 10 cm into the surface to obtain samples for a pyrolytic gas chromatograph. The arm also carried a camera for close-up observation of the soil, an alpha, proton and x-ray spectrometer for elemental analysis of the soil, a Mossbauer spectrometer to analyse the iron mineralogy of the soil, and a gas analyzer to identify any trace gases.

Both the balloon and the rover would ultimately be deleted, however. In 1990 the USSR was in financial distress, and the money for developing the Mars-94 mission arrived slowly and in smaller amounts than required. By April 1991 it was clear that the money was insufficient to address all of the ambitions of the mission. It became necessary to postpone the balloons and rovers to 1996, and send a simpler mission in

1994. The Mars-94 mission would now be a single launch of an orbiter with small landers similar to that of Mars 3 (which was successfully delivered to the surface, although it had failed immediately afterwards) and new penetrators developed by the Vernadsky Institute.

On New Year's Day 1992 the USSR was formally dissolved, Russia emerged as an independent state, and financial problems became acute. In the past, money had never been an issue in developing a planetary mission but it now became the pacing item. It was not delivered when required or fell short. Parts were not delivered by contractors. Work on Mars-94 declined into a stop-go affair depending on whether there was money and parts. In desperation the project asked for financial assistance from its international partners. In order to protect their investments in the mission, Germany and France sent $10 million in late 1993. An appeal was made to the US, but after the loss of the Phobos missions the Americans were suspicious of Russian capabilities and were nervous about investing in a foreign project that was in such a visibly dismal state. Besides, in August 1993 NASA had its first ever inflight loss of a planetary spacecraft when Mars Observer fell silent while preparing for Mars orbit insertion, and the agency was struggling to salvage its own program.

Fearing that the troubled Mars-94 project was in severe danger of delivering an ill-prepared spacecraft in 1994, the new Russian Space Agency (RSA) postponed the mission until 1996. The second spacecraft with the balloon and rover was slipped to 1998. The risk in this move was that money from the new and financially strapped government would dry up altogether, but the RSA gave the Mars-96 mission its full support and the government declared the project to be a high priority. If it were not for the international obligations and the Western currency involved, the mission just might have been canceled. It continued in the face of technical and financial issues. When the camera scan platform ran into technical difficulties the Russians proposed deleting it in favor of fixed mountings in order to save money. The Germans, who were building the cameras, became outraged, and in the end saved the scan platform by sending their own engineers to fix the problems. The Russian government did not send all the money that it had promised. The RSA pulled funds from lower priority missions, and more money – ultimately as much as $180 million more – was sent by Western partners to keep the project alive. Cancellation remained a possibility. The RSA went 80 million rubles into debt to complete the final integration and testing of the Mars-96 spacecraft in early 1996, by which time the Mars-98 mission with the balloon and rover had been canceled. Promised funding from the government just never arrived. The financial problems in the Russian space program were so bad that the ships in the tracking fleet were recalled to port and most of them sold off. One ship was made into a museum, and another was conscripted into the Ukrainian Black Sea naval fleet. The loss of these tracking vessels would eventually prove a serious problem to the project.

The situation at Baikonur in the summer of 1996 while preparing the mission for launch was atrocious. There were power shortages as utility bills went unpaid. Work was often done with heat from kerosene burners, light by candle, and labor without pay. With such a long protracted development under such adverse conditions, it took a heroic effort to get the Mars-96 spacecraft to the launch pad. Perhaps due to all the

Spacecraft launched

Spacecraft:	Mars-96 (M1 No.520)
Mission Type:	Mars Orbiter/Landers
Country/Builder:	USSR/NPO-Lavochkin
Launch Vehicle:	Proton-K
Launch Date/Time:	November 16, 1996 at 20:48:53 UT (Baikonur)
Outcome:	Launch failure, fourth stage misfired.

adversity in mounting this mission, it failed during the launch process and there was little hope that the Russians would be able to attempt another planetary mission for many years to come.

Mars-96 campaign objectives:

The Mars-96 mission was designed to use an orbiter, two small soft landers, and two penetrators to undertake a comprehensive investigation of the current state and past evolution of Mars by studying physical and chemical processes in the atmosphere, on the surface, and in the interior.

The scientific objectives of the orbiter included obtaining high resolution mapping and spectral imagery of the surface to study its geology, mineralogy and topography, to investigate the gravity field and crustal structure, and to monitor the climate. The spacecraft was also equipped to study the magnetic field, plasma characteristics, and magnetospheric structure. And it had instruments for astrophysical investigations of gamma-ray bursts, and both stellar and solar oscillations. The penetrators were to obtain images of the surface, undertake meteorological measurements, examine the physical, chemical, magnetic, and mechanical properties of the near-surface regolith, measure the water content of the soil, and measure seismic activity and the heat flow from the interior of the planet through the crust. The small landers were to study the vertical structure of the atmosphere and obtain images during the descent to assist in interpreting images taken on the surface, measure elemental, magnetic and oxidant composition of the soil, measure seismic activity, and monitor the local weather for its diurnal, seasonal, and annual variability.

The delays and problems with development had been extremely frustrating for the international participants, whose own budgets and schedules were heavily impacted. The failure of the launch and breakup of the spacecraft over the west coast of South America was the final straw for a Russian planetary exploration program that had been struggling against diminishing resources since the Phobos failures in 1988–1989 and the demise of the Soviet Empire in 1991.

The failures of the Phobos and Mars-96 missions were a heavy loss for planetary exploration. They were ambitious and complex missions aimed at comprehensively studying Mars and the larger of its two moons. They had more engineering systems, more observation platforms, more scientific instrumentation, and more subsidiary vehicles than any other missions in the history of planetary exploration. The array of measurements they were to have made was simply enormous. If these missions had been successful, the knowledge produced would have been astounding. They were also very international, complex, and expensive. They are missions the like of which will not be experienced in planetary exploration for a long time to come.

Spacecraft:

Orbiter:

The Mars-96 orbiter was a 3-axis stabilized spacecraft based on the Phobos design. A pressurized toroid at the base held the computer and most of the avionics, thermal regulation, communications, batteries and electronics for scientific instruments. The equipment tower was replaced with a flat deck on which equipment and instruments were mounted, including the solar panels and the entry systems for the two landers. The solar panels were larger, and extended out from opposite sides of the deck. They also carried low gain antennas and attitude control nozzles. Many subsystems and scientific apparatus were on the toroid below the deck level, including a pair of scan platforms for accurate and stable pointing of cameras and spectrometers. The high gain dish antenna extended off one side of the toroid, oriented perpendicular to the solar panels, and the medium gain antenna was on the opposite side. In this case the high gain was not steerable, and provided a communications rate of 130 kbits/s. The thermal control radiators, navigation and Sun and star sensors for the attitude control system were attached to the toroid, including the onboard propulsion system with its propellant tanks and thrusters, as for the Phobos spacecraft. The separable ADU propulsion system (now named the Fregat stage) was attached underneath, as before. The two penetrators were mounted on the ADU propellant tanks. The computers were more advanced and supplied by the Europeans who did not trust the Phobos computers after their poor performance. The orbiter carried more than two dozen instruments in addition to the landers and penetrators. Owing to the massive weight of the spacecraft, the Proton could not quite provide sufficient energy for the escape maneuver. After the Block D had released the spacecraft, the Fregat would fire to provide the final increment. The Fregat would perform the midcourse corrections, orbit insertion at Mars, and in-orbit maneuvers before being jettisoned.

Scan platforms were fairly new to Russian spacecraft. The 220.7 kg TPS 3-axis platform had its own computer control system, memory, thermal control, navigation camera, and a 53.5 kg payload of remote sensing instruments. The developers had difficulty achieving the stringent pointing and stability requirements, and when the Russians suggested deleting the platform and mounting the instruments on the body of the spacecraft, which would have severely limited the science objectives, German engineers were brought in to help to resolve the problems. The 74.2 kg PAIS 2-axis scan platform was simpler, and it carried instruments that had less stringent pointing requirements. The orbiter was to have deployed its landers on approach and then

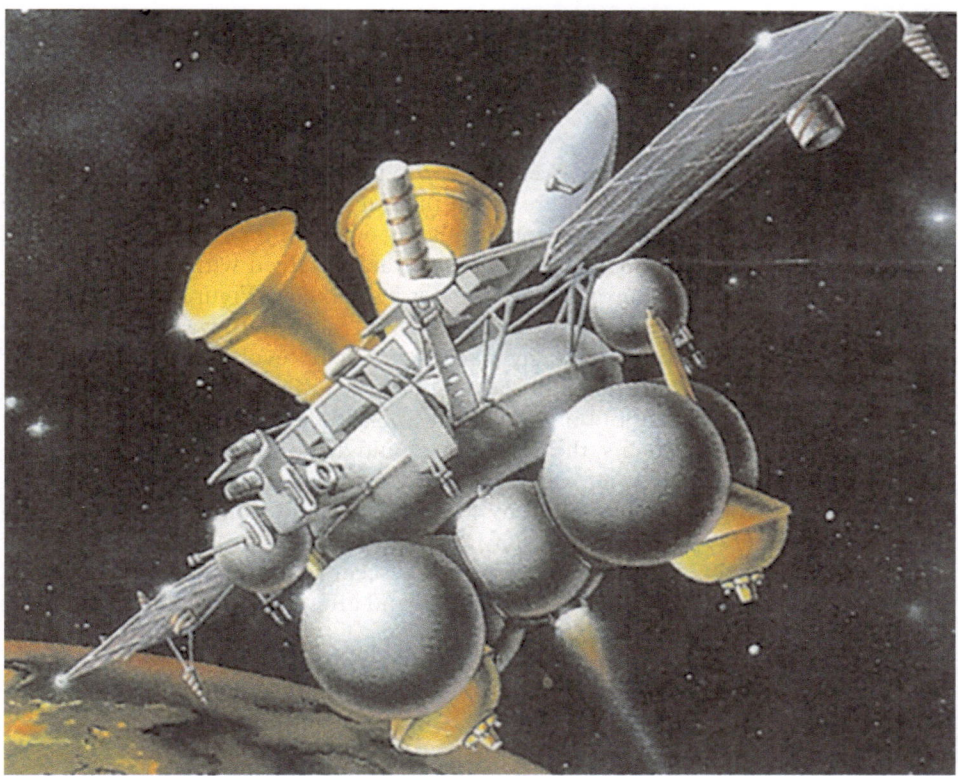

Figure 20.1 The Mars-96 spacecraft illustrating the toroidal base compartment with instrument deck above, Fregat propulsion stage below, small station entry canisters (gold) on top, penetrators (gold) nestled between propulsion tanks below (courtesy NPO-Lavochkin).

performed a deflection maneuver to reach the insertion point, whereupon it would put itself into an elliptical orbit. This would be shaped over the first month using several maneuvers to achieve an orbit in which the spacecraft would make four revolutions around the planet while Mars rotated seven times on its axis, because such orbits are generally quite stable. Once it had achieved an orbit with a periapsis of 300 km and a period of 43.09 hours, it would deploy the penetrators.

The Mars-96 spacecraft was 3.5 meters tall and 2.7 meters wide; 11.5 meters wide with its solar panels deployed.

Launch mass:	6,824 kg
Orbiter dry mass:	2,614 kg
Penetrators:	176 kg (2)
Landers:	241 kg (2)
Attachment structure:	283 kg (for landers and penetrators)
ADU dry mass:	490 kg

| *Fuel:* | 2,832 kg |
| *Hydrazine:* | 188 kg |

Landers:

The two landers or 'small stations' were similar to those used on the M-71 and M-73 missions but much smaller. They were about the same size as, but lighter than those of Luna 9 and 13. Each lander was approximately 60 cm in diameter and, including the 8 kg payload, was 30.6 kg. For entry, each was contained within a blunt-ended conical ablative aeroshell approximately 1 meter in diameter (Figure 20.3). The total mass of the lander and its entry system was 120.5 kg. Separation was to occur 4 or 5 days prior to orbit insertion, after being given a spin of 12 revolutions per minute for stability. Entry would begin at an altitude of about 100 km at 5.75 km/s and an angle of 11 to 21 degrees. After about 3 minutes, at an altitude in the range 19 to 44 km and a velocity of 200 to 320 m/s, the parachute would deploy. The aeroshell would be jettisoned 10 seconds later, and a 130 meter harness would unreel the lander. At about 18 to 4 km and a velocity of 20 to 40 m/s an airbag would inflate around the lander. This was designed to survive an impact at vertical and horizontal rates of about 20 m/s. Immediately upon contact after a descent lasting anything from 6 to 17 minutes, the lander would discard its parachute, and after rolling to a stop the airbag

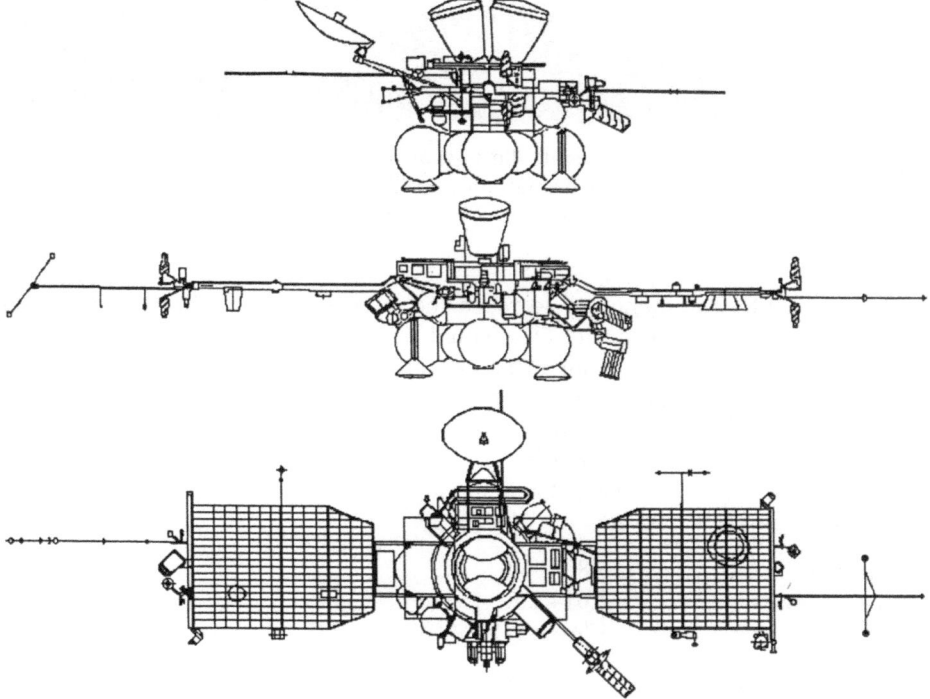

Figure 20.2 Mars-96 spacecraft line drawing (courtesy NPO-Lavochkin).

Figure 20.3 Mars-96 on the integration and test stand.

would split open at the seam and be discarded. In a manner similar to the Mars 3 lander four triangular petals would open, each extending about 30 cm from the central base. Three of these petals had springs to deploy instruments away from the lander.

On the surface the lander would draw power from two coffee-cup-sized 220 mW RTGs, a technology that had not previously been used in a Soviet mission but was intended for the rovers, supplemented by NiCd batteries. A lithium battery was used for the descent phase. The uplink and downlink at 2 and 8 kbits/s would use a UHF relay through the orbiter. There was no command capability; downlink was only to initiate transmission. The internal temperature of the lander would be maintained by a combination of insulation and heat from the RTGs and dedicated RHUs; 8.5 W of heater power was available. The expected operating life was about a local year.

In addition to the scientific payload, the landers carried a compact disk provided by The Planetary Society entitled 'Visions of Mars' which contained a compendium of knowledge about Mars.

Entry mass:	120.5 kg
Lander mass:	30.6 kg
Payload:	8 kg ($\sim$ 5 kg of science)

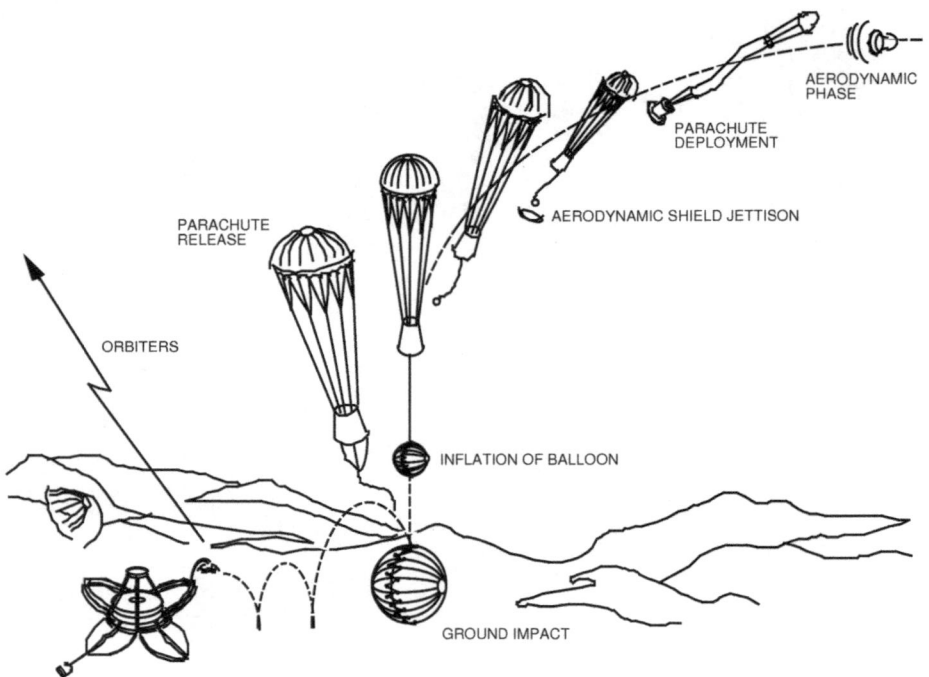

Figure 20.4 Small lander entry, descent and landing.

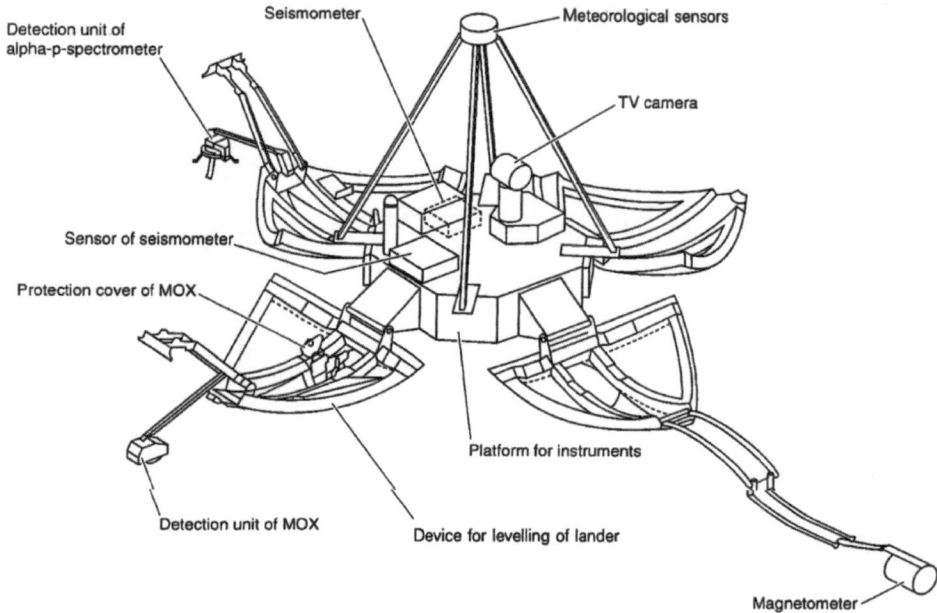

Seismometer

Meteorological sensors

Detection unit of
alpha-p-spectrometer

TV camera

Sensor of seismometer

Protection cover of MOX

Detection unit of MOX

Platform for instruments

Device for levelling of lander

Magnetometer

Figure 20.5 'Small Station' lander.

Figure 20.6 Test lander in a 'sandbox'.

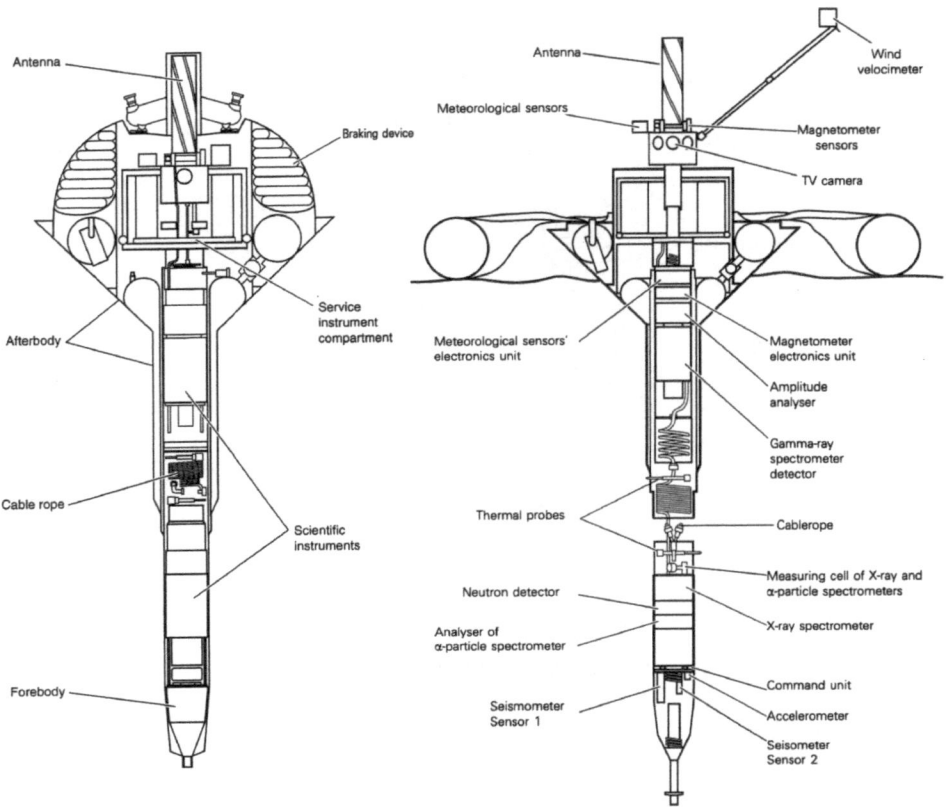

Figure 20.7 Mars-96 penetrators in flight (left) and deployed (right).

Penetrators:

The penetrators were cylinders 2 meters long with a pointed 12 cm diameter fore-body and a 17 cm diameter after-body that included a funnel-shaped tail section that broadened to 78 cm diameter. They drew power from a 0.5 W RTG and a 150 watt-hour lithium battery. A total of 4.5 kg of science instruments were distributed in the two sections. Each penetrator was to be released from orbit near apoapsis by pointing the spacecraft in the proper direction and spinning the entry system to 75 revolutions per minute for stabilization prior to release. At a safe distance the orbiter was to make a small diversion maneuver. The penetrator would then fire a solid rocket to reduce its velocity by 30 m/s, jettison this de-orbit motor, and inflate the ballute that was to decelerate the initial phase of its entry when it fell into the atmosphere about 21.5 hours later. Entry would be at a speed of 4.6 to 4.9 km/s and an angle of 12 degrees.

The penetrator would be slowed aerodynamically, then inflate an extension to the ballute in the after-body (Figure 20.8). After a 6 minute descent, it would hit the

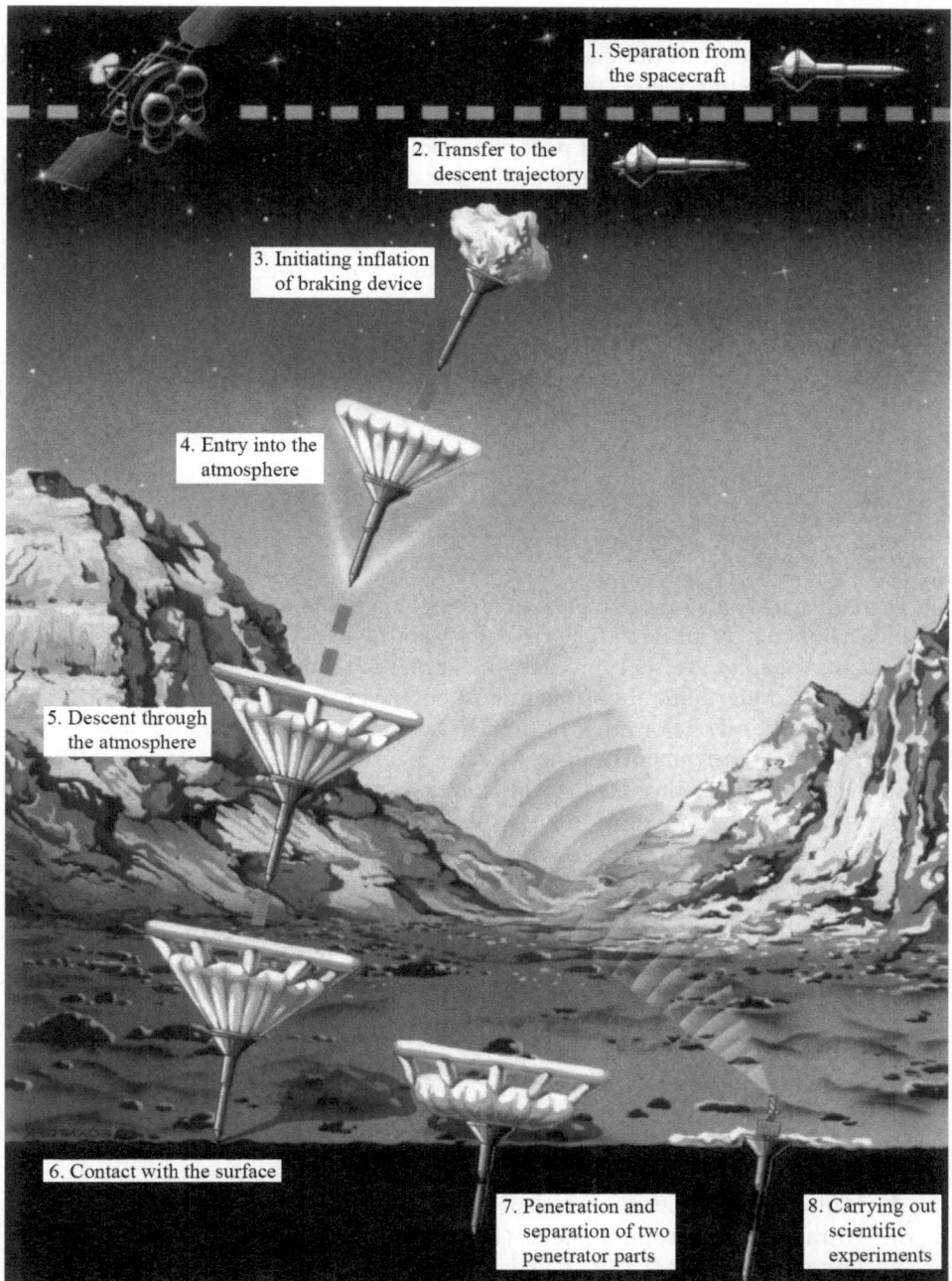

1. Separation from the spacecraft

2. Transfer to the descent trajectory

3. Initiating inflation of braking device

4. Entry into the atmosphere

5. Descent through the atmosphere

6. Contact with the surface

7. Penetration and separation of two penetrator parts

8. Carrying out scientific experiments

Figure 20.8 Penetrator entry, descent and landing scheme.

surface at about 75 m/s. The 500 G shock was to be cushioned by a fluid reservoir. The wide after-body was designed to stop at the surface as the fore-body separated and penetrated up to 6 meters into the ground, remaining attached by a coiled wire. The aft mast with the antenna, camera, magnetometer, and meteorological sensors was then to extend, and thermal probes would protrude into the soil.

One penetrator was to be sent to a site near one of the landers, and the other to a location at least 90 degrees away to facilitate the triangulation of seismic signals. Communications at 8 kbits/s would be feasible for about 5 to 6 minutes every 7 days through the transmitter on the after-body and would use the relay through the orbiter for uplink and downlink. The expected lifetime was a local year.

Penetrator mass w/engine:	88 kg
Entry mass:	45 kg
Science payload:	4.5 kg

Payload:

Orbiter:

The orbiter had twelve instruments to study the surface and atmosphere of Mars, seven instruments for plasma, fields, particles and ionospheric composition, and five instruments for solar and astrophysical research. And radio occultations during limb crossings would give data about the atmosphere. The instruments were on the sides of the spacecraft, on one or other of the two scan platforms, and on the solar panels. The three optical instruments of the ARGOS package were on the 3-axis TPS scan platform along with the navigation camera, and the SPICAM, EVRIS and PHOTON instruments were on the simpler 2-axis PAIS scan platform.

Remote sensing of the surface and atmosphere:

1. Multifunctional stereoscopic high resolution TV camera (ARGOS HRSC, Germany-Russia)
2. Wide-angle stereoscopic TV camera (ARGOS WAOSS, Germany-Russia)
3. Visible and infrared mapping spectrometer (ARGOS OMEGA, France-Russia-Italy)
4. Planetary infrared Fourier spectrometer (PFS, Italy-Russia-Poland-France-Germany-Spain)
5. Mapping radiometer (TERMOSKAN, Russia)
6. High resolution mapping spectrophotometer (SVET, Russia-USA)
7. Multi-channel optical spectrometer (SPICAM, Belgium-France-Russia)
8. Ultraviolet spectrophotometer (UVS-M, Russia-Germany-France)
9. Long wave radar (LWR, Russia-Germany-USA-Austria)
10. Gamma-ray spectrometer (PHOTON, Russia)
11. Neutron spectrometer (NEUTRON-S, Russia)
12. Quadrupole mass spectrometer (MAK, Russia-Finland)

The West Germans supplied the 21.4 kg high resolution HRSC camera and the East Germans the 8.4 kg wide angle WAOSS camera. After the reunification of that nation the instruments were combined into a single project. Each was a push-broom scanner using parallel linear arrays of 5,184 pixel CCDs. The narrow angle camera had nine arrays for multispectral, photometric, and stereoscopic imaging. The wide angle camera had three arrays for stereoscopic imaging. The best resolution would be 12 meters for the narrow angle camera and 100 meters for the wide angle camera. Because the altitude and velocity of the spacecraft would vary during the periapsis encounters in an elliptical orbit, the CCD integration times could be tailored over the long scans. The cameras on the TPS platform had an extensive 25.3 kg MORION-S onboard processing unit which incorporated a 21 kg solid-state memory system built in cooperation with ESA with a capacity of 1.5 gigabits to reduce the transmission requirements. These data acquisition resources were shared with other instruments. Also on the TPS platform was the 23.7 kg OMEGA visible and infrared mapping spectrometer to measure atmospheric composition and map surface composition.

The 28 kg TERMOSKAN instrument from Phobos was re-flown to measure the thermal properties of the regolith. The 12 kg SVET mapping spectrophotometer was to analyze the spectrum of the surface and aerosols. The 20 kg PHOTON gamma-ray spectrometer would map the elemental composition of the surface, and the 8 kg NEUTRON-S neutron spectrometer would determine ice and water abundances. The 35 kg LWR would probe the near-surface layer to measure vertical structure and ice deposits. The 25.6 kg PFS was to make atmospheric profiles of carbon dioxide, and measure atmospheric temperatures, winds, and aerosols. The 46 kg SPICAM would use solar and stellar occultations to produce vertical profiles of water vapor, ozone, oxygen, and carbon monoxide. The 9.5 kg UVS-M was to map atomic hydrogen, deuterium, oxygen and helium in the upper atmosphere of Mars and the interstellar medium. The 10 kg MAK mass spectrometer was to measure the composition and distribution of ions and neutrals in the upper atmosphere.

Space plasma and ionosphere:

1. Energy-mass ion spectrograph and neutral particle imager (ASPERA-C, Sweden-Russia-Finland-Poland-USA-Norway-Germany).
2. Fast omni-directional non-scanning energy-mass ion analyzer (FONEMA, UK-Russia-Czech Republic-France-Ireland).
3. Omnidirectional ionospheric energy-mass-spectrometer (DYMIO, France-Russia-Germany-USA).
4. Ionospheric plasma spectrometers (MARIPROB, Austria-Belgium-Bulgaria-Czech Republic-Germany-Hungary-Ireland-Russia-USA).
5. Electron analyzer and magnetometer (MAREMF, Austria-Belgium-France-Germany-Great Britain-Hungary-Ireland-Russia-USA).
6. Plasma wave instrument (ELISMA, France-Bulgaria-Great Britain-ESA-Poland-Russia- Ukraine).
7. Low-energy charged particle spectrometer (SLED-2, Ireland-Czech Republic-Germany-Hungary-Russia-Slovakia).

The 12.2 kg ASPERA instrument was to measure the energy distribution of ions and fast neutrals. The 10.7 kg FONEMA ion analyzer would measure the dynamics and structure of the upper atmosphere plasma. The 7.9 kg MARIPROB and 7.2 kg DYMIO instruments would complement these instruments. The 12.2 kg MAREMF instrument would analyze plasma electrons and it carried two fluxgate magnetometers to measure magnetic fields in interplanetary space and in orbit of Mars. The 12 kg ELISMA instrument was to measure plasma waves in the Martian environment. It consisted of three Langmuir probes and three search coil magnetometers. In addition to probing the surface of the planet the LWR radar was also to be used to measure the distribution of electrons in the ionosphere and how this interacted with the solar wind. The 3.3 kg SLED-2 instrument was to measure low energy cosmic rays during the interplanetary cruise and in the Mars environment.

Solar physics and astrophysics:

1. Precision gamma-ray spectrometer (PGS, Russia-USA).
2. Cosmic and solar gamma-ray burst spectrometer (LILAS-2, Russia-France).
3. Stellar oscillations photometer (EVRIS, France-Russia-Austria)
4. Solar oscillation photometer (SOYA, Ukraine-Russia-France-Switzerland)
5. Radiation dosage monitor (RADIUS-M, Russia-Bulgaria-Greece-USA-France-Czech Republic-Slovakia)

The 25.6 kg PGS gamma-ray spectrometer was intended to examine solar flares during the interplanetary cruise and then measure gamma-ray emission when in orbit around Mars. The 5 kg LILAS-2 instrument would locate celestial gamma-ray bursts in conjunction with several spacecraft in Earth orbit and Ulysses in deep space. The celestial sources were also to be examined by using Mars occultation observations. The 1 kg SOYA and 7.4 kg EVRIS photometers were to make helioseismology and astroseismology measurements respectively. SOYA was body mounted but EVRIS was on the 2-axis scan platform. The RADIUS-M radiation dosimeter was carried to obtain data pertinent to prospective future human Mars missions.

Landers:

Entry and descent:

1. Descent imager (DESCAM, France-Finland-Russia)
2. Three-axis accelerometer plus sensors for temperature and pressure (DPI, Russia)

The lander had a suite of sensors and a descent imager. DESCAM was mounted on the bottom of the lander to obtain imagery to provide context for the panorama that would be taken following landing. It had a 400 × 500 pixel CCD and was to be detached 2 minutes after landing, just before the airbag was discarded. The DPI had an accelerometer and temperature and pressure sensors to provide vertical profiles of temperature, pressure and density during entry and descent, and the dynamics of the landing. It was mounted outside the lander bay under one of the petal covers, so that descent data was convoluted with flow dynamics during descent.

Surface:

1. Panoramic camera on a central mast (PANCAM, Russia-France-Finland)
2. Meteorology instrument system on a 1-meter tall mast for temperature, pressure, humidity, wind, and optical depth (MIS, Finland-France-Russia)
3. Seismometer, magnetometer and inclinometer (OPTIMISM, France-Germany-Russia)
4. Alpha, proton and x-ray spectrometer for soil elemental analysis (APX, Germany-Russia-USA)
5. Oxidant sensor (MOX, USA-Russia)

At the top of the station shell, the lander carried a PANCAM scanning photometer camera similar to those on the earlier Mars landers to provide a 360 × 60 degree panorama composed of 6,000 × 1,024 pixels. A deployable overhead mast supported the MIS meteorology package with sensors for temperature, pressure and humidity, an ion anemometer for wind, plus an optical-depth sensor. The ODS optical sensor in this package was supplied by the French and measured direct Sun and scattered light at the zenith in three narrow bands at 270, 350, and 550 nm, and one broad band from 250 to 750 nm. The DPI package measured temperatures and wind velocities on the surface. Three of the petals contained instruments for deployment to the surface: the OPTIMISM instrument containing a seismometer, inclinometer and 3-axis fluxgate magnetometer; the APX backscatter analyzer to determine elemental abundances in the soil; and the MOX experiment. The latter was a colorimetric soil analyzer that had reactant spots sensitive to different types of oxidants. Supplied by the US it was developed in less than a year, weighed only 0.85 kg, and had its own power supply and data storage. Its function was to test the inference from the Viking landers that the soil was rich in oxidants and hence inimical to life.

Penetrators:

After-body above surface:

1. Television Camera (TVS, Russia)
2. Meteorological sensors for temperature, pressure, humidity, wind and opacity (MEKOM, Russia-Finland-USA)
3. Magnetometer (IMAP-6, Russia-Bulgaria)

After-body below surface:

1. Gamma-ray spectrometer for soil analysis (PEGAS, Russia)
2. Temperature sensor for heat flow (TERMOZOND Part 1, Russia)

Fore-body:

1. Seismometer for interior structure (KAMERTON, Russia-Great Britain)
2. Accelerometers for soil mechanics (GRUNT, UK-Russia)
3. Temperature sensor for heat flow (TERMOZOND Part 2, Russia)
4. Neutron detector for water detection (NEUTRON-P, Russia)
5. Alpha-proton spectrometer for soil analysis (ALPHA, Russia-Germany)
6. X-ray fluorescence spectrometer for soil analysis (ANGSTREM, Russia)

The GRUNT accelerometer in the fore-body was to measure the properties of the surface during impact and penetration. The KAMERTON seismometer would search for Martian activity. The TERMOZOND thermal probes would measure heat flow and provide data for thermal diffusivity and heat capacity. The gamma-ray, alpha, proton, neutron, and x-ray spectrometers would analyze the soil chemistry including its water content. Remaining above the surface on the after-body, the TVS 2,048 pixel linear camera would take a panoramic image of the site, the MEKOM package would monitor the temperatures and winds, and the IMAP-6 magnetometer would measure the local magnetic field.

Mission description:

Mars-96 was to have arrived at Mars in September 1997 on a direct trajectory about 10 months after launch. The small surface stations were to have been released 4 or 5 days out from the planet for direct atmospheric entry. The spacecraft would execute a deflection maneuver for its orbital insertion point. By this time the Mars ephemeris was very well known, so the complex optical navigation and close-in release of the M-71 and M-73 missions was not required. Three landing sites were selected for the landers, the two primary ones being at 41.31°N 153.77°W in Arcadia and at 32.48°N 163.32°W in Amazonia, with the backup at 3.65°N 193°W.

At insertion the spacecraft would perform a braking maneuver into an initial orbit of 500 × 52,000 km, and this would be reduced in stages to a 43.09 hour 7:4 Mars synchronous orbit at 106.4 degrees inclination with a 300 km periapsis.

The two penetrators were to have been deployed within 7 to 28 days, one targeted for Arcadia and the other at least 90 degrees away in Utopia Planitia to provide a good baseline for seismometry. The Fregat would be jettisoned after the deployment of the penetrators. Orbit maintenance would then be the task of the smaller onboard propulsion system.

At the start of the orbital mission communications sessions with the landers and penetrators were expected to be through the orbiter approximately once per day for 20 minutes each. One small orbit correction of about 1 to 2 m/s would be required each month to maintain visibility with the surface elements, whose nominal lifetime was to have been one local year or roughly two terrestrial years.

The spacecraft was launched on the optimal day of the window, November 16, 1996, and reached Earth orbit after the first firing of the Block D. If this stage had fired properly to initiate the escape maneuver, the Fregat propulsion system on the spacecraft would have provided the final increment required to reach Mars. It seems that either the Block D-2 did not fire or it shut down after only 20 seconds, perhaps owing to an inappropriate command from the spacecraft, which was in control of it. The logic of the situation then caused the spacecraft to separate and fire its Fregat as if to complete the escape maneuver. However, this burn left it in an 87 × 1,500 km elliptical orbit. With its periapsis inside the atmosphere the spacecraft was doomed. The Block D-2 stage entered the atmosphere at some time between 00:45 to 01:30 UT on November 17 and crashed into the Pacific between the Chilean coast and Easter Island. A day later, the spacecraft was spotted re-entering the atmosphere as a fireball over southern Chile, and is believed to have crashed in the Andes mountains of Chile near the border with Bolivia. It was carrying 270 grams of plutonium-238 in 18 pellets as part of the landers and penetrators. Designed to withstand heat and impact, these probably survived re-entry. Searches were made but the spacecraft was never found.

The failure occurred at the second ignition of the Block D-2 upper stage while the spacecraft was out of range of Russian ground stations. Owing to budget limitations the Russians had no tracking ships in the Pacific. The lack of telemetry data during critical parts of the escape phase of the mission precluded identification of the cause of the failure, and in particular whether it was due to failure of the Block D-2 upper stage or to a malfunction of the controlling spacecraft. It was an abysmal situation.

Results:

None.

21

The Soviet lunar and planetary exploration legacy

A HISTORICAL SYNOPSIS

The history of exploring the Solar System by spacecraft is short, spanning less than 42 years at the end of the 20th Century. Prior to January 1, 2001, there had been 182 launches. Of these, 89 were successful or partly successful, and three were in transit to their ultimate destinations. The exploration of the planets was dominated in the 20th Century by competition between the USSR and USA. Only five of the total of 182 missions were developed by other parties. It was not until 1985 that Europe and Japan launched their own deep space missions.

In the early years of the space race the USSR was usually first to achieve major feats at the Moon, Venus, and Mars. After the neck-to-neck race to the Moon in the 1960s, and its culmination with Apollo, the US, which had also had greater success with planetary missions, assumed the leading position in robotic exploration in the 1970s with unopposed successes at Mars, Mercury, and the outer Solar System. The USSR had no answer to the Mariner 9 Mars orbiter, the two Viking orbiters and landers at Mars, the Mariner 10 flybys of Mercury, or the Pioneer 10 and 11 and Voyager 1 and 2 missions to the outer planets – a realm where the Soviets were not technologically prepared to go. The US had conceded only Venus, where the Venera missions reigned supreme. At the beginning of the 1980s the Soviet program could be said to have won the competition at Venus, but lost it everywhere else. This trend changed with the Vega missions in the middle of that decade. The USSR vigorously participated in the International Halley Mission with the European Space Agency and Japan, and contributed two spacecraft as platforms for instruments from any country that wished to provide them. The Europeans and the USSR led this highly successful and precedent-setting enterprise for the Old Continent. The Americans of the New World were a minor player and did not even send a spacecraft to Halley for this first world-wide planetary exploration endeavor.

By the mid-1980s, the Soviets had seized the lead in planetary exploration from the Americans. The USSR gained a great deal of pride and prestige around the world from the Vega missions to Comet Halley, and decided as a matter of policy to open up future missions to international participation. They would be essentially Russian vehicles and Russian-led missions, but with instruments and scientific participation from around the world. The Americans were at a significant disadvantage in this situation, since they did not have the massive spacecraft to offer valuable instrument real estate to others and to compete with the Russians in this way. In addition, the US planetary program was suffering from a major decline in the 1980s starting with the administration of Ronald Reagan, who preferred a more direct competition with the Soviet Union.

The Vega campaign had been conceived as an almost entirely Soviet mission with some participation by the French, but was modified with international instruments, many from the Eastern Bloc, being added for the Halley intercept. The next Soviet planetary mission, Phobos, was internationalized earlier in its development, took this to a greater extreme, and had more instruments of Western origin. Mars-96 was the culmination of the international style of Soviet planetary missions, with instruments openly and broadly solicited from around the world and with a larger investment by Western countries including the US. It is supremely ironic that the Americans, who prided themselves on the openness of their space exploration program, remained far more xenophobic in their planetary exploration program than the Soviets, and were obliged to concede the lead in international planetary exploration missions to the Vega, Phobos and Mars-96 missions.

After the fall of the Soviet Union in 1991, attempts to form partnerships between the US and Russian planetary programs failed as the resources for further Russian planetary missions dried up in major national economic problems. The abysmal loss of Mars-96 created an international disaster, demoralized the national program, and embarrassed the post-Soviet Russian national government and its new space agency. Already beset with financial problems, Russia cut its investment in space science missions. At the end of the 20th Century, the Russian national program of robotic planetary exploration appeared to have been postponed indefinitely.

Unfortunate fate has been a bedfellow to Russian history for a millennium, and so it was for the Russian planetary exploration program just as it reached its peak in the late 1980s while that of the US was declining. A decade later there was no Russian planetary exploration program, the US program was revitalized, and the French, a bell-weather for international involvement in space science and a participant on Soviet missions since the early 1970s, were now making trips to Washington instead of to Moscow. But the hopes and dreams remained alive in Russia. After watching from the sidelines since 1996, and contributing primarily by offering launch services for cash, the Russians are just now emerging after a 15 year absence with the launch of a Phobos sample return mission scheduled for 2011 and with plans for a lunar orbiter/lander missions for later in the decade.

THE GOOD, THE BAD AND THE SAD

The Soviet and American enterprises to explore space were created as a by-product of the Cold War, specifically the development of intercontinental ballistic missiles and their modification to put spacecraft on interplanetary trajectories. While aiming their nuclear-tipped missiles at each other, the two opposed societies competed for the minds of the rest of the world by demonstrating their technological prowess by exploits in civilian space exploration. The Soviet space exploration program was not entirely divorced from the military, as it was in the US. As a result, Soviet robotic missions to the Moon and planets were cloaked in secrecy until the early 1980s, and only after the collapse of the USSR has reliable information become available on the full history of the Soviet lunar and planetary exploration program. The key leaders and institutions involved, and almost all decisions and events, were state secrets and unknown outside the closed circle of Soviet secrecy. Launches were not announced, and the Soviets rarely revealed the purpose of their spacecraft except for human missions where this could not be hidden. This policy hid embarrassing failures, and only when a success could be claimed was its purpose revealed.

The heavily cloaked Soviet robotic exploration program provided mystery and a challenge to the Americans. State secrecy concealed the fact that the Soviet robotic space exploration program was bolder, more innovative, and more tragic than any observers in the West could have imagined at the time. As each planetary launch window approached, the Americans would get very anxious about what spectacular the Soviets might be planning. The subliminal pressure to outdo the Soviets added to this anxiety, particularly in the first decade of the space race when the USSR always appeared to have the upper hand. Imagine the despair on the American side in the late 1960s if it had been known that the Soviets were planning landings on Mars in 1969 while the US was still conducting flybys. Over the long run, the Soviets were tragically jinxed at Mars, never gaining a true success despite expending enormous effort and resources on a resolute and more aggressive attack on the planet than was the case in the US. They were the first to launch at Mars in 1960, failed with their next generation spacecraft in the 1960s, fared poorly with their massive grand fleets in 1971 and 1973, fell tragically short with the Phobos missions in 1988, and ended abysmally with Mars-96. Yet many Soviet achievements endure – the lunar rovers and sample returns, the Vega missions, and the extensive and very successful in-situ exploration of Venus are all accomplishments that were never equaled by the US.

The story of the Soviet lunar and planetary exploration program is a tale of great adventure, excitement, suspense, and tragedy; a tale of courage and the patience to overcome obstacles and failure; a tale of fantastic accomplishment and debilitating loss; a tale of courage and enthusiasm to try the previously impossible. To carry it out they exhibited superb expertise in engineering design and development. They were very innovative in utilizing the technology available to produce engineering systems that accomplished the task. Their rocket engines are testimony to mastery of materials development and propulsion system engineering. Their innovative lunar mission design and return trajectories and their terminal optical navigation scheme for the M-71 and M-73 missions demonstrate excellence in celestial mechanics, navigation, and guidance and control. The automation of the midcourse maneuvers and optical navigation scheme for Mars were applied successfully well before the US even contemplated such complexity – a clear demonstration of superior skill in automation and software which unfortunately was to unravel in later missions. The success of any enterprise is ultimately the result of people, and the Soviet Union had excellent engineers, scientists and managers who faced immense difficulties with the heavy-handed, personality-driven, complex and entangled national system of control and supply, and succeeded thanks to an intense devotion to the space exploration enterprise and a strong sense of competition with America.

The successes of the Soviet robotic exploration program were achieved at a heavy price. The Luna, Venera and Mars programs all endured an enormous number of losses from launch vehicle and spacecraft failures – far more than would have been tolerated in a US program. Soviet persistence in pursuing their goals, particularly in the early years, would have appeared maniacal to an American. During one stretch between 1963 and 1965, the Soviets suffered eleven straight failures in attempting a lunar soft landing. Korolev had to exercise considerable political skill to save his lunar lander program after such a long string of disasters. This would have brought down an American program where no such commanding personality existed for the robotic program. The worst string of losses in an American robotic exploration program occurred roughly contemporaneously and was only about half this number. The Americans suffered six straight failures from 1961 to 1964 in their Ranger lunar impactor program. At one time they were very close to terminating both the program and its implementing organization. Thereafter the Americans never tolerated more than an occasional failure in their program, whereas the Soviets tolerated relatively large failure rates as a matter of course.

The poor reliability of Soviet rockets was the primary cause of failures until the mid-1970s, but it was these very same rockets that enabled the Soviet Union to be so bold in executing their program. The Molniya was capable of lifting many times the weight of American rockets. It was produced in quantity, and readily available from the military on short notice and at no apparent cost. This characteristic was essential since the Soviets, in contrast to the Americans, tested their spacecraft by flying them – resulting in many more attempts to launch Soviet spacecraft than American: 106 versus 51 through 1996. This may have been a consequence of the readier access to Soviet launch vehicles before cost became an issue, but in any case Soviet engineers also lacked discipline in ground testing. They rushed their designs through assembly with insufficient system test time in low quality clean facilities with loose ground test procedures. This showed in the poor performance of their spacecraft in flight. By the end of 1965, they had lost all four of their spacecraft launched to Venus, both of their spacecraft launched to Mars, and five of nine lunar spacecraft. In that same time, they lost an additional 24 of their 39 missions to launch vehicle failures, which was not only a very large number in itself but also a huge percentage loss from an American point of view. The situation improved with time, but failures continued to plague the program. The inflight failure rate dropped from over 70% through 1965 to 39% by 1976, and the launch vehicle failure rate dropped from over 60% to 48% over that same period of time. After 1976 the inflight failure rate fell to 10% and the launch vehicle failure rate dropped to 9%.

The absence of strong ground test discipline was symptomatic of a weakness in systems engineering. The Americans learned their skill in this discipline through the trouble-plagued Ranger lunar program in the early 1960s, and rarely had an inflight failure afterwards. The Soviets were much slower in applying this discipline, and suffered continuing inflight failures. Their problems were exacerbated by a handicap in electronics technology. Decades after vacuum electronics had become standard in Western spacecraft the Soviets continued to fly pressurized spacecraft – in fact, right up to Mars-96. One impetus for continuing this practice was that Soviet rockets were so large that the mass penalty of old electronics was not a major consideration, but another problem was that Soviet industry did not produce vacuum qualified complex electronic systems for its exploration missions. The reliability and operating lifetime of Soviet space avionics systems were a problem throughout the program and were a principal reason why the USSR never attempted a mission to the outer planets. Their Mars spacecraft were for some reason particularly prone to inflight avionics failures from start to finish.

The sad part of the story is the disappearance of Russia from the scene after the fiasco of Mars-96. This has been a great loss of vision, enterprise and expertise in robotic space exploration. The Soviet enterprise was born as part of the Cold War, and seemingly expired with it. After 1991 the Russian space program turned almost exclusively to humans-in-orbit. The Academy of Sciences had a great deal of trouble acquiring funds to keep its robotic space exploration program alive. After Mars-96 failed and government interest in robotic space exploration plummeted, it increased its investment in human spaceflight and partnership with the US in the International Space Station. Now, after a long hiatus, the Russians are reviving their robotic space exploration program with the Phobos-Grunt mission.

A NEW BEGINNING RISES FROM A CHERISHED LEGACY

The Phobos-Grunt spacecraft embraces the latest in space technology and erases the tradition of pressurized planetary spacecraft. The launcher will be the Zenit-Fregat. The Zenit rocket is a legacy of the Energiya-Buran program, and the Fregat stage is a legacy from the Phobos and Mars-96 missions. For Phobos-Grunt to achieve the desired interplanetary trajectory, a burn of the spacecraft's engine will be required after the Fregat burns out. The spacecraft will make the midcourse maneuvers, orbit insertion, and orbital maneuvers to rendezvous with and ultimately land on Phobos, the larger of the two Martian moons. Tradition also survives in the boldness of Russia's return to robotic exploration: not just a modest step but an ambitious sample return mission, which seems appropriate for a program amongst whose unanswered legacies is a sample return from the Moon. Phobos-Grunt also continues the legacy of international cooperation, because it will carry the first Chinese Mars spacecraft, Yinghuo 1, and release it into orbit around the planet.

The main goal of the Phobos-Grunt mission is to return a sample from Phobos to Earth for in-depth laboratory studies to help to answer key questions concerning the origin and evolution of the Solar System. The payload also includes instruments for navigation and for studying the Martian environment (television, space plasma and magnetic field detectors, and a dust particle detector) and instruments to study the surface of Phobos following the landing. The latter include a panoramic camera, gas

Figure 21.1 Phobos-Grunt spacecraft. At bottom is the Fregat propulsion stage, then (in turn) adapter ring, spacecraft-lander system, Earth return system and entry capsule (courtesy NPO-Lavochkin).

chromatograph, gamma-ray spectrometer, neutron spectrometer, laser time-of-flight mass spectrometer, secondary ion mass spectrometer, infrared spectrometer, thermal detector, long-wave subsurface radar and a seismometer. The robotic manipulator to be used for sampling carries a micro-television camera, an alpha, proton and x-ray spectrometer, and a Mossbauer spectrometer.

The Soviet lunar and planetary exploration program in the 20th Century left a legacy of scientific results and new knowledge. It is difficult to remember how little we knew about the Moon and planets at the beginning of the space age in 1957, and how much we have learned as a result of sending out spacecraft. Table 21.1 provides a summary of the exploration milestones achieved by the Soviet program, most of which occurred during the first 15 years of the space age. Soviet scientists also made many scientific discoveries in the course of their missions. The early Luna missions established that whereas the near side of the Moon is dominated by the dark maria, the far side is dominated by the bright highlands, an interesting dichotomy that has yet to be adequately explained. Lunar mass concentrations were first discovered by Soviet spacecraft. Much of what we know of the atmosphere and surface of Venus comes from the Venera missions. And while the Soviets were thwarted at Mars, they were first to successfully land (albeit the lander failed after a few seconds), made the first in-situ measurements in the Martian atmosphere, the early discoveries about its ionosphere, and the lack of an intrinsic magnetic field.

The Phobos-Grunt mission represents a hope that Russia will resume a lunar and planetary exploration enterprise with the same boldness, innovation, and persistence that they demonstrated in the first 40 years of space exploration.

Table 21.1 'Firsts' in lunar and planetary exploration by Soviet spacecraft

Lunar missions		
First spacecraft to escape Earth's gravity	Luna 1	1959, January 2
First spacecraft to fly by the Moon	Luna 1	1959, January 4
First spacecraft to impact another celestial body	Luna 2	1959, September 14
First photographs of the far side of the Moon	Luna 3	1959, October 6
First lunar lander	Luna 9	1966, February 3
First lunar orbiter	Luna 10	1966, April 3
First circumlunar mission with Earth return	Zond 5	1968, September 20
First robotic sample return mission	Luna 16	1970, September 21
First robotic rover (Lunokhod 1)	Luna 17	1970, November 17
Venus missions		
First launch attempt to Venus	1VA No.1	1961, February 4
First spacecraft to impact another planet	Venera 3	1966, March 1
First planetary entry probe	Venera 4	1967, October 18
First planetary lander	Venera 7	1970, December 15
First Venus orbiter	Venera 9	1975, October 22
First photographs from the surface of a planet	Venera 9	1975, October 22
First radar imagery of Venusian surface	Venera 15	1983, October 10
First planetary balloon	Vega 1	1985, June 11
First comet distant flyby	Vega 1	1986, March 6
Mars missions		
First planetary launch attempt	1M No.1	1960, October 10
First spacecraft to impact Mars	Mars 2	1971, November 27
First lander on Mars (failed after landing)	Mars 3	1971, December 2
First atmospheric probe of Mars (lost at landing)	Mars 6	1973, March 12

Appendices

APPENDIX A. EARLY SPACECRAFT 'TAIL NUMBERS'

Russian spacecraft were given "tail numbers" during construction. Luna spacecraft were given the designation Ye followed by a number indicating the design series and a second number indicating the serial number of the particular spacecraft under construction, i.e. Ye-3 No.2 was the second spacecraft built in the third design series of lunar spacecraft. Sometimes a letter was attached to indicate a modification to the original design, such Ye-2A No.1. After successful translunar injection, the spacecraft were renamed "Luna".

The designation scheme for planetary spacecraft was somewhat different. The early 1960-1961 spacecraft were simply designated 1M or 1V for the first design series of Mars or Venus spacecraft. The next generation were a common design for both Mars and Venus and were designated as follows:

Example: 3MV-1 No.3

First number:	Serial number of design (3rd major design series)
Second set of letters:	Spacecraft targets (MV = Mars, Venus common design)
Third number:	Mission modification number:
	1 – Venus Entry Mission
	2 – Venus Flyby Mission
	3 – Mars Entry Mission
	4 – Mars Flyby Mission
Fourth number:	Serial number of vehicle (No.3, or third to be built)

The spacecraft were renamed after successful departure from earth orbit as "Venera" (for Venus) or "Mars" spacecraft.

A few 3MV planetary spacecraft were built for engineering test flights and were given "1A" designations including a failed Mars test flight 3MV-1A No.2 on Nov 11, 1963, and a failed Venus test flight 3MV-1A No.4A on Feb 19, 1964. Oddly, the Zond 3 Mars spacecraft, 3MV-4 No.3, was not given a "1A" designation, but carried out a successful flyby test at the Moon before failing to reach Mars distance.

Three 3MV Mars spacecraft, one entry probe (3MV-3 No.1) and two flyby spacecraft (3MV-4 No.4 and No.6) that missed their launch window in 1964 were modified as Venus spacecraft and launched in 1965. Their original construction as Mars missions accounts for their anomalous tail numbers.

There is confusion in the literature over the tail numbers assigned by OKB-1 to the early Luna, Venera and Mars spacecraft before Lavochkin assumed responsibility. The most authoritative original source for Mars and Venus spacecraft is Chertok. The most authoritative secondary source for all spacecraft is Siddiqi's Deep Space Chronicle. There remain some inconsistencies between these and other sources in the literature. We have attempted to reconcile all these sources to the extent possible through communications with both Asif Siddiqi and Timothy Varfolomeyev, and on this basis have chosen to use the tail number designations given in Chertok.

APPENDIX B. USSR LUNAR AND PLANETARY SPACECRAFT FAMILIES

Launch date	L/V	Mass (kg)	Builder	Spacecraft	Mission name	Mission type	Result
Luna			Soviet Lunar Missions				
Ye-1 series (OKB-1)							
Sep 23, 1958	Luna	~360	OKB-1	Ye-1 No.1		Lunar Impactor	fb
Oct 11, 1958	Luna	~360	OKB-1	Ye-1 No.2		Lunar Impactor	fb
Dec 4, 1958	Luna	~360	OKB-1	Ye-1 No.3		Lunar Impactor	fb
Jan 2, 1959	Luna	361.3	OKB-1	Ye-1 no.4	Luna 1	Lunar Impactor	ft
Jun 18, 1959	Luna	~390	OKB-1	Ye-1A No.5		Lunar Impactor	fb
Sep 12, 1959	Luna	390.2	OKB-1	Ye-1A No.7	Luna 2	Lunar Impactor	s
Ye-2,3 series (OKB-1)							
Oct 4, 1959	Luna	278.5	OKB-1	Ye-2A No.1	Luna 3	Circumlunar Flyby	s
Apr 15, 1960	Luna	?	OKB-1	Ye-3 No.1		Circumlunar Flyby	fu
Apr 19, 1960	Luna	?	OKB-1	Ye-3 No.2		Circumlunar Flyby	fb
Ye-6 Series (OKB-1)							
Jan 4, 1963	Molniya	1,420	OKB-1	Ye-6 No.2	[Sputnik 25]	Lunar Lander	fi
Feb 3, 1963	Molniya	1,420	OKB-1	Ye-6 No.3		Lunar Lander	fb
Apr 2, 1963	Molniya	1,422	OKB-1	Ye-6 No.4	Luna 4	Lunar Lander	fc
Mar 21, 1964	Molniya-M	~1,420	OKB-1	Ye-6 No.6		Lunar Lander	fu
Apr 20, 1964	Molniya-M	~1,420	OKB-1	Ye-6 No.5		Lunar Lander	fu
Mar 12, 1965	Molniya	~1,470	OKB-1	Ye-6 No.9	Cosmos 60	Lunar Lander	fi
Apr 10, 1965	Molniya	~1,470	OKB-1	Ye-6 No.8		Lunar Lander	fu
May 9, 1965	Molniya-M	1,476	OKB-1	Ye-6 No.10	Luna 5	Lunar Lander	ft
Jun 8, 1965	Molniya-M	1,442	OKB-1	Ye-6 No.7	Luna 6	Lunar Lander	fc
Oct 4, 1965	Molniya	1,506	OKB-1	Ye-6 No.11	Luna 7	Lunar Lander	ft
Dec 3, 1965	Molniya	1,552	OKB-1	Ye-6 No.12	Luna 8	Lunar Lander	ft

Ye-6 series (NPO-L)

Jan 31, 1966	Molniya-M	1,538	NPO-L	Ye-6M No.202/13	Luna 9	Lunar Lander	s
Mar 1, 1966	Molniya-M	~1,580	NPO-L	Ye-6S No.204	Cosmos 111	Lunar Orbiter	fi
Mar 31, 1966	Molniya-M	1,582	NPO-L	Ye-6S No.206	Luna 10	Lunar Orbiter	s
Aug 24, 1966	Molniya-M	1,640	NPO-L	Ye-6LF No.101	Luna 11	Lunar Orbiter	s
Oct 22, 1966	Molniya-M	1,620	NPO-L	Ye-6LF No.102	Luna 12	Lunar Orbiter	s
Dec 21, 1966	Molniya-M	1,620	NPO-L	Ye-6M No.205/14	Luna 13	Lunar Lander	s
May 16, 1967	Molniya-M	~1,700	NPO-L	Ye-6LS No.111	Cosmos 159	Lunar Orbiter Test Flight	fu
Feb 7, 1968	Molniya-M	~1,700	NPO-L	Ye-6LS No.112		Lunar Orbiter	fu
Apr 7, 1968	Molniya-M	1,700	NPO-L	Ye-6LS No.113	Luna 14	Lunar Orbiter	s

Ye-8 series (NPO-L)

Feb 19, 1969	Proton-D	~5,700	NPO-L	Ye-8 No.201		Lunar Lander/Rover	fu
Jun 14, 1969	Proton-D	~5,700	NPO-L	Ye-8-5 No.402		Lunar Sample Return	fu
Jul 13, 1969	Proton-D	5,667	NPO-L	Ye-8-5 No.401	Luna 15	Lunar Sample Return	ft
Sep 23, 1969	Proton-D	~5,700	NPO-L	Ye-8-5 No.403	Cosmos 300	Lunar Sample Return	fu
Oct 22, 1969	Proton-D	~5,700	NPO-L	Ye-8-5 No.404	Cosmos 305	Lunar Sample Return	fu
Feb 6, 1970	Proton-D	~5,700	NPO-L	Ye-8-5 No.405		Lunar Sample Return	fu
Sep 12, 1970	Proton-D	5,727	NPO-L	Ye-8-5 No.406	Luna 16	Lunar Sample Return	s
Nov 10, 1970	Proton-D	5,660	NPO-L	Ye-8 No.203	Luna 17	Lunar Lander/Rover	s
Sep 2, 1971	Proton-D	5,750	NPO-L	Ye-8-5 No.407	Luna 18	Lunar Sample Return	ft
Sep 28, 1971	Proton-D	5,700	NPO-L	Ye-8LS No.202	Luna 19	Lunar Orbiter	s
Feb 14, 1972	Proton-D	5,750	NPO-L	Ye-8-5 No.408	Luna 20	Lunar Sample Return	s
Jan 8, 1973	Proton-D	5,700	NPO-L	Ye-8 No.204	Luna 21	Lunar Lander/Rover	s
May 29, 1974	Proton-D	5,700	NPO-L	Ye-8LS No.206	Luna 22	Lunar Orbiter	s
Oct 28, 1974	Proton-D	5,795	NPO-L	Ye-8-5M No.410	Luna 23	Lunar Sample Return	ft
Oct 16, 1975	Proton-D	~5,800	NPO-L	Ye-8-5M No.412		Lunar Sample Return	fu
Aug 9, 1976	Proton-D1	5,795	NPO-L	Ye-8-5M No.413	Luna 24	Lunar Sample Return	s

Zond

Soviet lunar test missions

Sep 27, 1967	Proton-D	~5,375	TsKBEM	7K-L1 No.4L		Circumlunar/Return	fb
Nov 22, 1967	Proton-D	~5,375	TsKBEM	7K-L1 No.5L		Circumlunar/Return	fu
Mar 2, 1968	Proton-D	5,375	TsKBEM	7K-L1 No.6L	Zond 4	Lunar Distance/Return	ft

Launch date	L/V	Mass (kg)	Builder	Spacecraft	Mission name	Mission type	Result
Apr 22, 1968	Proton-D	~5,375	TsKBEM	7K-L1 No.7L		Circumlunar/Return	fu
Sep 14, 1968	Proton-D	5,375	TsKBEM	7K-L1 No.9L	Zond 5	Circumlunar/Return	s
Nov 10, 1968	Proton-D	5,375	TsKBEM	7K-L1 No.12L	Zond 6	Circumlunar/Return	ft
Jan 20, 1969	Proton-D	~5,375	TsKBEM	7K-L1 No.13L		Circumlunar/Return	fu
Feb 21, 1969	N-1	6,900	TsKBEM	7K-L1S No.3S		Orbiter/Return	fb
Jul 3, 1969	N-1	6,900	TsKBEM	7K-L1S No.5L		Orbiter/Return	fb
Aug 7, 1969	Proton-D	5,375	TsKBEM	7K-L1 No.11	Zond 7	Circumlunar/Return	s
Oct 20, 1970	Proton-D	5,375	TsKBEM	7K-L1 No.14	Zond 8	Circumlunar/Return	s
Nov 23, 1972	N-1	9,500	TsKBEM	7K-LOK No.6A		Orbiter/Return	fb

Mars

Soviet Mars missions

1M series (OKB-1)

Launch date	L/V	Mass (kg)	Builder	Spacecraft	Mission name	Mission type	Result
Oct 10, 1960	Molniya	650	OKB-1	1M No.1		Mars Flyby	fu
Oct 14, 1960	Molniya	650	OKB-1	1M No.2		Mars Flyby	fu

2MV combination Mars-Venus series (OKB-1)

Launch date	L/V	Mass (kg)	Builder	Spacecraft	Mission name	Mission type	Result
Oct 24, 1962	Molniya	~900	OKB-1	2MV-4 No.3		Mars Flyby	fi
Nov 1, 1962	Molniya	893.5	OKB-1	2MV-4 No.4	Mars 1	Mars Flyby	fc
Nov 4, 1962	Molniya	1,097	OKB-1	2MV-3 No.1		Mars Atm/Surf Probe	fi

3MV combination Mars-Venus series (OKB-1)

Launch date	L/V	Mass (kg)	Builder	Spacecraft	Mission name	Mission type	Result
Nov 11, 1963	Molniya	~800	OKB-1	3MV-1A No.2	Cosmos 21	Test Flight	fi
Nov 30, 1964	Molniya	950	OKB-1	3MV-4 No.2	Zond 2	Mars Flyby	fc
Jul 18, 1965	Molniya	960	OKB-1	3MV-4 No.3	Zond 3	Test Flight with Lunar Flyby	p

NPO-L Proton series

Launch date	L/V	Mass (kg)	Builder	Spacecraft	Mission name	Mission type	Result
Mar 27, 1969	Proton-D	4,850	NPO-L	M-69 No.521		Mars Orbiter	fu
Apr 2, 1969	Proton-D	4,850	NPO-L	M-69 No.522		Mars Orbiter	fb
May 10, 1971	Proton-D	4,549	NPO-L	M-71 No.170	Cosmos 419	Mars Orbiter	fi
May 19, 1971	Proton-D	4,650	NPO-L	M-71 No.171	Mars 2	Mars Orbiter/Lander	p

Date	Vehicle	Mass	Designation	Name	Mission	
May 28, 1971	Proton-D	4,650	M-71 No.172	Mars 3	Mars Orbiter/Lander	p
Jul 21, 1973	Proton-D	3,440	M-73 No.52S	Mars 4	Mars Orbiter	ft
Jul 25, 1973	Proton-D	3,440	M-73 No.53S	Mars 5	Mars Orbiter	p
Aug 5, 1973	Proton-D	4,470	M-73 No.50P	Mars 6	Mars Flyby/Lander	p
Aug 9, 1973	Proton-D	4,470	M-73 No.51P	Mars 7	Mars Flyby/Lander	ft
Jul 7, 1988	Proton-D2	6,220	1F No.101	Phobos 1	Mars Orbiter/Phobos Landers	fc
Jul 12, 1988	Proton-D2	6,220	1F No.102	Phobos 2	Mars Orbiter/Phobos Landers	p
Nov 16, 1996	Proton-D2	6,828	M1 No.520	Mars 96	Mars Orbiter/Landers	fu

Venera/Vega

Soviet Venus missions

1VA Series (OKB-1)

Date	Vehicle	Mass	Designation	Name	Mission	
Feb 4, 1961	Molniya	~645	1VA No.1		Venus Impactor	fi
Feb 12, 1961	Molniya	643.5	1VA No.2	Venera 1	Venus Impactor	fc

2MV combination Mars-Venus series (OKB-1)

Date	Vehicle	Mass	Designation	Name	Mission	
Aug 25, 1962	Molniya	1,097	2MV-1 No.3		Venus Atm/Surf Probe	fi
Sep 1, 1962	Molniya	1,097	2MV-1 No.4		Venus Atm/Surf Probe	fi
Sep 12, 1962	Molniya	~890	2MV-2 No.1		Venus Flyby	fi

3MV combination Mars-Venus series (OKB-1)

Date	Vehicle	Mass	Designation	Name	Mission	
Feb 19, 1964	Molniya-M	~800	3MV-1A No.4A		Test Flight	fu
Mar 27, 1964	Molniya-M	948	3MV-1 No.5	Cosmos 27	Venus Atm/Surf Probe	fi
Apr 2, 1964	Molniya-M	948	3MV-1 No.4	Zond 1	Venus Atm/Surf Probe	fc
Nov 12, 1965	Molniya-M	963	3MV-4 No.4	Venera 2	Venus Flyby	ft
Nov 16, 1965	Molniya-M	958	3MV-3 No.1	Venera 3	Venus Atm/Surf Probe	fc
Nov 23, 1965	Molniya-M	~960	3MV-4 No.6	Cosmos 96	Venus Flyby	fi

NPO-L Molniya series

Date	Vehicle	Mass	Designation	Name	Mission	
Jun 12, 1967	Molniya-M	1,106	1V No.310	Venera 4	Venus Atm/Surf Probe	s
Jun 17, 1967	Molniya-M	~1,100	1V No.311	Cosmos 167	Venus Atm/Surf Probe	fi
Jan 5, 1969	Molniya-M	1,138	2V No.330	Venera 5	Venus Atm/Surf Probe	s

Launch date	L/V	Mass (kg)	Builder	Spacecraft	Mission name	Mission type	Result
Jan 10, 1969	Molniya-M	1,138	NPO-L	2V No.331	Venera 6	Venus Atm/Surf Probe	s
Aug 17, 1970	Molniya-M	1,180	NPO-L	3V No.630	Venera 7	Venus Atm/Surf Probe	s
Aug 22, 1970	Molniya-M	~1,180	NPO-L	3V No.631	Cosmos 359	Venus Atm/Surf Probe	fi
Mar 27, 1972	Molniya-M	1,184	NPO-L	3V No.670	Venera 8	Venus Atm/Surf Probe	s
Mar 31, 1972	Molniya-M	~1,180	NPO-L	3V No.671	Cosmos 482	Venus Atm/Surf Probe	fi
NPO-L Proton series							
Jun 8, 1975	Proton-D	4,936	NPO-L	4V-1 No.660	Venera 9	Venus Orbiter/ Lander	s
Jun 14, 1975	Proton-D	5,033	NPO-L	4V-1 No.661	Venera 10	Venus Orbiter/ Lander	s
Sep 9, 1978	Proton-D1	4,450	NPO-L	4V-1 No.360	Venera 11	Venus Flyby/Lander	s
Sep 14, 1978	Proton-D1	4,461	NPO-L	4V-1 No.361	Venera 12	Venus Flyby/Lander	s
Oct 30, 1981	Proton-D1	4,363	NPO-L	4V-1M No.760	Venera 13	Venus Flyby/Lander	s
Nov 4, 1981	Proton-D1	4,363	NPO-L	4V-1M No.761	Venera 14	Venus Flyby/Lander	s
Jun 2, 1983	Proton-D1	5,250	NPO-L	4V-2 No.860	Venera 15	Venus Orbiter	s
Jun 7, 1983	Proton-D1	5,300	NPO-L	4V-2 No.861	Venera 16	Venus Orbiter	s
Dec 15, 1984	Proton-D1	4,924	NPO-L	5VK No.901	Vega 1	Venus Balloon & Lander / Halley Flyby	s / s
Dec 21, 1984	Proton-D1	4,926	NPO-L	5VK No.902	Vega 2	Venus Balloon & Lander / Halley Flyby	s / s

1. Mass column lists mass at launch
2. Result Codes:

fb	booster failure
fu	upper stage failure
fi	interplanetary trajectory injection failure
fc	failure in transit during cruise
ft	failure at the target
p	partial success
s	success

APPENDIX C1. USSR LUNAR MISSION RECORD

Robotic spacecraft missions

Successes		Post-launch failures		Launch failures	
Luna 1 Impactor [partial]	1959	Luna 4 Lander	1963	Ye-1 No.1 Impactor	1958
Luna 2 Impactor	1959	Luna 5 Lander	1965	Ye-1 No.2 Impactor	1958
Luna 3 Circumlunar	1959	Luna 6 Lander	1965	Ye-1 No.3 Impactor	1958
Zond 3 Flyby	1965	Luna 7 Lander	1965	Ye-1A No.5 Impactor	1959
Luna 9 Lander	1966	Luna 8 Lander	1965	Ye-3 No.1 Circumlunar	1960
Luna 10 Orbiter	1966	Luna 15 Sample Return	1969	Ye-3 No.2 Circumlunar	1960
Luna 11 Orbiter [imager failed]	1966	Luna 18 Sample Return	1971	Ye-6 No.2 Lander (Sputnik 25)	1963
Luna 12 Orbiter	1966	Luna 23 Sample Return	1974	Ye-6 No.3 Lander	1963
Luna 13 Lander	1966			Ye-6 No.5 Lander	1964
Luna 14 Orbiter	1968			Ye-6 No.6 Lander	1964
Luna 16 Sample Return	1970			Ye-6 No.8 Lander	1965
Luna 17 Lander/Rover	1970			Ye-6 No.9 Lander	1965
Luna 19 Orbiter	1971			Ye-6S No.204 Orbiter (Cosmos 111)	1966
Luna 20 Sample Return	1972			Ye-6LS No.111 Orbiter Test (Cosmos 159)	1967
Luna 21 Lander/Rover	1973			Ye-6LS No.112 Orbiter	1968
Luna 22 Orbiter	1974			Ye-8 No.201 Lander/Rover	1969
Luna 24 Sample/Return	1976			Ye-8-5 No.402 Sample Return	1969
				Ye-8-5 No.403 Sample Return (Cosmos 300)	1969
				Ye-8-5 No.404 Sample Return (Cosmos 305)	1969
				Ye-8-5 No.405 Sample Return	1970
				Ye-8-5M No.412 Sample Return	1975

Dates are for launch

Automated tests of lunar Soyuz spacecraft

Successes		Post-launch failures		Launch failures	
Zond 5 Circumlunar/Return	1968	Zond 4 Lunar Distance/Return	1968	7K-L1 No.4L Zond Circumlunar	1967
Zond 7 Circumlunar/Return	1969	Zond 6 Circumlunar/Return	1968	7K-L1 No.5L Zond Circumlunar	1967
Zond 8 Circumlunar/Return	1970			7K-L1 No.7L Zond Circumlunar	1968
				7K-L1 No.13L Zond Circumlunar	1969
				7K-L1S No.3S Zond Orbiter/Return	1969
				7K-L1S No.5L Zond Orbiter/Return	1969
				7K-LOK No.6A Soyuz Orbiter/Return	1972

Dates are for launch

APPENDIX C2. USA ROBOTIC LUNAR MISSION RECORD

Successes		Post-launch failures		Launch failures	
Pioneer 4 Flyby [partial]	1959	Ranger 3 Hard Lander	1962	Pioneer 0 Orbiter	1958
Ranger 7 Impactor	1964	Ranger 4 Hard Lander	1962	Pioneer 1 Orbiter	1958
Ranger 8 Impactor	1965	Ranger 5 Hard Lander	1962	Pioneer 2 Orbiter	1958
Ranger 9 Impactor	1965	Ranger 6 Impactor	1964	Pioneer 3 Flyby	1958
Surveyor 1 Lander	1966	Surveyor 2 Lander	1966	Atlas-Able 4 Orbiter	1959
Lunar Orbiter 1	1966	Surveyor 4 Lander	1967	Atlas-Able P-3 Orbiter	1959
Lunar Orbiter 2	1966			Atlas-Able P-30 Orbiter	1960
Lunar Orbiter 3	1967			Atlas-Able P-31 Orbiter	1960
Lunar Orbiter 4	1967			Ranger 1 Deep Space Test*	1961
Lunar Orbiter 5	1967			Ranger 2 Deep Space Test*	1961
Surveyor 3 Lander	1967				
Surveyor 5 Lander	1967				
Surveyor 6 Lander	1967				
Surveyor 7 Lander	1968				
Clementine Orbiter	1994				
Lunar Prospector Orbiter	1998				
Lunar Reconnaissance Orbiter	2009				

Dates are for launch

* test launch

APPENDIX D1. USSR MARS MISSION RECORD

Partial successes		Post-launch failures		Launch failures	
Mars 2 Orbiter [lander failed]	1971	Mars 1 Flyby	1962	1M No.1 Flyby	1960
Mars 3 Orbiter [lander failed]	1971	Zond 2 Flyby	1964	1M No.2 Flyby	1960
Mars 5 Orbiter [short lived]	1973	Zond 3* [success at the Moon]	1965	2MV-4 No.3 Flyby (Sputnik 22)	1962
Mars 6 Flyby/Lander [descent data only]	1973	Mars 4 Orbiter	1973	2MV-3 No.1 Probe (Sputnik 24)	1962
Phobos 2 Orbiter/Landers [failed at Phobos]	1988	Mars 7 Flyby/Lander	1973	3MV-1A No.2 Probe* (Cosmos 21)	1963
		Phobos 1 Orbiter/Landers	1988	M69-1 Orbiter	1969
				M69-2 Orbiter	1969
				M71-S Orbiter	1971
				Mars 96 Orbiter/Landers	1996

Dates are for launch   * test launch

APPENDIX D2. USA MARS MISSION RECORD

Successes		Post-launch failures		Launch failures	
Mariner 4 Flyby	1964	Mars Observer Orbiter	1992	Mariner 3 Flyby	1964
Mariner 6 Flyby	1969	Mars Climate Orbiter	1998	Mariner 8 Orbiter	1971
Mariner 7 Flyby	1969	Mars Polar Lander/Penetrators	1999		
Mariner 9 Orbiter	1971				
Viking 1 Orbiter/Lander	1975				
Viking 2 Orbiter/Lander	1975				
Mars Global Surveyor Orbiter	1996				
Mars Pathfinder Lander	1996				
Mars Odyssey Orbiter	2001				
Spirit Rover	2003				
Opportunity Rover	2003				
Phoenix Lander	2005				
Mars Reconnaissance Orbiter	2007				

Dates are for launch

APPENDIX E1. USSR VENUS MISSION RECORD

Successes		Post-launch failures		Launch failures	
Venera 4 Atm/Surf Probe [lost in atm]	1967	Venera 1 Impactor	1961	1VA No.1 Impactor (Sputnik 7)	1961
Venera 5 Atm/Surf Probe [imploded]	1969	Zond 1 Atm/Surf Probe	1964	2MV-1 No.3 Atm/Surf Probe (Sputnik 19)	1962
Venera 6 Atm/Surf Probe [imploded]	1969	Venera 2 Flyby	1965	2MV-1 No.4 Atm/Surf Probe (Sputnik 20)	1962
Venera 7 Atm/Surf Probe	1970	Venera 3 Atm/Surf Probe	1965	2MV-2 No.1 Flyby (Sputnik 21)	1962
Venera 8 Atm/Surf Probe	1972			3MV-1A No.4A Atm/Surf Probe*	1964
Venera 9 Orbiter/Lander	1975			3MV-1 No.5 Atm/Surf Probe (Cosmos 27)	1964
Venera 10 Orbiter/Lander	1975			3MV-4 No.6 Flyby (Cosmos 96)	1965
Venera 11 Flyby/Lander [imager failed]	1978			1V No.311 Atm/Surf Probe (Cosmos 167)	1967
Venera 12 Flyby/Lander [imager failed]	1978			3V No.631 Atm/Surf Probe (Cosmos 359)	1970
Venera 13 Flyby/Lander	1981			3V No.671 Atm/Surf Probe Cosmos 482)	1972
Venera 14 Flyby/Lander	1981				
Venera 15 Orbiter	1983				
Venera 16 Orbiter	1983				
Vega 1 Flyby/Lander/Balloon	1984				
Vega 2 Flyby/Lander/Balloon	1984				

Dates are for launch

* test launch

APPENDIX E2. USA VENUS MISSION RECORD

Successes		Post-launch failures	Launch failures	
Mariner 2 Flyby	1962	(None)	Mariner 1 Flyby	1962
Mariner 5 Flyby	1967			
Mariner 10 Flyby	1973			
Pioneer 12 Orbiter	1978			
Pioneer 13 Bus/Probes(3)	1978			
Magellan Orbiter	1989			
Galileo Flyby	1989			
Cassini Flyby	1997			

Dates are for launch

APPENDIX F. SPACE EXPLORATION MILESTONES IN THE 20TH CENTURY

Lunar missions

First lunar mission attempt	Pioneer 0	1958, August 17
First spacecraft to escape Earth's gravity	Luna 1	1959, January 2
First spacecraft to fly by the Moon	Luna 1	1959, January 4
First to impact another celestial body	Luna 2	1959, September 14
First photographs of the lunar farside	Luna 3	1959, October 6
First lunar lander	Luna 9	1966, February 3
First lunar orbiter	Luna 10	1966, April 3
First image of Earth from the Moon	Lunar Orbiter 1	1966, August 23
First liftoff from the Moon	Surveyor 6	1967, November 17
First circumlunar mission and Earth return	Zond 5	1968, September 20
First piloted circumlunar mission	Apollo 8	1968, December 24
First piloted landing	Apollo 11	1969, July 20
First robotic sample return mission	Luna 16	1970, September 21
First robotic lunar rover (Lunokhod 1)	Luna 17	1970, November 17
First piloted lunar rover	Apollo 15	1971, July 30

Mercury missions

First mission to Mercury (multiple flyby)	Mariner 10	1974, March 29

Venus missions

First launch attempt to Venus	1VA No.1	1961, February 4
First successful mission to Venus (flyby)	Mariner 2	1962, December 14
First spacecraft to impact another planet	Venera 3	1966, March 1
First successful planetary entry probe	Venera 4	1967, October 18
First successful planetary lander	Venera 7	1970, December 15
First spacecraft to use gravity assist (Venus)	Mariner 10	1974, February 5
First successful Venus orbiter	Venera 9	1975, October 22
First photographs from a planetary surface	Venera 9	1975, October 22
First radar imagery of Venusian surface	Venera 15	1983, October 10
First successful planetary balloon	Vega 1	1985, June 11

Mars missions

First planetary launch attempt (Mars)	1M No.1	1960, October 10
First successful mission to Mars (flyby)	Mariner 4	1965, July 15
First planetary orbiter (Mars)	Mariner 9	1971, November 14
First spacecraft to impact Mars	Mars 2	1971, November 27
First landing on Mars	Mars 3	1971, December 2
First Mars atmospheric probe	Mars 6	1973, March 12
First successful Mars lander	Viking 1	1976, July 20
First images from the surface of Mars	Viking 1	1976, July 20
First planetary rover (Sojourner)	Mars Pathfinder	1997, July 5

Small bodies missions

First comet plasma tail flythrough (G-Z)	ICE	1985, September 11
First comet nucleus distant flyby (Halley)	Vega 1	1986, March 6
First comet nucleus close flyby (Halley)	Giotto (ESA)	1986, March 14
First asteroid flyby (Gaspra)	Galileo	1991, October 29
First asteroid orbiter (Eros)	NEAR	2000, February 14
First asteroid lander (Eros)	NEAR	2001, February 12

Outer planet missions

First spacecraft through the asteroid belt	Pioneer 10	1973
First Jupiter flyby	Pioneer 10	1973, December 3
First Saturn flyby	Pioneer 11	1979, September 1
First spacecraft to leave the Solar System	Pioneer 10	1983, June 13
First Uranus flyby	Voyager 2	1986, January 24
First Neptune flyby	Voyager 2	1989, August 25
First Jupiter orbiter	Galileo	1995, December 7
First Jupiter probe	Galileo	1995, December 8
First Saturn orbiter	Cassini	2004, July 1
First Titan probe	Huygens (ESA)	2005, January 14

Code

Soviet Missions
US and European Missions

APPENDIX G. TIMELINE OF PLANETARY EXPLORATION MISSIONS IN THE 20TH CENTURY

Launch	L/V	Source	Mission name (Spacecraft)	Target	Mission type	Result	Description
1958							
17-Aug	TA	US (ARPA)	Pioneer 0 (Able 1)	Moon	Orbiter	fb	Booster exploded
23-Sep	R7E	USSR	Ye-1 No.1	Moon	Impactor	fb	First stage destroyed
11-Oct	TA	US (ARPA)	Pioneer 1 (Able 2)	Moon	Orbiter	fu	Reached 115,000 km altitude
11-Oct	R7E	USSR	Ye-1 No.2	Moon	Impactor	fb	First stage destroyed
8-Nov	TA	US (ARPA)	Pioneer 2 (Able 3)	Moon	Orbiter	fu	Third stage failure, reached only 1,550 km altitude
4-Dec	R7E	USSR	Ye-1 No.3	Moon	Impactor	fb	Second stage premature engine shutdown
6-Dec	J2	US (ABMA)	Pioneer 3	Moon	Flyby	fu	Reached 107,500 km altitude
1959							
2-Jan	R7E	USSR	Luna 1 (Ye-1 No.4)	Moon	Impactor	ft	Missed moon by 5,965 km
3-Mar	J2	US (ABMA)	Pioneer 4	Moon	Flyby	p	Low injection velocity, flew past Moon at 60,030 km
18-Jun	R7E	USSR	Ye-1A No.5	Moon	Impactor	fb	Second stage guidance failure
12-Sep	R7E	USSR	Luna 2 (Ye-1A No.7)	Moon	Impactor	s	First successful lunar impact 14 September
24-Sep	AA	US (NASA)	Atlas-Able 4 (Pioneer)	Moon	Orbiter	fb	Pad explosion during test
4-Oct	R7E	USSR	Luna 3 (Ye-2A No.1)	Moon	Circumlunar Flyby	s	Circled moon, first far side images
26-Nov	AA	US (NASA)	P-3 (Pioneer)	Moon	Orbiter	fu	Shroud collapsed during launch destroying spacecraft
1960							
15-Apr	R7E	USSR	Ye-3 No.1	Moon	Circumlunar Flyby	fu	Third stage malfunction
19-Apr	R7E	USSR	Ye-3 No.2	Moon	Circumlunar Flyby	fb	First stage disintegrated
25-Sep	AA	US (NASA)	P-30 (Pioneer)	Moon	Orbiter	fu	Launch failed, second stage malfunction
10-Oct	R7M	USSR	1M No.1	Mars	Flyby	fu	Third stage failure, did not achieve Earth orbit
14-Oct	R7M	USSR	1M No.2	Mars	Flyby	fu	Third stage failure, did not achieve Earth orbit
15-Dec	AA	US (NASA)	P-31 (Pioneer)	Moon	Orbiter	fb	Booster exploded
1961							
4-Feb	R7M	USSR	Sputnik 7 (1VA No.1)	Venus	Impactor	fi	Fourth stage failure, stranded in low Earth orbit
12-Feb	R7M	USSR	Venera 1 (1VA No.2)	Venus	Impactor	fc	Failed in transit, carried medallions in small entry probe
23-Aug	AAB	US (NASA)	Ranger 1 (P-32)	Moon	Deep Space Test	fi	Upper stage failed 2nd burn

Date							
18-Nov	AAB	US (NASA)	Ranger 2 (P-33)	Moon	Deep Space Test	fi	Upper stage failed 2nd burn
1962							
26-Jan	AAB	US (NASA)	Ranger 3 (P-34)	Moon	Hard Lander	ft	Terminal maneuver failed, missed Moon by 37,745 km
23-Apr	AAB	US (NASA)	Ranger 4 (P-35)	Moon	Hard Lander	fc	Computer failed in Earth orbit, impacted lunar farside
22-Jul	AAB	US (NASA)	Mariner 1 (P-37)	Venus	Flyby	fu	Launch vehicle failed
25-Aug	R7M	USSR	Sputnik 19 (2MV-1 No.3)	Venus	Atm/Surf Probe	fi	Fourth stage engine failed, stranded in Earth orbit
27-Aug	AAB	US (NASA)	Mariner 2 (P-38)	Venus	Flyby	s	First successful planetary mission, flew by Venus on 14 Dec
1-Sep	R7M	USSR	Sputnik 20 (2MV-1 No.4)	Venus	Atm/Surf Probe	fi	Fourth stage failed, stranding vehicle in low Earth orbit
12-Sep	R7M	USSR	Sputnik 21 (2MV-2 No.1)	Venus	Flyby	fi	Third and fourth stages failed, stranded in low Earth orbit
18-Oct	AAB	US (NASA)	Ranger 5 (P-36)	Moon	Hard Lander	fc	Power and control system failed, passed Moon at 724 km
24-Oct	R7M	USSR	Sputnik 22 (2MV-4 No.3)	Mars	Flyby	fi	Fourth stage exploded
1-Nov	R7M	USSR	Mars 1 (2MV-4 No.4)	Mars	Flyby	fc	Failed in transit on 21 Mar 1963
4-Nov	R7M	USSR	Sputnik 24 (2MV-3 No.1)	Mars	Atm/Surf Probe	fi	Fourth stage failed, stranding vehicle in low Earth orbit
1963							
4-Jan	R7My	USSR	Sputnik 25 (Ye-6 No.2)	Moon	Lander	fi	Fourth stage failed to ignite, stranding vehicle in Earth orbit
3-Feb	R7My	USSR	Ye-6 No.3	Moon	Lander	fb	Launch vehicle veered off-course, did not reach Earth orbit
2-Apr	R7My	USSR	Luna 4 (Ye-6 No.4)	Moon	Lander	fc	Navigation system failed, missed Moon by 8,500 km
11-Nov	R7M	USSR	Cosmos 21 (3MV-1A No.2)	Mars	Test Flight	fi	Test launch, fourth stage failed, left in Earth orbit
1964							
30-Jan	AAB	US (NASA)	Ranger 6 (Ranger A/P-53)	Moon	Impactor	ft	Impacted Moon on 2 Feb, but cameras failed to operate
19-Feb	R7M'	USSR	3MV-1A No.4A	Venus	Test Flight	fu	Third stage engine exploded, did not reach Earth orbit
21-Mar	R7M'y	USSR	Ye-6 No.6	Moon	Lander	fu	Third stage engine failure, did not reach Earth orbit
27-Mar	R7M'	USSR	Cosmos 27 (3MV-1 No.5)	Venus	Atm/Surf Probe	fi	Fourth stage engine did not ignite, did not leave Earth orbit
2-Apr	R7M'	USSR	Zond 1 (3MV-1 No.4)	Venus	Atm/Surf Probe	fc	Failed in transit

Launch	L/V	Source	Mission name (Spacecraft)	Target	Mission type	Result	Description
20-Apr	R7M'y	USSR	Ye-6 No.5	Moon	Lander	fu	Upper stage failures, did not reach Earth orbit
28-Jul	AAB	US (NASA)	Ranger 7 (Ranger B/P-54)	Moon	Impactor	s	First completely successful US lunar mission
5-Nov	AAD	US (NASA)	Mariner 3 (Mariner-64C)	Mars	Flyby	fu	Shroud failed to jettison properly, damaging spacecraft
28-Nov	AAD	US (NASA)	Mariner 4 (Mariner-64D)	Mars	Flyby	s	First successful Mars mission on 15 Jul 1965
30-Nov	R7M	USSR	Zond 2 (3MV-4 No.2)	Mars	Flyby	fc	Communications failed in transit after one month
1965							
17-Feb	AAB	US (NASA)	Ranger 8 (Ranger C)	Moon	Impactor	s	Returned 7,137 photos of Sea of Tranquility
12-Mar	R7My	USSR	Cosmos 60 (Ye-6 No.9)	Moon	Lander	fi	Fourth stage failed to ignite, left in Earth orbit
21-Mar	AAB	US (NASA)	Ranger 9 (Ranger D)	Moon	Impactor	s	Returned photos before impact
10-Apr	R7M	USSR	Ye-6 No.8	Moon	Lander	fu	Third stage engine failure, did not reach Earth orbit
9-May	R7M'	USSR	Luna 5 (Ye-6 No.10)	Moon	Lander	ft	Guidance and retro-rockets malfunctioned, crashed
8-Jun	R7M'	USSR	Luna 6 (Ye-6 No.7)	Moon	Lander	fc	Mid-course maneuver failed, missed the Moon
18-Jul	R7M	USSR	Zond 3 (3MV-4 No.3)	Moon	Flyby	s	Photographed the lunar farside on 20 Jul, then
				Mars	Test Flight	fc	communications lost before reaching Mars distance
4-Oct	R7M'	USSR	Luna 7 (Ye-6 No.11)	Moon	Lander	ft	Crashed into the Ocean of Storms near Kepler
12-Nov	R7M'	USSR	Venera 2 (3MV-4 No.4)	Venus	Flyby	ft	Communications failed during Venus flyby
16-Nov	R7M'	USSR	Venera 3 (3MV-3 No.1)	Venus	Atm/Surf Probe	fc	Communications failed 17 days before arrival
23-Nov	R7M'	USSR	Cosmos 96 (3MV-4 No.6)	Venus	Flyby	fi	Upper stage failures, did not leave Earth orbit
3-Dec	R7M	USSR	Luna 8 (Ye-6 No.12)	Moon	Lander	ft	Crashed into the Ocean of Storms near Galilaei
1966							
31-Jan	R7M'	USSR	Luna 9 (Ye-6M No.202/13)	Moon	Lander	s	First lunar lander on 3 Feb, returned pictures from the surface
1-Mar	R7M'	USSR	Cosmos 111 (Ye-6S No.204)	Moon	Orbiter	fi	Fourth stage failed, stranded in Earth orbit
31-Mar	R7M'	USSR	Luna 10 (Ye-6S No.206)	Moon	Orbiter	s	First successful lunar orbiter on 3 Apr
30-May	AC	US (NASA)	Surveyor 1 (Surveyor-A)	Moon	Lander	s	First US lunar lander successful 2 Jun
10-Aug	AAD	US (NASA)	Lunar Orbiter 1 (LO-A)	Moon	Orbiter	s	First US lunar orbiter successful on 14 Aug
24-Aug	R7M'	USSR	Luna 11 (Ye-6LF No.101)	Moon	Orbiter	p	Lunar orbit photo and science mission, no images returned
20-Sep	AC	US (NASA)	Surveyor 2 (Surveyor-B)	Moon	Lander	ft	Crashed southeast of Copernicus
22-Oct	R7M'	USSR	Luna 12 (Ye-6LF No.102)	Moon	Orbiter	s	Lunar orbit photo and science mission
6-Nov	AAD	US (NASA)	Lunar Orbiter 2 (LO-B)	Moon	Orbiter	s	Lunar orbital photographic mapper for Apollo
21-Dec	R7M'	USSR	Luna 13 (Ye-M No.205/14)	Moon	Lander	s	Lunar surface science and images

1967

Date		Country	Mission	Target	Type		Description
5-Feb	AAD	US (NASA)	Lunar Orbiter 3 (LO-C)	Moon	Orbiter	s	Lunar photographic mapper for Apollo
17-Apr	AC	US (NASA)	Surveyor 3 (Surveyor-C)	Moon	Lander	s	Lunar surface science and images
4-May	AAD	US (NASA)	Lunar Orbiter 4 (LO-D)	Moon	Orbiter	s	Lunar photographic mapper for Apollo
16-May	R7M'	USSR	Cosmos 159 (Ye-6LS No.111)	Moon	Orbiter Test Flight	fu	Fourth stage burn insufficient for very high Earth orbit
12-Jun	R7M'	USSR	Venera 4 (1V No.310)	Venus	Atm/Surf Probe	s	First planetary atmospheric probe, did not reach surface
14-Jun	AAD	US (NASA)	Mariner 5 (Mariner-67E)	Venus	Flyby	s	Flew by Venus at 3,990 km on 19 Oct
17-Jun	R7M'	USSR	Cosmos 167 (1V No.311)	Venus	Atm/Surf Probe	fi	Failed to depart low Earth orbit
14-Jul	AC	US (NASA)	Surveyor 4 (Surveyor-D)	Moon	Lander	ft	Lost contact minutes before landing on 17 Jul
1-Aug	AAD	US (NASA)	Lunar Orbiter 5 (LO-E)	Moon	Orbiter	s	Scientific photographic mapping
8-Sep	AC	US (NASA)	Surveyor 5 (Surveyor-E)	Moon	Lander	s	Lunar surface science and images
27-Sep	PrD	USSR	7K-L1 No.4L	Moon	Circumlunar/Return	fb	Test of Soyuz lunar craft in Earth orbit, booster failed
7-Nov	AC	US (NASA)	Surveyor 6 (Surveyor-F)	Moon	Lander	s	Lunar surface science and images
22-Nov	PrD	USSR	7K-L1 No.5L	Moon	Circumlunar/Return	fu	Test of lunar Soyuz in Earth orbit, 2nd stage failure

1968

Date		Country	Mission	Target	Type		Description
7-Jan	AC	US (NASA)	Surveyor 7 (Surveyor-G)	Moon	Lander	s	Lunar surface science and imagery
7-Feb	R7M'	USSR	Ye-6LS No.112	Moon	Orbiter	fu	Third stage terminated early at 524 sec, ran out of propellant
2-Mar	PrD	USSR	Zond 4 (7K-L1 No.6L)	Moon	Lunar Distance Test Flight	ft	Soyuz lunar craft self-destructed before touchdown
7-Apr	R7M'	USSR	Luna 14 (Ye-6LS No.113)	Moon	Orbiter	s	Mapped lunar gravity field.
22-Apr	PrD	USSR	7K-L1 No.7L	Moon	Circumlunar/Return	fu	Second stage shutdown
14-Sep	PrD	USSR	Zond 5 (7K-L1 No.9L)	Moon	Circumlunar/Return	s	First circumlunar flight and return to Earth on 21 Sep
10-Nov	PrD	USSR	Zond 6 (7K-L1 No.12L)	Moon	Circumlunar/Return	ft	Crashed on landing after circumlunar flight.
21-Dec	S5	US (NASA)	Apollo 8 (CSM103)	Moon	Orbiter	s	First human mission to the Moon entered orbit on 24 Dec

1969

Date		Country	Mission	Target	Type		Description
5-Jan	R7M'	USSR	Venera 5 (2V No330)	Venus	Atm/Surf Probe	s	Returned in-situ atmospheric science, did not reach surface
10-Jan	R7M'	USSR	Venera 6 (2V No.331)	Venus	Atm/Surf Probe	s	Returned in-situ atmospheric science, did not reach surface
20-Jan	PrD	USSR	7K-L1 No.13L	Moon	Circumlunar/Return	fu	Test of Soyuz lunar craft in Earth orbit, upper stage failures

Launch	L/V	Source	Mission name (Spacecraft)	Target	Mission type	Result	Description
19-Feb	PrD	USSR	Ye-8 No.201	Moon	Lander/Rover	fu	Shroud failure, vehicle disintegrated
21-Feb	N1	USSR	7K-L1S No.3S	Moon	Orbiter/Return	fb	First N-1 launch, first stage failed in flight
25-Feb	AC	US (NASA)	Mariner 6 (Mariner-69F)	Mars	Flyby	s	Returned 75 images during flyby
27-Mar	AC	US (NASA)	Mariner 7 (Mariner-69G)	Mars	Flyby	s	Returned 126 images during flyby
27-Mar	PrD	USSR	M-69 No.521	Mars	Orbiter	fu	Planned atmospheric probe deleted, third stage exploded
2-Apr	PrD	USSR	M-69 No.522	Mars	Orbiter	fb	Planned atmospheric probe deleted, booster exploded
18-May	S5	US (NASA)	Apollo 10 (CSM106/LM4)	Moon	Orbiter	s	Tested lunar lander in lunar orbit
14-Jun	PrD	USSR	Ye-8-5 No.402	Moon	Sample Return	fu	Fourth stage failed to ignite
3-Jul	N1	USSR	7K-L1S No.5L	Moon	Orbiter/Return	fb	Second N-1 launch, first stage exploded at liftoff
13-Jul	PrD	USSR	Luna 15 (Ye-8-5 No.401)	Moon	Sample Return	ft	Crashed during landing attempt on 21 July after 52 orbits
16-Jul	S5	US (NASA)	Apollo 11 (CSM107/LM5)	Moon	Orbiter/Lander	s	First human landing on the Moon 20 July, 1969
7-Aug	PrD	USSR	Zond 7 (7K-L1 No.11)	Moon	Circumlunar/Return	s	Returned successfully to Earth on 14 Aug
23-Sep	PrD	USSR	Cosmos 300 (Ye-8-5 No.403)	Moon	Sample Return	fi	Fourth stage failed to ignite for translunar injection
22-Oct	PrD	USSR	Cosmos 305 (Ye-8-5 No.404)	Moon	Sample Return	fi	Fourth stage misfire
14-Nov	S5	US (NASA)	Apollo 12 (CSM108/LM6)	Moon	Piloted Orbiter/Lander	s	Successful precision landing 156 meters from Surveyor 3
1970							
6-Feb	PrD	USSR	Ye-8-5 No.405	Moon	Sample Return	fu	Second stage premature shutdown at 127 sec mark
11-Apr	S5	US (NASA)	Apollo 13 (CSM109/LM7)	Moon	Orbiter/Lander	fc	Support module exploded en route, astronauts returned safely
17-Aug	R7M'	USSR	Venera 7 (3V No.630)	Venus	Atm/Surf Probe	s	First successful planetary lander
22-Aug	R7M'	USSR	Cosmos 359 (3V No.631)	Venus	Atm/Surf Probe	fi	Fourth stage misfired, failed to depart low Earth orbit
12-Sep	PrD	USSR	Luna 16 (Ye-8-5 No.406)	Moon	Sample Return	s	First robotic lunar sample return
20-Oct	PrD	USSR	Zond 8 (7K-L1 No.14)	Moon	Circumlunar/Return	s	Returned successfully to Earth on 27 Oct
10-Nov	PrD	USSR	Luna 17 (Ye-8 No.203)	Moon	Lander/Rover	s	First lunar rover, Lunokhod 1
1971							
31-Jan	S5	US (NASA)	Apollo 14 (CSM110/LM8)	Moon	Orbiter/Lander	s	Landed in Fra Mauro on 5 Feb
9-May	AC	US (NASA)	Mariner 8 (Mariner-71H)	Mars	Orbiter	fu	Centaur stage failure
10-May	PrD	USSR	Cosmos 419 (M-71 No.170)	Mars	Orbiter	fi	Fourth stage failed to re-ignite, left in low Earth orbit

Date		Country	Mission	Target	Type		Notes
19-May	PrD	USSR	Mars 2 (M-71 No.171)	Mars	Orbiter/Lander	p	Orbiter successful, lander crashed but first artifact on Mars
28-May	PrD	USSR	Mars 3 (M-71 No.172)	Mars	Orbiter/Lander	p	Orbiter successful, lander failed after 20 sec
30-May	AC	US (NASA)	Mariner 9 (Mariner-71I)	Mars	Orbiter	s	First successful Mars and planetary orbiter on 13 Nov
26-Jul	S5	US (NASA)	Apollo 15 (CSM112/LM10)	Moon	Orbiter/Lander/Rover	s	First human lunar rover, deployed lunar sub-satellite
2-Sep	PrD	USSR	Luna 18 (Ye-8-5 No.407)	Moon	Sample Return	ft	Lost communications during landing attempt
28-Sep	PrD	USSR	Luna 19 (Ye-8LS No.202)	Moon	Orbiter	s	Lunar orbital photography and gravity field mapping
1972							
14-Feb	PrD	USSR	Luna 20 (Ye-8-5 No.408)	Moon	Sample Return	s	Returned samples on 25 Feb
2-Mar	AC	US (NASA)	Pioneer 10 (Pioneer-F)	Jupiter	Flyby	s	First Jupiter flyby, first to leave Solar System
27-Mar	R7M'	USSR	Venera 8 (3V No.670)	Venus	Atm/Surf Probe	s	55 min descent and 63 min on the surface
31-Mar	R7M'	USSR	Cosmos 482 (3V No.671)	Venus	Atm/Surf Probe	fi	Fourth stage misfired, failed to depart low Earth orbit
16-Apr	S5	US (NASA)	Apollo 16 (CSM113/LM11)	Moon	Orbiter/Lander/Rover	s	Carried rover and deployed sub-satellite
23-Nov	N1	USSR	7K-LOK No.6A	Moon	Orbiter/Return	fb	Fourth N-1 launch, booster exploded in flight
7-Dec	S5	US (NASA)	Apollo 17 (CSM114/LM12)	Moon	Orbiter/Lander/Rover	s	Final Apollo mission, first and last to carry a scientist
1973							
8-Jan	PrD	USSR	Luna 21 (Ye-8 No.204)	Moon	Lander/Rover	s	Deployed Lunokhod 2 rover
5-Apr	AC	US (NASA)	Pioneer 11 (Pioneer-G)	Jupiter	Flyby	s	Flew past Jupiter on 4 Dec 1974
				Saturn	Flyby	s	Flew past Saturn on 1 Sep 1979
21-Jul	PrD	USSR	Mars 4 (M-73 No.52S)	Mars	Orbiter	ft	Failed to achieve Mars orbit
25-Jul	PrD	USSR	Mars 5 (M-73 No.53S)	Mars	Orbiter	p	Failed early, completed only 22 orbits
5-Aug	PrD	USSR	Mars 6 (M-73 No.50P)	Mars	Flyby/Lander	p	Returned descent data, but no communication from surface
9-Aug	PrD	USSR	Mars 7 (M-73 No.51P)	Mars	Flyby/Lander	ft	Entry system failed, flew past Mars
3-Nov	AC	US (NASA)	Mariner 10 (Mariner-73J)	Venus	Flyby	s	Successful flyby of Venus
				Mercury	Flyby	s	Three successful sequential flybys
1974							
29-May	PrD	USSR	Luna 22 (Ye-8LS No.206)	Moon	Orbiter	s	Orbital photography and surface elemental composition
28-Oct	PrD	USSR	Luna 23 (Ye-8-5M No.410)	Moon	Sample Return	ft	Sampler damaged in landing, no return

Launch	L/V	Source	Mission name (Spacecraft)	Target	Mission type	Result	Description
1975							
8-Jun	PrD	USSR	Venera 9 (4V-1 No.660)	Venus	Orbiter/ Lander	s	First B/W images of Venus at the surface
14-Jun	PrD	USSR	Venera 10 (4V-1 No.661)	Venus	Orbiter/ Lander	s	Same science as Venera 9
20-Aug	T3EC	US (NASA)	Viking 1 (Viking-B)	Mars	Orbiter/Lander	s	First successful Mars lander deployed from orbit
9-Sep	T3EC	US (NASA)	Viking 2 (Viking-A)	Mars	Orbiter/Lander	s	Deployed second successful lander
16-Oct	PrD	USSR	Ye-8-5M No.412	Moon	Sample Return	fu	Fourth stage failed
1976							
9-Aug	PrD1	USSR	Luna 24 (Ye-8-5M No.413)	Moon	Sample Return	s	Landed 18 Aug and returned core sample
1977							
20-Aug	T3EC	US (NASA)	Voyager 2 (Voyager B)	Jupiter	Flyby	s	Explored Jupiter system on 9 Jul 1979
				Saturn	Flyby	s	Explored Saturn system on 25 Aug 1981
				Uranus	Flyby	s	Explored Uranus system on 24 Jan 1986
				Neptune	Flyby	s	Explored Neptune system on 25 Aug 1989
5-Sep	T3EC	US (NASA)	Voyager 1 (Voyager A)	Jupiter	Flyby	s	Explored Jupiter system on 5 Mar 1979
				Saturn	Flyby	s	Explored Saturn system on 12 Nov 1980
1978							
20-May	AC	US (NASA)	Pioneer 12 (Pioneer Venus 1)	Venus	Orbiter	s	Conducted atmospheric science and planetary radar mapping
8-Aug	AC	US (NASA)	Pioneer 13 (Pioneer Venus 2)	Venus	Entry Bus/Atm Probes	s	Kamikaze bus, one large, three small atmosphere probes
12-Aug	D	US (NASA)	Int'l Comet Explorer (ICE)	Comet G-Z	Flyby	s	ISEE-3 diverted to Giaccobini-Zinner for first comet flyby
9-Sep	PrD1	USSR	Venera 11 (4V-1 No.360)	Venus	Flyby/Lander	s	Atmospheric science on descent, surface science failed
14-Sep	PrD1	USSR	Venera 12 (4V-1 No.361)	Venus	Flyby/Lander	s	Atmospheric science on descent, surface science failed
1979			(No Missions)		(No Missions)		
1980			(No Missions)		(No Missions)		

Date		Agency	Mission	Target	Type		Description
1981							
30-Oct	PrD1	USSR	Venera 13 (4V-1M No.760)	Venus	Flyby/Lander	s	First color imagery from Venus surface
4-Nov	PrD1	USSR	Venera 14 (4V-1M No.761)	Venus	Flyby/Lander	s	Same science as Venera 13
1982			(No Missions)				
1983							
2-Jun	PrD1	USSR	Venera 15 (4V-2 No.860)	Venus	Orbiter	s	Radar mapper covered northern hemisphere down to 30°N
7-Jun	PrD1	USSR	Venera 16 (4V-2 No.861)	Venus	Orbiter	s	Radar mapper with same coverage as Venera 16
1984							
15-Dec	PrD1	USSR	Vega 1 (5VK No.901)	Venus	Flyby/Lander/Balloon	s	Deployed lander and balloon at Venus on 11 Jun 1985
				Halley	Flyby	s	Flew by Halley at 8,890 km 6 Mar 1986
21-Dec	PrD1	USSR	Vega 2 (5VK No.902)	Venus	Flyby/Lander/Balloon	s	Deployed lander and balloon at Venus on 15 Jun 1985
				Halley	Flyby	s	Flew by Halley at 8,030 km 9 Mar 1986
1985							
7-Jan	M3S2	Japan (ISAS)	Sakigake (MS-T5)	Halley	Flyby	s	Very distant Halley flyby on 8 Mar 1986
2-Jul	Ar1	ESA	Giotto	Halley	Flyby	s	Close Halley flyby at 596 km on 14 Mar 1986
				Comet G-S	Flyby	s	Flew by Grigg-Skjellerup on 10 Jul 1992
18-Aug	M3S2	Japan (ISAS)	Suisei (Planet-A)	Halley	Flyby	s	Distant Halley flyby at 151,000 km on 14 Mar 1986
1986			(No Missions)				
1987			(No Missions)				
1988							
7-Jul	PrD2	USSR	Phobos 1 (1F No.101)	Mars	Orbiter/Phobos Lander	fc	Lost en route due to a command error on 1 Sep

Launch	L/V	Source	Mission name (Spacecraft)	Target	Mission type	Result	Description
12-Jul	PrD2	USSR	Phobos 2 (1F no.102)	Mars	Orbiter/Phobos Lander	p	Entered Mars orbit, but failed just prior to Phobos encounter
1989							
4-May	STS	US (NASA)	Magellan	Venus	Orbiter	s	Radar mapper, high resolution coverage of entire planet
18-Oct	STS	US (NASA)	Galileo	Venus	Flyby	s	Feb 10, 1990
				Earth	Flyby	s	Dec 8, 1990
				Gaspra	Flyby	s	Oct 29, 1991
				Earth	Flyby	s	Dec 8, 1992
				Ida	Flyby	s	Aug 28, 1993
				Jupiter	Atm Probe	s	Successful entry on 7 Dec 1996
				Jupiter	Orbiter	s	Orbited Jupiter system for nearly 7 years
1990							
24-Jan	M3S2	Japan (ISAS)	Hiten -	Moon	Multiple Flybys	s	Lunar flybys, lunar orbit 15 Feb 92, impacted 10 Apr 93
			Hagoromo (MUSES-A)	Moon	Orbiter	ft	Deployed from Hiten, but no communications received
1991			(No Missions)		(No Missions)		
1992							
25-Sep	T3C	US (NASA)	Mars Observer	Mars	Orbiter	ft	Propulsion system exploded 3 days prior to arrival
1993			(No Missions)		(No Missions)		
1994							
25-Jan	T2G	US (DoD) and (NASA)	Clementine	Moon	Orbiter	s	Conducted global photographic and spectral mapping, then
				Geographos	Flyby	fc	departed lunar orbit and failed en-route to asteroid

Date	Code	Agency	Mission	Target	Type		Details
1995 (No Missions)			(No Missions)				
1996							
17-Feb	D2	US (NASA)	Near Earth Asteroid Rendezvous	Eros	Orbiter	s	Flyby of Mathilda en route, first to orbit an asteroid
7-Nov	D2	US (NASA)	Mars Global Surveyor (MGS)	Mars	Orbiter	s	First aerobraking to achieve close mapping orbit
16-Nov	PrD2	Russia	Mars 96 (M1 No.520)	Mars	Orbiter/Landers/Penetrators	fu	Fourth stage failure, re-entered atmosphere
4-Dec	D2	US (NASA)	Mars Pathfinder	Mars	Lander/Rover	s	Landed on 4 Jul 1997
						s	First Mars rover "Sojourner" deployed from lander on 6 Jul
1997							
15-Oct	T4BC	US (NASA) and (ESA)	Cassini-Huygens	Saturn	Orbiter	s	Orbiter with Titan probe, arrived 1 July 2004
				Titan	Atm/Surf Probe	s	Dispatched 25 Dec 2004, entry and landing 14 Jan 2005
				Venus	Flyby	s	Apr 26, 1998
				Venus	Flyby	s	Jun 24, 1999
				Earth	Flyby	s	Aug 17, 1999
				Jupiter	Flyby	s	Dec 1, 2000
1998							
7-Jan	A2	US (NASA)	Lunar Prospector	Moon	Orbiter	s	UV/Vis and IR mapper entered orbit on 11 Jan
3-Jul	M5	Japan (ISAS)	Nozomi (Planet-B)	Mars	Orbiter	ft	Failed to orbit and flew by Mars
24-Oct	D2	US (NASA)	Deep Space 1 (DS-1)	Asteroid	Test Flight	p	Flew by asteroid 1992KD and Comet Borrelly
			EPOXI	Hartley 2	Flyby	s	Flyby of Comet Hartley 2
11-Dec	D2	US (NASA)	Mars Climate Orbiter	Mars	Orbiter (MCO)	ft	Failed to achieve Mars orbit, impacted atmosphere
1999							
3-Jan	D2	US (NASA)	Mars Polar Lander (MPL)	Mars	Lander	ft	Failed during entry
	D2	US (NASA)	Deep Space 2 (DS-2)	Mars	Penetrators	ft	Both lost with lander
7-Feb	D2	US (NASA)	Stardust	Wild-2	Sample Return	s	Flyby of Comet Wild-2, returned coma samples
			Stardust-NExT	Temple 1	Flyby	s	Flyby of Temple 1 post Deep Space impact

Launch	L/V	Source	Mission name (Spacecraft)	Target	Mission type	Result	Description
2000			(No Missions)		(No Missions)		

Launch Vehicle Codes:

R7E	Luna
R7M	Molniya
R7M'	Molniya, upgraded
y	suffix: modified for spacecraft control of upper stages
PrD	Proton with Block D restartable upper stage
PrD1	Proton with Block D-1 (or DM) restartable upper stage
PrD2	Proton with Block D-2 restartable upper stage
N1	N1-L3 Soviet Moon launcher similar to the US Saturn V
SF	Soyuz-Frigat
TA	Thor-Able
J2	Juno 2 (modified Jupiter-C)
AA	Atlas-Able
AAB	Atlas-Agena B
AAD	Atlas-Agena D
AC	Atlas-Centaur
A5	Atlas V Centaur
T2G	Titan 2G
T3EC	Titan 3E Centaur
T3C	Titan 3C
T4BC	Titan 4B Centaur
D	Delta
D2	Delta 2
A2	Athena 2
S5	Saturn V
STS	Space Shuttle with Interim Upper Stage (IUS)
Ar1	ESA Ariane 1
Ar5	ESA Ariane 5
M3S2	ISAS M3S2
M5	ISAS M5

Outcome Codes:

fb	booster failure
fu	upper stage failure
fi	interplanetary trajectory injection failure
fc	failure in transit during cruise
ft	failure at the target
p	partial success
s	success
e	enroute

APPENDIX H. USSR LUNAR AND PLANETARY PROBE LOCATIONS

USSR Venus Probe Locations

Arrival	Mission	Latitude	Longitude	Last P	Last T	Last Alt	Day/Nite	Images?	Description
01-Mar-66	Venera 3	20°S-20°N	60-80°E	-	-	-	night	N	failed prior to entry
18-Oct-67	Venera 4	019°N	038°E	7.2 bar[1]	535K	24 km[2]	night	N	93 minutes on descent
16-May-69	Venera 5	003°S	018°E	27 bar	600K	18 km[2]	night	N	53 minutes on descent
17-May-69	Venera 6	005°S	023°E	27 bar	567K	22 km[2]	night	N	51 minutes on descent
15-Dec-70	Venera 7	005°S	351°E	~92 bar[3]	~738K	surface	night	N	on surface for 23 min
22-Jul-72	Venera 8	010.7°S	335.25°E	93 bar	743K	surface	day	N	on surface for 63 min
22-Oct-75	Venera 9	031.01°N	291.64°E	85 bar	728K	surface	day	Y	on surface for 53 min
25-Oct-75	Venera 10	015.42°N	291.51°E	91 bar	737K	surface	day	Y	on surface for 65 min
25-Dec-78	Venera 11	014°S	299°E	91 bar	728K	surface	day	N	on surface for 95 min
21-Dec-78	Venera 12	007°S	294°E	92 bar	729K	surface	day	N	on surface for 110 min
01-Mar-82	Venera 13	007.55°S	303.69°E	89.5 bar	738K	surface	day	Y	on surface for 127 min
05-Mar-82	Venera 14	013.055°S	310.19°E	93.5 bar	743K	surface	day	Y	on surface for 57 min
11-Jun-85	Vega 1	007.2°N	177.8°E	95 bar	740K	surface	night	N	on surface for 56 min
15-Jun-85	Vega 2	006.45°S	181.08°E	91 bar	735K	surface	night	N	on surface for 57 min

(1) Pressure data terminated earlier than temperature data.
(2) Altitude at Last T using the Venus Int'l Ref. Atmosphere.
(3) Calculated. Pressure data not transmitted.

USSR Mars probe locations

Arrival	Mission	Latitude	Longitude	Day/Nite	Images?	Description
27-Nov-71	Mars 2 Lander	044.2°S	313.2°W	day	Y	crashed
02-Dec-71	Mars 3 Lander	045°S	158°W	day	Y	failed shortly after landing
12-Mar-73	Mars 6 Lander	023.90°S	019.42°W	day	Y	failed just at landing

442

USSR lunar probe locations

Arrival	Mission	Latitude	Longitude	Day/Nite	Images?	Description
14-Sep-59	Luna 2	29.1°N	0.0°E	day	N	first lunar impact
5-Dec-65	Luna 5	1.6°S	335°E	day	Y	crashed lander
10-Jul-65	Luna 7	9.8°N	312.2°E	day	Y	crashed lander
06-Dec-65	Luna 8	9.1°N	296.7°E	day	Y	crashed lander
03-Feb-65	Luna 9	7.08°N	295.63°E	day	Y	first lander
24-Dec-66	Luna 13	18.87°N	297.95°E	day	Y	lander
21-Jul-69	Luna 15	17°N	60°E	day	Y	crashed sample return
20-Sep-70	Luna 16	0.68°S	56.30°E	day	Y	first robotic sample return
17-Nov-70	Luna 17	38.25°N	325.00°E		N	lander
	(Lunokhod 1)	38.32°N	324.99°E		Y	final rover location
11-Sep-71	Luna 18	3.57°N	56.50°E	day	Y	crashed sample return
21-Feb-72	Luna 20	3.53°N	56.55°E	day	Y	sample return
15-Jan-73	Luna 21	26.92°N	30.45°E	day	N	lander
	(Lunokhod 2)	25.84°N	30.90°E		Y	final rover location
06-Nov-74	Luna 23	12.68°N	62.28°E	day	N	damaged sample return
18-Aug-76	Luna 24	12.75°N	62.20°E	day	N	sample return

Bibliography

BOOKS AND MONOGRAPHS

Almas, I., Horvath, A.: *Space Travel Encyclopedia*. Akademiai Kiado and Zrinyi Katonai Kiado, Budapest (1981)

Anon.: *Pervye photografii obratnoi storony Luny*. Izdatelstvo AN SSSR, Moskva (1959)

Anon.: *Atlas obratnoi storony Luny*. Izdatelstvo AN SSSR, Moskva (1960)

Anon.: *Pervye panoramy lunnoi poverhnosti*. Nauka, Moskva (1966)

Ball, A.J., Garry, J.R.C., Lorenz, R.D., Kerzhanovich, V.V.: *Planetary Landers and Entry Probes*. Cambridge Univ. Press, Cambridge (2007)

Barsukov, V.I., Basilevsky, A.T., Volkov, V.P., Zharkov, V.N. eds.: *Venus Geology, Geochemistry, and Geophysics, Research results from the USSR*. The University of Arizona Press, Tucson, London (1992)

Bougher, S.W., Hunten, D.M., and Phillips, R.J., eds.: *Venus II*. The University of Arizona Press, Tucson, Arizona (1997)

Burns, J. A., and Matthews M. S., eds.: *Satellites*. The University of Arizona Press, Tucson, Arizona (1986)

Fimmel, R.O., Colin, L., Burgess E.: *Pioneering Venus*. NASA SP-518 (1995)

Gatland, K. W.: *Robot Explorers*. Blanford Press, London (1972)

Chertok, B.Ye.: *Rakety i liudi*, 2nd ed. Mashinostroenie, Moskva (1999)

Chertok, B.Ye.: *Rakety i liudi Fili, Podlipki, Tyuratam*. Mashinostroenie, Moskva (1999)

Chertok, B.Ye.: *Rakety i liudi Gorjachie dni holodnoi voiny*. Mashinostroenie, Moskva (1999)

Chertok, B.Ye.: *Rakety i liudi Lunnaya gonka*. Mashinostroenie, Moskva (1999)

Glushko, V. P. ed.: *Kosmonavtika entsiklopediya*. Sovetskaja entsiklopediya, Moskva (1985)

Grewing, M., Praderie, F., Reinhard R.,eds.: *Exploration of Halley's Comet*. Springer-Verlag, Berlin, Heidelberg, New York, London, Paris, Tokyo (1987)

Harvey, B.: *Russian Planetary Exploration.* Springer-Praxis, Chichester (2007)

Harvey, B.: *Soviet and Russian Lunar Exploration.* Springer-Praxis, Chichester (2007)

Hunten, D.M., Colin L., Donahue T.M., and Moroz V.I., eds.: *Venus.* The University of Arizona Press, Tucson, Arizona (1983)

Harford, J.: *Korolev: How one Man Masterminded the Soviet Drive to Beat America to the Moon.* John Wiley & Sons, New York (1997)

Johnson, N. L.: *Handbook of Soviet Lunar and Planetary Exploration. Amer. Astronau. Soc. Publ.,* Volume 47 science and technology series (1979)

Keldysh, M.V. and Marov, M.Ya. *Kosmichesckiye Issledovaniya.* Nauka, Moskva (1981)

Kieffer, H.H., Jakosky, B.M., Snyder, C.W. and Mattheus, M. S., eds.: *Mars.* The University of Arizona Press, Tucson (1992)

Kuzmin, A.D., and Marov, M. Ya.: *Physics of the planet Venus.* Nauka, Moskow (1974)

Marov, M.Ya., and Grinspoon, D.: *The Planet Venus.* Yale University Press (1997)

Moroz, V.I.: *Physics of the planet Mars.* Nauka, Moskow (1978)

Perminov, V.G.: *The difficult road to Mars, a brief history of Mars exploration in the Soviet Union.* NASA NP-1999-251-HQ (1999)

Reinhardt, R. ed.: *Space missions to Halley's Comet.* ESA SP-1066 (1986)

Sagdeev, R.Z.: *The making of the Soviet scientist.* John Wiley, New York, Chichester, Brisbane, Toronto, and Singapore (1994)

Semenov, Yu.P., ed.: *Raketno-kosmicheckaya korporatsiya Energiya imeni S.P. Koroleva.* RKK Energiya (1996)

Serebrennikov, V.A, Voitik, V.L., Shevalev I.L. et al., eds.: *NPO imeni S.A. Lavochkina, Na zemle, v nebe i v kosmose.* Voennyi Parad, Moskva (1997)

Siddiqi, A.A.: *Challenge to Apollo: the Soviet Union and the space race, 1945-1974.* NASA SP 2000-4408 (2000)

Siddiqi, A.A.: *Deep Space Chronicle: A Chronology of Deep Space and Planetary Probes, 1958-2000.* NASA SP-2002-4524 (2002)

Sidorenko, A.V. ed.: *Poverkhnost Marsa.* Nauka, Moskva (1980)

Surkov, Yu. A.: *Exploration of Terrestrial Planets from Spacecraft,* 2nd ed. Wiley-Praxis, Chichester (1997)

Ulivi, P. with Harland, D. M.: *Lunar Exploration: Human Pioneers and Robotic Surveyors.* Springer-Praxis, Chichester (2004)

Ulivi, P. with Harland, D. M.: *Robotic Exploration of the Solar System, Part 1. The Golden Age 1957-1982.* Springer-Praxis, Chichester (2007)

Ulivi, P. with Harland, D. M.: *Robotic Exploration of the Solar System, Part 2. Hiatus and Renewal 1983-1996.* Springer-Praxis, Chichester (2009)

Vinogradov, A.P. ed.: *Peredvizhnaia laboratoria na Lune Lunokhod-1.* Nauka, Moskva (1971)

Vinogradov, A.P. ed.: *Lunnyi grunt iz Morja Izobiliia.* Nauka, Moskva (1974)

JOURNAL AND BOOK ARTICLES

Anonymous: 'Jodrell Bank Tracks Zond 2'. *Flight International* 1965, 303 (1965)

Anonymous: 'Mars from Orbit'. *Spaceflight* 14, 68-70 and 118-120 (1972)

Avduesvsky, V.S., Marov, M.Ya., and Rozhdestvensy, M.K.: 'The Model of the Atmosphere of the Planet Venus Based on the Results of Measurements Mad by the Soviet Automatic Interplanetary Station Venera 4'. In: Jastrow, R. and Rasool, S.I. eds. *The Venus Atmosphere*. Gordon and Breach, New York-London-Paris (1969)

Avduesvsky, V.S., Marov, M.Ya., and Rozhdestvensy, M.K.: 'The Tentative Model of the Atmosphere of the Planet Venus based on the results of measurements of probes Venera 5 and Venera 6'. *J.Atm. Sci.* 27, No.4 (1970)

Avduevsky, V.S., Kerzhanovich, V.V., Marov, M.Ya., et al.: 'Soft Landing of Venera 7 on the Venus Surface and Preliminary Results of Investigations of the Venus Atmosphere'. *J. Atm. Sci.* 28, 263-269 (1971)

Avduesvsky, V.S., Akim, E.L., Kerzhanovich, V.V., Marov, M.Ya., et al.: 'Atmosphere of Mars in the site of "Mars 6" landing (Preliminary results)'. *Kosmich. Issled.* 13, 21-32 (1975)

Day, D.: 'Mission Impossible: the Kidnapping of Lunik 5'. *Quest* 5, 55 (1996)

Deryugin, V.A. et al.: 'Vega-1 and Vega-2 Spacecraft. Operation of Landers in Venusian Atmosphere'. *Cosmic Research* 25, 494-498 (1987)

Govorchin, G.G.: 'The Soviets in Space – An Historical Survey'. *Spaceflight* 7, 74-82 (1965)

Huntress, W. T., Moroz, V. I., Shevalev, I.: 'Robotic Planetary Exploration Missions of the 20th Century'. *Space Science Reviews* 107, 541-649 (2003)

Kerzhanovich, V.V., and Marov, M.Ya.: 'On the wind-velocity measurements from Venera Spacecraft data'. *Icarus* 30, 320 (1977)

Klaes, L.: 'The Robot Explorers of Venus'. *Quest* 8:1, 24-36 (2000), *Quest* 8:2, 24-31 (2000) and *Quest* 8:3, 17-23 (2000)

Kremnev, R.S., et al.: 'VEGA Balloon System and Instrumentation'. *Science* 231, 1408-1411 (1986)

Kremnev, R.S., et al.: 'The VEGA balloons: a tool for studying atmosphere dynamics on Venus'. *Soviet Astronomy Letters* 12, 7-9 (1986)

Lantratov, K.: '25 Years from Lunokhod-1'. *Novosti kosmonavtiki* 23, 79-83 (1995) and 24, 70-79 (1995)

Lantratov, K.: 'To Mars!'. *Novosti kosmonavtiki* 20, 53-72 (1996) and 21, 41-51 (1996)

Lantratov, K., Hendrickx, B.: 'Mars-69: the Forgotten Mission to the Red Planet'. *Quest* 7, 26-31 (1999)

Linkin, V., Harri, A.-M., Lipatov, A., et al.: 'A sophisticated lander for scientific exploration of Mars: scientific objectives and implementation of the Mars-96 Small Station'. *Planetary and Space Science* 46, 717-737 (1998)

Maksimov, G.Yu.: 'Construction and Testing of the First Soviet Automatic Interplanetary Stations'. *History of Rockets and Astronautic* 20, 233-246 (1997)

Marov, M.Ya.: 'Model of the Venus Atmosphere'. *Soviet Academy of Sci. Doklady*, 196 (No.1), 67 (1971)

Marov, M.Ya., Avduesvsky, V.S., Borodin, N.F., Kerzhanovich, V.V., Rozhdest-vensky, M.K., et al.: 'Preliminary results on the Venus atmosphere from the Venera 8 descent module'. *Icarus* 20, 407 (1973).

Marov, M.Ya., and Petrov, G.I.: 'Investigations of Mars from the Soviet automatic stations Mars 2 and 3'. *Icarus* 19, 163 (1973)

Marov, M.Ya., and Moroz, V.I.: 'Preliminary results of investigation carried out with automatic stations Venera 9 and Venera 10'. *Kosmich. Issled.* 14, 651-654 (1976)

Marov, M.Ya.: 'The Atmosphere of Venus: Venera Data'. *Fundamentals of Cosmic Physics* 5, 46 (1979)

Marov, M.Ya.: 'Scientific and Technical Strategy for Planetary Exploration'. *Space Policy* 10, No.1, 32 (1994)

Marov, M.Ya.: 'Missions to Mars: An Overview and a Perspective at the Turn of the Century'. In: *Astronomical and Biochemical Origins and the Search for Life in the Universe* (eds. C.B. Cosmovici, S. Bowyer, and D. Werthimer). Editrice Compositori (1997)

Marov, M.Ya.: 'Strategy for Solar System studies: A view from Russia'. In: *Solar System Plasma Physics* (eds. F. Mariani and N.F. Ness). Editrice Compositori (1997)

Moroz, V.I.: 'Preliminary results of research conducted aboard the Soviet Mars-4, Mars-5, Mars-6 and Mars-7 planetary probes'. *Kosmich. Issled.* 13, 3-8 (1975)

Moroz, V.I.: 'Spectra and spacecraft'. *Planetary and Space Science* 49, 173-190 (2001)

Moroz, V.I., Huntress, W.T., Shevalev, I.: 'Planetary Missions of the 20th Century'. *Kosmich. Issled.* 40, 451 (2002)

Moroz, V.I., Ksanfomality, L.V.: 'Preliminary Results of Astrophysical Observations of Mars from Mars-3'. *Icarus* 17, 408-422 (1972)

Oja, H.: 'Soviet Mars Landers'. *Spaceflight* 15, 242-245 (1973)

Petrov, B.N.: 'Space Research in the USSR and the Venera 4 Experiment'. *Spaceflight* 11, 171-173 (1969)

Rocard, F. et al.: 'French Participation in the Soviet Phobos Mission'. *Acta Astronautica* 22, 261-7 (1990)

Sagdeev, R.Z., Blamont J., Galeev A.A., Moroz V.I., Shapiro V.D., Shevchenko V.I., Szego K.: 'Vega spacecraft encounters with comet Halley'. *Nature* 321, 259-262 (1986)

Sagdeev, R.Z., Linkin, V.M., Blamont, J.E., Preston, R.A.: 'The Vega Venus Balloon Experiment'. *Science* 231, 1407-1408 (1986)

Sagdeev, R.Z., Moroz, V.I.: 'Venera 13 and Venera 14'. *Sov. Astron. Lett.* 8, 209-211 (1982)

Sagdeev, R.Z., Zakharov A.V.: 'Brief history of the Phobos mission'. *Nature* 341, 581-584 (1989)

Sagdeev, R.Z.: 'A brief history of the expedition to Phobos'. *Sov. Astron. Lett.* 16, 125-128 (1990)

Siddiqi, A.A.: 'First to the Moon'. *JBIS* 51, 231-238 (1998), additional comments by T. Varfolomeyev *JBIS* 52, 157-160 (1999)

Siddiqi, A.A., Hendrickx, B., Varfolomeyev, T.: 'The Tough Road Traveled: A New Look at the Second Generation Luna Probes'. *JBIS* 53, 319-356 (2000)

Snyder, C.W., Moroz, V.I.: 'Spacecraft exploration of Mars'. In *Mars* (H. H. Kieffer, B. M. Jakosky, C. W. Snyder and M. S. Mattheus eds.) The University of Arizona Press, Tucson (1992)

Vakhnin, V.M.: 'A Review of the Venera 4 Flight and Its Scientific Program'. *J. Atm. Sci.* 25, 533-534 (1968)

Varfolomeyev, T.: 'The Soviet Venus Programme'. *Spaceflight* 35, 42-43 (1993)

Varfolomeyev, T.: 'The Soviet Mars Programme'. *Spaceflight* 35, 230-231 (1993)

Varfolomeyev, T.: 'Soviet Rocketry that Conquered Space, Part 1: From First ICBM to Sputnik Launcher'. *Spaceflight* 37, 260-263 (1995)

Varfolomeyev, T.: 'Soviet Rocketry that Conquered Space, Part 2: Space Rockets for Lunar Probes'. *Spaceflight* 38, 49-52 (1996)

Varfolomeyev, T.: 'Soviet Rocketry that Conquered Space, Part 3: Lunar Launchings for Impact and Photography'. *Spaceflight* 38, 206-208 (1996)

Varfolomeyev, T.: 'Soviet Rocketry that Conquered Space, Addendum: Launch Vehicle Designations'. *Spaceflight* 38, 317-318 (1996)

Varfolomeyev, T.: 'Sputnik Era Launches'. *Spaceflight* 39, 331-332 (1997)

Varfolomeyev, T.: 'Soviet Rocketry that Conquered Space, Part 4: The Development of a Four-Stage Launcher 1958-1960'. *Spaceflight* 40, 28-30 (1998)

Varfolomeyev, T.: 'Soviet Rocketry that Conquered Space, Part 5: The First Planetary Probe Attempts 1960-1964'. *Spaceflight* 40, 85-88 (1998)

Varfolomeyev, T.: 'Soviet Rocketry that Conquered Space, Part 6: The Improved Four-Stage Launch Vehicle 1964-1972'. *Spaceflight* 40, 181-184 (1998)

Varfolomeyev, T.: 'Soviet Rocketry that Conquered Space, Part 9: Launchers for an Early Circumlunar Program'. *Spaceflight,* 41 207-210 (1999)

Vinogradov A.P., ed.: 'Cosmochemistry of Moon and planets'. *Proceedings of US-Soviet conference*, Nauka, Moskva (1975)

Wotzlaw, S., Kasmann, F.C.W., Nagel, M.: 'Proton – Development of a Russian Launch Vehicle'. *JBIS*, 51, 2-18

Zaitsev, Y.: 'The Successes of Phobos-2'. *Spaceflight* 31, 374-377 (1989)

JOURNAL SPECIAL ISSUES AND DEVOTED SECTIONS

Mars 2,3: *Icarus* 18 No.1, entire issue (1973)

Mars 4,5,6,7: *Kosmich. issled.* 13, 3-130 (1975)

Phobos 1,2: *Nature* 341, 581-618 (1989) and *Planet. Sp. Sci.* 39, 1-399 (1991)

Vega Halley flyby: *Nature* 321, 259-366 (1986)

Vega Venus balloons: *Science* 231, 1407-1425 (1986)

Vega 1,2: *Kosmich. issled.* 25, 643-958 (1987)

Venera 9,10: *Kosmich. issled.* 14, 651-877 (1976)

Venera 11,12: *Kosmich. issled.* 17, 646-829 (1979)

Venera 13,14: *Kosmich. issled.* 21, 147-319 (1983)

Venera 15,16: *Kosmich. issled.* 23, 179-267 (1985)

WEBSITES

Aerospaceguide *http://www.aerospaceguide.net*
Planetary Sciences at NSSDC *http://nssdc.gsfc.nasa.gov/planetary*
Encyclopedia Astronautica *http://www.astronautix.com*
Energia Site *http://www.energia.ru/english/index.html*
Grand Tour! A Planetary Exploration Page *http://utenti.lycos.it/paoloulivi*
History of Space Exploration *http://solarviews.com/eng/history.htm*
IKI Space Research Institute *http://www.iki.rssi.ru/*
Don P. Mitchell *http://www.mentallandscape.com*
N1 Photo Clearinghouse *http://www.personal.psu.edu/faculty/g/h/ghb1/n1/n1synth.htm*
NPO Lavochkin *http://www.laspace.ru/rus/index.php*
Russian Mission Histories *http://vsm.host.ru*
Soviet Web Space – Siddiqi *http://faculty.fordham.edu/siddiqi/sws/index.html*
Sven's Space History *http://www.svengrahn.pp.se/histind/histind1.htm*
Zak's Spaceweb Site *http://www.russianspaceweb.com/index.html*
Zarya – Soviet and Russian Space Programmes *http://www.zarya.info*
Zheleznyakov's Encyclopedia *http://www.cosmoworld.ru/spaceencyclopedia*

Index